OXFORD PAPERBACK REFERENCE

A Dictionary of
Geography

Susan Mayhew is a geography teacher and the
author of *Masterstudies Geography*. Her current
research interests are in feminist and historical
geography with particular reference to the
reification of concept through place.

Oxford Paperback Reference

The most authoritative and up-to-date reference books for both students and the general reader.

Abbreviations
ABC of Music
Accounting
Allusions
Archaeology
Architecture
Art and Artists
Art Terms
Astronomy
Better Wordpower
Bible
Biology
British History
Buddhism*
Business
Card Games
Catchphrases*
Celtic Mythology
Chemistry
Christian Art
Christian Church
Classical Literature
Computing
Dance
Dates
Dynasties of the World
Earth Sciences
Ecology
Economics
Engineering*
English Etymology
English Folklore
English Grammar
English Language
English Literature
English Place-Names
Euphemisms*
Everyday Grammar
Finance and Banking
First Names
Food and Drink
Food and Nutrition
Foreign Words and Phrases
Geography
Handbook of the World
Humorous Quotations
Idioms
Irish Literature
Jewish Religion
Kings and Queens of Britain
Language Toolkit
Law
Linguistics
Literary Quotations
Literary Terms
Local and Family History
London Place-Names

Mathematics
Medical
Medicines
Modern Design
Modern Quotations
Modern Slang
Music
Nursing
Opera
Philosophy
Phrase and Fable
Physics
Plant-Lore
Plant Sciences
Pocket Fowler's Modern
  English Usage
Political Biography
Political Quotations
Politics
Popes
Proverbs
Psychology
Quotations
Reverse Dictionary
Rhyming*
Rhyming Slang*
Sailing Terms
Saints
Science
Scientists
Shakespeare
Ships and the Sea
Slang
Sociology
Spelling
Statistics
Superstitions
Surnames
Synonyms and Antonyms
Theatre
Twentieth-Century Art
Twentieth-Century Poetry
Twentieth-Century World
  History
Weather
Weights, Measures, and Units
Who's Who in the Classical
  World
Who's Who in the Twentieth
  Century
World History
World Mythology
World Religions
Writers' Dictionary
Zoology

*forthcoming

A Dictionary of
Geography

SECOND EDITION

SUSAN MAYHEW

Oxford   New York

OXFORD UNIVERSITY PRESS

OXFORD
UNIVERSITY PRESS

Great Clarendon Street, Oxford OX2 6DP

Oxford University Press is a department of the University of Oxford.
It furthers the University's objective of excellence in research, scholarship,
and education by publishing worldwide in

Oxford New York

Athens Auckland Bangkok Bogotá Buenos Aires Calcutta
Cape Town Chennai Dar es Salaam Delhi Florence Hong Kong Istanbul
Karachi Kuala Lumpur Madrid Melbourne Mexico City Mumbai
Nairobi Paris São Paulo Singapore Taipei Tokyo Toronto Warsaw

First published 1992 as *The Concise Oxford Dictionary of Geography*
First edition (with Anne Penny) 1992
Second edition 1997

British Library Cataloguing in Publication Data

Data available

Library of Congress Cataloging in Publication Data

A dictionary of geography.—2nd ed. / edited by Susan Mayhew.
(Oxford paperback reference)
Rev. ed. of: The concise Oxford dictionary of geography / Susan Mayhew and Ann Penny. 1992.
1. Geography—Dictionaries.   I. Mayhew, Susan.   II. Mayhew, Susan.   Concise Oxford
dictionary of geography.
G63.M39   1997   910'.3—dc21   96–52361

ISBN 0-19-280034-5

20 19 18 17 16 15 14 13 12 11

Printed in Great Britain by
Clays Ltd, St Ives plc

How to use this book

This dictionary provides coverage in one volume of the terms used in both human and physical geography. There are over 6000 definitions across the following fields: cartography, surveying, remote sensing, statistics, meteorology, climatology, biogeography, ecology, simple geology, soils, geomorphology, population, migration, settlement, agriculture, industry, transport, development, and diffusion.

Headwords are printed in bold type and appear in alphabetical order. However, some entries contain further definitions. These have been included under the headword to avoid unnecessary repetition and to indicate some of the wider applications of the headword. Many entries have an asterisk. This points the reader to cross-references. In a very few cases the cross-reference as indicated does not have the exact wording as in the entry, but is close enough to make further reference possible.

References to geographical authors include surname, initials and date. The use of these details in a card or electronic library catalogue will yield the book referred to.

A

aa *See* *block lava.

abîme An upright or near vertical shaft in an area of *karst geomorphology.

abiotic Not living, non-biological, usually describing factors in an ecosystem such as atmospheric gases, inorganic salts, mineral soil particles, and water. This expression is also used to describe the chemical and physical factors, such as salinity and humidity, which influence organisms. An **abiotic environment** is one without any life.

ablation Loss of snow and ice from a glacier by, for example, melting and evaporation. Ablation also results from *sublimation, the *calving of icebergs, and *avalanches. In temperate and subpolar regions melting is the most important process in ablation, whereas in the Antarctic the most important ablation process is calving. The rate of loss varies with the meteorological factors of air temperature, relative humidity, wind speed, and *insolation, together with *aspect and the nature of the surface. The **ablation sub-system** is the zone of a glacial system between the *firn line and the snout where there is an annual net loss of ice since annual ablation exceeds annual *accumulation. The zone of net ice loss from a glacier is the **ablation zone**.

At the edges of the glaciers, where ablation has occurred, large quantities of *debris are released and accumulate to form **ablation moraines (ablation till)**.

Abney level A surveying instrument which can measure angles to within 10 seconds of an arc.

aborigine A member of an indigenous people existing in a land before invasion or colonization by another race. This term is especially used for the original inhabitants of Australia.

abrasion The grinding away of bedrock by fragments of rock which may be incorporated in ice (**glacial abrasion**), water (**marine abrasion, fluvial abrasion**), or wind (**aeolian abrasion**). In *fluvial environments, the main agent of abrasion is the *bed load. The mass of solid material removed varies with the size, density, and velocity of the particles, and the density of the *vector bearing these particles. Ice ceases to be an effective agent for abrasion when the weight of the ice is thick enough to bring about *plastic flow. Abrasion is an alternative term for *corrasion. *See* *striations.

abrasion platform *Wave-cut platform.

abrasion terrace A former *wave-cut platform, now above sea level because of either tectonic uplift of the mainland or eustatic lowering of sea level. Abrasion terraces are thus indicative of *emerging coastlines.

abscissa The horizontal, or *x* axis, of a graph. Where a *causal factor or an *independent variable can be clearly defined, it is recorded along the abscissa.

absenteeism A failure to show up for work.

absolute drought In the UK, this is a period of 15 days, on none of which more than 0.25 mm of rain falls. National definitions vary with climate; in Libya, droughts are recognized only after two years without rain. These arbitrary definitions give no indication of the impact of drought. *See* *Palmer Drought Severity Index (PDSI).

absolute humidity The density of the water vapour present in a mixture of air and water vapour, that is, the ratio of the mass of water vapour to the volume occupied by the mixture, usually measured in grams per cubic centimetre. Cold air cannot contain as much water vapour as warm air, so cold air has a lower absolute humidity than warm air. *See also* *relative humidity.

absolute plate motion The movement of a *crustal *plate in relation to a fixed point, such as a *hot spot.

absolute zero The lowest temperature theoretically obtainable: $-273.15\ °C$.

absorption The process by which a material or system takes in another material or system. *See also* *adsorption.

abstraction The selection and conceptualization of a phenomenon, or some aspect of it. Abstraction is an essential part of *model building where some aspect or part of the real world is extracted and simplified. Unfortunately, during the process of simplification, so much information has to be jettisoned that the resulting model may have very limited success.

Other abstractions are based on *idealism; models are made of an 'ideal type' such as the Latin American city model. The problem is that it is quite possible to construct very different ideal types of the same phenomenon.

There are different **levels of abstraction**: global, national, societal, class, and so on.

abundance The total number of individuals of a certain species present in an area. Abundance is generally estimated by using one or more of a variety of sampling methods (such as capture–recapture) and may vary according to competition, predation, and resources. *See also* *diversity.

abyssal At great depths; over 3000 m below sea level; thus, **abyssal plain**—the deep sea floor with a gradient of less than 1 in 10 000—and **abyssal deposits**. **Abyssal hills** are hills of 50–250 m which interrupt the abyssal plain. The word abyssal for a rock has now been replaced by *plutonic.

abysso- At great depths. Hence **abyssopelagic zone**; that part of deep lakes, oceans, or seas characterized by specific pelagic organisms (forms of plankton and nekton which inhabit open water), and **abyssobenthic zone**; the bottom of a deep lake, ocean, or sea. *See* *benthos.

accelerated For accelerated soil erosion, *see* *soil erosion.

accelerator A factor which increases the momentum of a boom or slump in an economy so that a small change in demand, for example, leads to a greater industrial growth or decline.

acceptable-dose limit The highest safe level of an introduced substance; the maximum level at which the substance poses no health hazards to the environment in which it is used. Currently, the EU has ruled that nitrate levels in the water supply should not exceed 50 mg. nitrate/litre of water.

accessibility 1. The ease of approach to one location from other locations. This may be measured in terms of the distance travelled, the cost of travel, or the time taken. In *network analysis, accessibility may be expressed by measures of *connectivity. Accessibility can be calculated by using a framework known as an **accessibility matrix**. Consider the five towns in the diagram below.

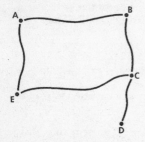

FIGURE 1: *Accessibility*

In the matrix derived from it the number of roads used to travel from each town in turn to each of the others is recorded. The total is noted in the column marked row sum. The town with the lowest total, C, is the most accessible because it needs the fewest roads to reach the others.

TOWN	A	B	C	D	E	ROW SUM
A	—	1	2	3	1	7
B	1	—	1	2	2	6
C	2	1	—	1	1	5
D	3	2	1	—	2	8
E	1	2	1	2	—	6

See also *node, *link.

2. The ease of access to a resource or service.

In the case of both resources and services, it is possible to distinguish between **physical accessibility** where a resource is within reach, and **social accessibility** whereby the individual actually has the means to reach the resource or location. For example, it is argued that class structures, income, age, educational background, gender, or race can limit people's access to services. *See also* *location.

accordant Complying with. Thus, **accordant drainage** is *drainage which has evolved in conformity with the underlying geological structure, so domes show a *radial pattern, and *trellised patterns develop on gently dipping scarplands, for example.

The **law of accordant junctions (Playfair's law)** states that tributaries join a stream or river at the same elevation as that of the larger watercourse; thus there is no sudden 'drop' in the level of the tributary. This means that tributaries are *graded to the level of the junction.

Accordant summits are hill or mountain tops of approximately the same elevation. The presence of accordant summits has been seen as confirmation of the theory of the *cycle of erosion.

accreting margin *See* *constructive margin.

accretion **1.** The growth of land by the offshore deposition of sediment. Accretion is most active in *estuaries, particularly within the Tropics. *Spits and *tombolos are features of accretion.

2. The increase in size of a continent by the addition of terranes (**accretion terranes**).

3. The growth of a landform by the addition of deposits; *seif dunes grow by accretion.

4. The increase in size of particles by additions to the exterior, as in the formation of *hailstones.

acculturation The adaptation to, and adoption of, a new culture. This may occur simultaneously as two cultures meet but more often occurs as an immigrant group takes to the behaviour patterns and standards of the receiving group. A major example is the acceptance of American norms by the millions of European immigrants who arrived in the USA in the first half of the twentieth century. *See also* *assimilation, *charter group.

accumulated temperature From a specific date, the length of time for which mean daily temperatures have been above, or below, a stated temperature; the total time for which temperatures varied from that standard. It is used, for example, to judge the fitness of a climate for a particular crop by showing how long temperatures are above the minimum required for that crop. *See also* *degree day.

accumulation **1.** The input of ice to a glacier. Observations from the Decade glacier, Baffin Island, suggest that accumulation is greatest in shaded upland areas. The **accumulation zone**, or **accumulation sub-system**, is that part of a glacier between the *firn line and the source where the

input of snow, firn, and ice exceeds losses by *ablation. The lower limit of
the accumulation zone is at the *equilibrium line.

2. The reinvestment of surplus value in the form of capital in order to
increase that capital. Accumulation is a key feature of *capitalism because,
in order to remain in business, the capitalist has not only to preserve the
value of the capital raised but also to add to the value of that capital. The
effects of this 'imperative of accumulation' are, on the one hand, a class of
capitalists who own the *means of production, and, on the other hand, a
class of workers who sell their labour to the capitalists; an economic
system that has had profound societal implications. A further effect of the
imperative of accumulation is the necessity for technical change and
economic growth. Among others, David Harvey (1982), argues that the
obligatory accumulation of the capitalist system has been responsible for
*uneven development.

Economists have classified two **regimes of accumulation**: *Fordism and
*post-Fordism.

acid A substance containing hydrogen ions which can be neutralized by
an *alkali. The *pH of acid is below 7. *See* *acid rain, *acid soil.

The term 'acid' as applied to rocks has an entirely different meaning. *See*
*acid rocks.

acid rain When *fossil fuels are burned, dioxides of sulphur and nitrogen
are released into the air. When inhaled, these dry deposits can lead to
breathing problems. Industrial development, particularly in the mid-
latitudes of the northern hemisphere, has been responsible for the
emission of increasing quantities of such atmospheric pollutants, which
can travel large distances, generally being carried eastwards by the
prevailing westerly winds; the *OECD estimated in the early 1990s that
85% of the sulphur dioxide deposition on the Nordic countries was
'imported'. The primary sources were the former USSR, the reunified
Germany, and Poland.

The pollutants dissolve in atmospheric water particles to form acid
rain.

Any form of atmospheric water, such as rain, dew, or snow, with a
*pH of less than 5.6 is properly termed **acid precipitation**. (Note that the
pH scale is logarithmic, so that an increase of one point represents a
tenfold increase in acidity.) When the concentration of sulphur dioxide
reaches 0.2 p.p.m., acid precipitation is toxic to vegetation. Extensive
damage has been reported, for example, in the Black Forest, Germany.
Humans are at risk when the concentration rises above 1 p.p.m.

Within soils, acidification seems to limit bacterial activity, displace
nutrient ions by hydrogen ions, and liberate toxic heavy metals such as
aluminium and lead, which may contaminate drinking water. High levels
of aluminium in lake water in Scandinavia have been linked to acid
emissions from the UK, and have caused the destruction of aquatic flora
and fauna. Aquatic ecosystems seem to react more rapidly than terrestrial
systems to acidification. Acid precipitation may also attack building-stone
containing calcium and magnesium carbonates.

acid rock In geology, an *igneous rock containing more than 60% of silica or silicates by weight, over 10% of this being free quartz. Examples include *granite, granodiorites, and rhyolites. Compare with *basic rock.

acid soil A soil with a *pH of less than 7. Acidity in a soil may be due to the *leaching out of *cations when *precipitation exceeds *evapotranspiration. The cations are replaced by hydrogen ions. Other factors promoting acidity in a soil include the nature of the vegetation, and thus the *humus, and the acidity of the parent rock. Examples of acid soils are *podzols and *brown earths.

acidification A soil-forming process whereby the presence of organic acids (developed through the incorporation of humus into the soil) results in increasing hydrogen ion concentration. In humid temperate forest regions, acidification transforms *brown-earth soils into acid brown earths.

acre A unit of area, defined in British law as 4840 square yards (about 0.4 hectares).

actinometer A device measuring the intensity of electromagnetic radiation (radiant energy); usually that of the sun. It is used to record *insolation at the earth's surface.

action space The area in which an individual moves and makes decisions about her or his life, including, for example, shopping, studying, or working; the set of places which an individual is aware of. See *mental map.
Localities which are well known by individuals are more often chosen as the places in which to base their activities; the decision-maker evaluates all the locations within the relevant action space by accrediting each with *place utility but, if none of these sites offers adequate utility, then that individual will extend the action space by *search behaviour. Put in simple language, this means that a person will look for a suitable site for activity within the area best known to him or her. All possible sites are rated, according to how satisfactory they are. If none of the sites within the action space comes up to scratch, the decision-maker will widen the search area, become familiar with a new area, and thus extend the action space. *See also* *activity space.

active layer Also known as *annually thawed layer, depth of thaw*, and *depth to permafrost*, this is the highly mobile layer of soil, subject to periodic thawing, located above the *permafrost in *tundra regions such as Spitzbergen or Alaska, and ranging in depth from a few centimetres to 3 m. The thickness of the layer depends on factors such as slope angle and aspect, drainage, rock and/or soil type, depth of snow cover, and ground-moisture conditions. Thawing may occur daily or only in summer. On refreezing, the active layer may expand, especially if *silt-sized particles predominate.
The mobility of the active layer is due to the restricted nature of *tundra vegetation which does little to bind together this surface zone, so that it moves on slopes as gentle as 2°, and clearance of vegetation will increase the depth and mobility of the active layer. *See* *thermokarst.

Many *periglacial processes occur in the active layer, such as *frost
heaving, *frost thrusting, *ice wedging, *gelifluction, and the formation of
*patterned ground; most can cause havoc to buildings. *See also* *mollisol.

active margin A type of destructive *plate margin characterized by
*ocean trenches, *earthquakes, *andesitic *volcanic chains, and young
*fold mountains. Active margins are alternatively known as Pacific
margins after the active margin found where the Nasca *plate collides
with the South American plate.

active volcano A volcano known to have erupted in recent times, or
which is likely to erupt. Examples include Mt. St Helens, USA, and Etna,
Sicily.

activity allocation model A type of planning model used to determine
the location of activity within an area.
    The first stage is to make as accurate a forecast as possible of future
population, industry, trade, housing, and so on. This may be done through
*extrapolation, or through the use of methods such as *economic base
theory. Future needs are indicated—the simplest way is to consider separate
sub-models for each land use, such as residential, industrial, retail—and the
planner then allocates new developments to the most suitable points in
the area, often using the *gravity model. It is then possible to model the
flows of people within the area which result from this planning to see how
the various sub-models fit together.

activity index A measure of the extent to which a local authority
develops policies to attract industry. Such policies may range from the
provision of sites to the construction of factories.

activity rate The percentage of people of working age who are actually
employed. This may be calculated for a region or a nation. A low activity
rate indicates high unemployment.

activity segregation The *spatial separation of the sexes during the
working day. An extreme example was the traditional mining village,
where the mines were an exclusively male preserve and the kitchen an
exclusively female one. *See also* *feminist geography.

activity space The space we live in from day to day; that part of *action
space with which an individual interacts on a daily basis. There seems to
be a hierarchy of activity spaces for most people, increasing in spatial
extent from family space, to *neighbourhood, to economic space, and then
to urban space. With movement up the hierarchy, the individual's
knowledge of the space becomes less comprehensive. *See also* *mental
map.

actuarial data *Demographic statistics having a bearing on calculation
of risk for births and deaths.

adaptation Any change in the structure or functioning of an organism
that makes it better suited to its environment.

adaptive radiation A surge of evolution from an original ancestral form as new forms 'fan out', adapting over time to new niches. The classic example must be the fourteen Galapagos finches examined by Darwin, all presumably descended from a common ancestral species, but each of which had a different mode of life.

additional worker hypothesis The view that a rise in unemployment leads to a rise in the working population. When the major wage earner becomes unemployed, other members of the family who were not in paid work now seek employment in order to sustain the household. This might be valid if part-time work is recognized.

adiabat A line plotted on a *thermodynamic diagram, usually on a *tephigram, showing as a continuous sequence the temperature and pressure states of a parcel of air with changing height. Dry adiabats show temperature change at the dry *adiabatic lapse rate.

adiabatic change, adiabatic process A change in temperature, pressure, or volume, involving no transfer of energy to or from another material or system. In an adiabatic process, compression is accompanied by warming, and expansion by cooling. An **adiabatic temperature change** thus results from a pressure change. The speed at which the temperature of rising air falls with altitude is the **adiabatic lapse rate**. Dry, rising air expands with height. The energy needed for this expansion comes from the air itself in the form of heat.

The resulting change in temperature is expressed in the equation:

$$\frac{Dt}{dz} = \frac{-g}{Cp}$$

where Dt is the temperature change, g is the acceleration due to gravity, dz is the height change, and Cp is the specific heat of the air parcel. This change is the **dry adiabatic lapse rate (DALR)**: 9.84 °C/1000 m. The temperature change sustained by any parcel of dry air is calculated using Poisson's equation.

If the rising air becomes saturated to *dew point, condensation of vapour will begin. This condensation is accompanied by the release of *latent heat, which partly offsets the cooling with height, so that the rate of cooling of moist air—the **saturated adiabatic lapse rate (SALR)**—is lower than the DALR. In the lower *troposphere, the vapour content of air is high so the latent heat of condensation is high; SALRs may be as low as 5 °C/1000 m. In the cold, dry, high *troposphere, though, there is little vapour ready for condensation, so the SALR may be close to the DALR. Quantitative expressions of the SALR are therefore quite complex. See also *sublimation.

Adiabatic changes rarely occur in the *stratosphere because this layer experiences very little vertical atmospheric motion.

adiabatic chart See *aerological diagram.

adit A tunnel driven horizontally into a hillside for the purpose of mining. Such shafts were common in the early development, for example, of the South Wales coalfield.

administrative principle The principle advanced by Christaller (1933; trans. C. W. Baskin 1966) which proposes that, in a region with a highly developed system of central administration, settlement is so arranged that one major centre administers six centres of lesser rank, each of which, in turn, oversees a further six centres. The number of settlements at progressively lower levels, starting with the highest in rank, thus follows the sequence 1, 7, 49, 343 . . . This hierarchy is known as $k = 7$.

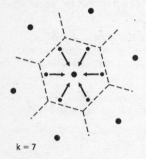

$k = 7$

FIGURE 2: *Administrative principle*

adobe Bricks of sun-dried earth or clay; **adobe houses** are made from such bricks. The term has been extended to include alluvial clay and *playa clay, which is often found in dry, desert lake-beds, such as those in the south-west of the USA.

adret The sunny slope of a hill or valley side. In the northern hemisphere adret slopes face south; in the southern hemisphere they face north. Adret slopes are warmer because they receive more *insolation. This can have important implications for land use; along the Rhine rift valley, for example, the adret slopes are terraced for vines, while the north-facing *ubac slopes are largely forested. Similarly, in the Alps, villages are generally located on adret slopes.

adsorption In *soil science, the addition of ions or molecules to the electrically charged surface of a particle of clay or humus. In this way, minerals such as potassium, sodium, magnesium, or calcium become bonded to soil particles. The term **adsorption complex** is given to those soil particles which can absorb ions or molecules.

advanced economy, economically advanced economy A synonym for a more developed country, generally defined as having a per capita GNP in excess of $10 000 per annum (1995) and an agricultural workforce of less than 6% of the working population.

advanced gas-cooled reactor, AGR A nuclear reactor where the heat of the reaction is transmitted to carbon dioxide under pressure. The hot gas is then used to heat water to make the steam which turns turbines.

advection The movement of a quality, such as heat, or humidity, due to the flow of the fluid possessing that property. In meteorology, the term is usually applied to the *horizontal* transfer of heat (compare with *convection).

  Advective transfers occur in air streams, in ocean currents, and in surface *run-off. They redress, in part, the imbalance of *insolation between the tropics and the poles. For **advection fog**, *see* *fog.

adventitious population People living in the countryside by choice and not because they are employed in rural occupations.

aeolian Of the wind. Aeolian processes include the erosion, transport, and deposition of material by wind, and work best when vegetation cover is sparse, or absent. **Aeolian landforms** include *zeugen (pedestal rocks) and *yardangs, both of which are desert features.

aeration zone In the *hydrological cycle, the zone between the soil-moisture zone and the capillary zone immediately above the water-table.

aerobic Referring to any living organism which depends on atmospheric oxygen for the release of energy from foodstuffs during the process of respiration. *Compare with* *anaerobic.

aerological diagram A chart plotting the factors which determine the movement of air. Variations of temperature, pressure, dry and saturated *adiabatic lapse rates, and *saturated mixing ratio lines are plotted against height in a *tephigram. An adiabatic chart may be used to predict the *convective condensation level.

aerology The study of the air.

aerosol Very precisely, a suspension of droplets or particles in a gas; and, even more precisely, of those particles with a maximum diameter of 1 μm. Thus, *fog and *mist are aerosols.

  In meteorology, the term is often used to describe the particles suspended within the gas, such as minute fragments of sea-salt, dust (particularly silicates), organic matter, and smoke. Such aerosols enter the atmosphere by natural processes such as *vulcanicity, and by human agency like the burning of *fossil fuels. Aerosols absorb heat and may act as *condensation nuclei.

aesthetic landscape The landscape in terms of taste and beauty. Similar economic systems, like wheat farming on the Prairies of Canada and the Great Plains of the USA, can give rise to different aesthetic landscapes. Some landscapes are actually created according to current taste, as exemplified in the parklands of English country houses.

afforestation The planting of trees. Tree-planting can stabilize soils by increasing *interception and reducing *run-off, reduce flooding through the reduction of silting, improve soil fertility (in *agroforestry schemes), provide timber and firewood, and counteract possible *global warming through the absorption of the increasing levels of atmospheric carbon dioxide.

aftershock A smaller tremor or series of tremors commonly occurring after an *earthquake, as the regional stresses are redistributed. Aftershocks usually decrease in frequency over time, but can continue for a period of months, and generally add to the damage caused by the original earthquake.

age dependency The dependency on the contributions of those in full employment of those members of the population too young or too old to be employed full-time; in the UK, those below 16 or above 65 are reckoned to be age-dependent. While the 36% of Sweden's age-dependent population are evenly split between under 15 year olds and over 64 year olds, 35% of Turkey's 38% dependent population are under 15. The respective figures for Kenya are 52% dependent, of whom 50% are children (1989 data). As population growth slackens in developed societies and as health care improves, there are large numbers of retired people whose accumulated savings have been eroded by inflation and who cannot, therefore, support themselves. The ageing of the population means that fewer workers are obliged to support increasing numbers of old people. *See* *dependency ratio.

age of towns scheme A scheme of town classification advanced by Griffith Taylor, who saw towns developing and changing through time, from infantile towns with a haphazard distribution of shops and houses with no factories, through juvenile and adolescent towns with increasing levels of land-use zoning, to mature towns.

age structure The composition of a nation by age groups. Three types of structure have been identified: the West European with fewer than 30% of children and 15% of old people; the North American type with 35–40% of children and 10% of old people; and the Brazilian type with 45–55% of children and 4–8% of old people. The type of age structure has a profound effect on the future of the nation, with old *age dependency at one extreme and explosive population growth at the other.

age–sex pyramid A set of two *histograms set on a vertical axis and back to back which depicts the numbers of the two sexes in different age groups. Males are usually on the left with females on the right. The youngest group is at the base; the oldest group at the apex. Actual figures, or percentages of the age groups may be used.
The importance of such diagrams is that they can show, in pictorial form, the varying population structures of different types of society. For example, in many less developed countries the pyramid will have a very wide base; these are termed **progressive pyramids** because they suggest future population growth. In contrast, the **regressive pyramids** characteristic of Western nations, where there are fewer children and more old people, are more cylindrical; population is likely to decrease. In

between the two are **stationary pyramids** which show a balance between old and young and no population growth.

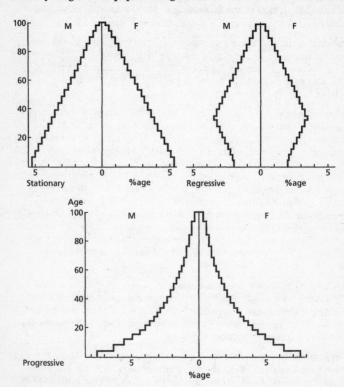

FIGURE 3: *Age–sex pyramids*

ageism Prejudice towards and/or discrimination against people because of their age.

agency Another term for action, as in 'human agency'.

agglomerate A coarse-grained volcanic rock composed of sharp or sub-angular fragments of lava, set in a fine matrix. Some agglomerates are *pyroclastic, having formed during a volcanic eruption; others are deposits from *mud flows or *lahars.

agglomeration 1. The concentration of activities, usually industries, near to each other, for example in a specialized industrial region, such as 'Silicon Fen' Cambridgeshire, UK, or in a large town or city. *See* *agglomeration economies, *transactional analysis.
    2. In meteorology, the process by which cloud droplets grow by assimilating other droplets. *See* *precipitation.

agglomeration economies For an industry, the benefits of locating in a densely peopled and highly industrialized situation. The market is large, but concentrated in a relatively small area. Transport costs are therefore low, so that many specialized industries can evolve, since local demand is sufficiently high. Further benefits arise because of *functional linkages, and through *external economies. However, when land costs and congestion are high, these agglomeration economies may be offset to such an extent that *decentralization begins to set in. *See also* *concentration and centralization.

aggradation The deposition of unconsolidated deposits by *aeolian, *marine, or *fluvial processes, when the quantity or calibre of the load is greater than the competence of the transporting medium to carry it, or as a result of *mass movement.
  *Alluvial deposits are major features of aggradation, and detailed analyses of **aggradation deposits**, such as the East Anglian Fens, using *stratigraphical and *palynological methods, have enabled geomorphologists to throw light on post-glacial climatic change.

aggregate A group of soil particles held together by electrostatic forces, *polysaccharide gums, and *cementation by carbonates and iron oxides.

aggregate data Statistics which relate to broad classes, groups, or categories, so that it is not possible to distinguish the properties of individuals within those classes, groups, or categories.

aggregate travel model A method of estimating the total distances travelled in serving a market from a choice of locations. The model tries to find the point at which the distance covered is at a minimum, using the formula

$$A_i = \sum_{j=1}^{n} Q_j T_{ij}$$

where A_i is the aggregate distance travelled to serve the market from the factory site at i, Q_j is the expected volume of sales in the market j, and T_{ij} is either the distance between i and j or the transport cost between the two. Market size, Q, could be taken as proportional to per capita income, or to the volume of retail sales in the area.
  The formula is applied to a number of competing locations, and the results may be used to map a *cost surface—a contour map of the costs of serving a market.

aggressivity In *karst geomorphology, the ability of water to dissolve calcium carbonate. Water has this ability when atmospheric carbon dioxide is dissolved in it, forming a weak acid.

agrarian Describing an agricultural system which combines *horticulture with the rearing of animals.

agribusiness Large agricultural operations which are run like an industry. A single business can be concerned with the whole of agricultural output: the ownership of land, the agricultural process, the manufacture of agricultural machinery, the processing of the product, and its shipment. This is typical of agribusinesses in the USA; European equivalents are not generally as all-embracing. An agribusiness is characterized by very large production units, and considerable vertical and horizontal *integration. For example, a firm producing frozen vegetables sets up contracts with farmers and also owns the company which provides their contract labour and sells them fertilizer. Management tends to be by administrators and accountants rather than farmers because the farms may be only a minor part of the business.

agricultural density The density of the agricultural population per unit area of farmed land. It is, however, rather difficult to define and isolate the 'agricultural' population, as many rural inhabitants are engaged in essentially *urban occupations. *See* *adventitious population.

agricultural geography The study of spatial patterns in agricultural activity. Major themes include variations in agricultural activity within the main *biomes, the delimitation of agricultural regions, the study of agriculture as a system, and the classification of agricultural systems, usually with reference to the terms: intensive/extensive; commercial/subsistence; shifting/sedentary and pastoral/arable/mixed. Some agricultural geographers are concerned with the way in which agricultural systems change with levels of development.

Other major concerns are: increasing agricultural output by such means as *land reform, the *green revolution and *appropriate technology, the diffusion of agricultural innovation, and the impact of agriculture on *ecosystems.

Important theories of agricultural geography include *von Thünen's land use model, incorporating the concept of *economic rent, and the *Böserup theory of population increase.

agricultural location theory An attempt to explain the pattern of agricultural land use in terms of costs, distance, and prices.

One explanation is concerned with the effect of the city on rural land use. It predicts that circular zones of different agricultural production will surround the city. Zones may also occur beside a coastal strip, or along both sides of a transport artery. A pattern of agricultural land use zoning may occur from a unit as small as a single farm to one as large as a subcontinent. The classic theory of agricultural location is that of *von Thünen.

agricultural revolution A period of rapid change in agriculture, usually associated with increases in output. One such change took place during the Agricultural Revolution (although the timing and use of this term is disputed) in England from 1750 onwards, when the use of rotations, nitrogen-fixing crops, such as legumes and clover, and mixed farming with manuring caused output to rise. These changes were helped by major reforms in land tenure; large 'common' fields were divided and much common land was fenced during the enclosures of the eighteenth century.

Technology also improved, with new machinery for sowing, harvesting, and threshing.

The introduction in China in 1979 of the Responsibility System which gave individual households plots for cultivation, negotiated production contracts for each household, and encouraged specialization, is an example of a more modern agricultural revolution, as is the *green revolution.

agricultural system Any method of farming may be seen as part of an agricultural system. 'Inputs' include seeds, water, pesticides, herbicides, fertilizer, and livestock which are introduced to the 'plant': the buildings, machinery, and land. The 'output' is the produce of the farm.

agroforestry Any agricultural system which incorporates the planting or encouragement of trees on land where animals are raised or crops grown. Thus, agroforestry includes *slash-and-burn, or the use of shade trees.

The planting of trees may aid farmers since tree roots can bind soil and limit soil erosion, deep-rooted trees can tap new nutrient sources, leguminous trees can fix atmospheric nitrogen and improve soil fertility, leaf litter can add organic matter, and tree cover can moderate temperatures. In addition, trees may provide food, fodder, firewood, and timber.

agro-town A town of up to 20 000 people with agriculture as the main economic activity. As such they are not truly *urban settlements, although they may be the only significant settlements in the region. Agro-towns occur in parts of Mediterranean Europe and may find their counterparts in the Yoruba regions of West Africa.

aid The provision of resources from developed to less developed countries. This is usually from the Western democracies to the Third World and, more recently, to Eastern Europe. Aid may take the form of finance or credit, or other forms such as expertise, education and training, and advanced or *intermediate technology. **Bilateral aid** is aid from a donor to a recipient country, while **multilateral aid** is provided by a group of countries. **Emergency aid** is short-term aid, generally given as a response to disasters, while **structural aid** is given to promote long-term *development.

Ostensibly, the provision of aid is to encourage development but donors of **tied aid** are rewarded, perhaps by interest payments from the receiving nation, access to new markets, or by political allegiance. Furthermore, agencies such as the International Monetary Fund or the World Bank may impose *structural adjustment of the economy as a condition of receiving aid. The 'charities', or Non-Governmental Organizations (NGOs), also provide aid but with fewer conditions and more emphasis on development.

aiguille French for pyramidal peak or *horn. Its sharply pointed nature is often due to *nivation and *frost wedging. Perhaps the most famous example is Mont Blanc.

air frost See *frost.

air mass An area of the *atmosphere extending for hundreds of kilometres that, horizontally, has generally uniform properties, especially of temperature and *humidity, and with similar vertical variations of temperature and pressure throughout. Air masses obtain these attributes from their areas of origin, known as source regions, which confer uniform properties to the overlying air. They become less uniform with movement into different areas, and with the effects of *wind shear. Air masses are classified first by the source region: over land, **continental air mass, c**, over sea, **maritime air mass, m**; and secondly by the latitude of the source area: the arctic, **A**, the poles, **P**, and the tropics, **T**. These two combine to distinguish most air masses, such as **cA—continental arctic, mP—maritime polar, mT—maritime tropical**, and so on. Some include **equatorial, E**, and **monsoon, M**, in classifications. *See also* *secondary air masses.

air parcel An abstract volume of air. A simple forecast of the likely behaviour of an air parcel is made by assigning the relevant qualities to it such as humidity or temperature and then predicting its movements according to the laws of physics.

air pollution The presence in the earth's atmosphere of man-caused, or man-made, contaminants which may adversely affect property, or the lives of plants, animals, or humans. Common air pollutants include: carbon dioxide, carbon monoxide, lead, nitrogen oxides, ozone, smoke, and sulphur dioxide. In 1988 the EC adopted a directive to limit emissions of sulphur dioxide, nitrogen oxides and dust from power stations above a certain capacity, and the most rigorous environmental protection targets in the 1990s have been set by the six founding states of the EC, together with Denmark. *See* *ambient air standard. *See also* *pollution.

air pressure *See* *atmospheric pressure.

air–sea interaction The transfer of heat, momentum, solids, liquids, and gases between the atmosphere and the oceans. These transfers occur because the atmosphere and the oceans constitute a single system in terms of matter, forces, heat, and other forms of energy. Thus, small processes like wind stress and *evaporation are of essential importance in the *general circulation of the atmosphere and oceanic circulations. Similarly, larger processes, such as *planetary winds, affect oceanic circulation. Ocean temperatures are critical in determining *depression tracks.

Aitken nuclei *See* *nucleus.

alas In a *periglacial landscape, a large, steep-sided, flat-bottomed depression which can be several kilometres across. Lakes often occur in alases, which may then develop into *pingos. Alases are *thermokarst features, and, in Siberia, may take up more than one third of the landscape. It is suggested that they form in a sequence running from the formation of *baydjarakhs to *duyodas, to alas depressions.

Alaska current A warm *ocean current, an offshoot of the North Pacific current, washing the southern coasts of Alaska, and keeping ports such as Valdez ice-free.

albedo That proportion of solar radiation reflected from any surface, such as clouds (meagre from tenuous stratus cloud, but up to 80% from thick strato-cumulus), or bare rock. Characteristic values are:

	%
fresh snow	40–70
dry sand	35–45
wet sand	20–30
tarmac	5–10
grassland	10–20
coniferous forest	5–15
deciduous forest	10–20
crops	15–25

Lighter, whiter bodies have higher albedos than darker, blacker bodies. The total albedo of the earth is about 35%.

alcove A semi-circular, steep-sided cavity on a rock outcrop, caused by water erosion, especially spring-sapping.

Aleutian current A cold ocean current of the north Pacific.

Aleutian trench The ocean trench at the boundary of the North American and Pacific plates. The *subduction of the Pacific plate changes west–east from normal to oblique, so that this boundary becomes a *transform fault.

alfisol A soil order of the *US soil classification. Alfisols are young, acid soils, with a clay-rich B horizon, commonly occurring beneath deciduous forests. *See* *brown earth.

algae A large and diverse group of simple plants that contain chlorophyll and can therefore photosynthesize. Algae live in aquatic habitats or in moist regions inland.

algal bloom The dense spread of algae which results from changes in the chemistry and/or temperature of lake water. Blooms may be green or red, and are most common in spring or early summer, when primary production outstrips the growth of the consumer organisms. The addition of phosphorus as a result of pollution from fertilizers is an important factor in the growth of algal blooms. The bloom will disappear if the input of fresh phosphorus ceases. *See* *eutrophication.

algorithm A set of calculations used to solve a problem, a formula. In the context of computers, it is a step-by-step method of solving a problem, usually backed up by a mathematical proof.

*Social geographers often use a series of steps to construct a new variable from a set of other variables, for example, they might work out an index of housing quality by combining the variables of housing density, occupation density, and the existence in the house of all amenities such as central heating, sanitation, and so on.

Alguhas current A warm *ocean current off the coast of southern and south-eastern South Africa.

alienation The estrangement, or separation of individuals from one another, so that they do not identify with each other as a group. Thus, Karl Marx (1844, 1975) argued that it was employers who enjoyed the products of industry, rather than the workers who actually made the goods; under capitalism, workers become appendages of the machine they use rather than whole people.

The psychological state of alienation is said to include isolation, powerlessness, and meaninglessness, and, not surprisingly, has been linked with state-owned industries just as much, if not more than, with capitalism, since no identifiable person owns a state industry.

The concept of alienation has been used to explain the rise of urban problems such as violent football fans, street gangs, and alcohol and drug addiction.

alkali A substance which neutralizes, or is neutralized by, *acid. Alkalis are generally oxides, hydroxides, or compounds such as ammonia which dissolve in water to form hydroxide ions. Alkalis have a *pH of more than 7.

alkaline rock *See* *basic rock.

alkaline soil Any soil, such as a *rendzina, which has a *pH above 7. This alkalinity usually reflects a high concentration of carbonates, notably those of sodium and calcium.

Alleghanian orogeny An *orogeny extending from the Early Carboniferous to the late Permian, caused by a collision between North America and Africa, affecting an area from Alabama to Newfoundland, and resulting in the formation of the Appalachian Mountains.

Allen's rule J. A. Allen (1871) suggested that in many species, extremities such as ears and tails are larger in varieties living in a hot climate than in related types found in cold climates. This enables animals living in hot climates to dissipate excess heat through these extremities, while animals in cold climates minimize heat loss. Some writers contest the validity of this 'rule'.

Allerød A milder phase of less intense glaciation in north-west Europe, occurring during the last *glacial of the *Pleistocene and possibly lasting from about 12 000 BP to 11 000 BP, although the dates are disputed. Pollen records suggest that birch was widespread.

allochthonous An allochthonous rock was formed of materials brought from outside, for example from sediments transported by ice; *till is an example. *Compare with* *autochthonous.

allogenic Caused externally. Thus, an **allogenic stream** is one—for example, the R. Nile—fed from outside the local area, where precipitation and run-off are sufficient to generate flow. In ecology, **allogenic succession** is an alteration in a plant *succession brought about by some external factor, such as a volcanic eruption, rather than by the organisms themselves.

alluvial Referring to *fluvial contexts, processes and products, so that an **alluvial channel** is a river channel cut in *alluvium; it is self-forming, in that its form reflects the *load and discharge of the river rather than the constraints of a *bedrock.

An **alluvial fan** is fan-shaped landform composed of *alluvium dropped by a river after it loses momentum as it enters a broad valley from a narrow, upland course. The fan shape is the result of the river swinging back and forth over the *aggradation deposits while the apex is fixed at the point where the river emerges from the uplands. It is the terrestrial equivalent of a *delta.

An **alluvial cone** is similar in origin, but tends to have a steeper slope. Cones tend to grow from seasonal streams in semi-arid or arid areas, such as the western flank of the Sangre de Cristo range, Colorado. **Alluvial fill** would appear to be another term for *alluvium, and the process of *sedimentation from *fluvial channels may also be described as **alluvial filling**. This generally comes about in one of four ways:

■ when the *discharge decreases,
■ when the amount of *load increases,
■ when the width/depth ratio of the channel changes,
■ when a change in *base level lowers the gradient.

Alluvial flood plains are formed by rivers with marked variations in their seasonal flow. During periods of low water, most of the alluvial plain is dry, but when water levels are high the entire width of the flood plain may be covered. Coarse material is transported short distances during times of flood, temporarily deposited, and then picked up during the next flood.

alluvium A general term for all deposits laid down by present-day rivers, especially at times of flood. Alluvium is characterized by:

■ sorting, so that coarser alluvium is found in the upper course of rivers and finer in the lower courses
■ *stratification, such that coarse material in river *bars is overlain by finer material
■ structures such as *current bedding.

Alluvial landforms include *alluvial cones and fans, *deltas, river bed materials, and *flood plains. Alluvial deposits are fertile, and can contain minerals, such as the alluvial gold and diamonds of West Africa. There is a tendency to restrict the term alluvium to fine-grained deposits such as *silt or silty clays.

Alonso model An explanation of urban land use and land values, developed by William Alonso (1964). It is grounded on the concept of *bid rents whereby the urban land user seeks central locations, but is willing to accept a location further from the city centre if rents are lower in compensation. The use that can extract the greatest return from a site will be the successful bidder. To this basis, Alonso, in a study of housing, added the quantity of land required, and variations in the amount of disposable income used on land and transport costs on one hand, and on all goods and services on the other. If the amount of goods and services is held constant, the price of land should decrease with increasing distance from the centre. The well-off will choose to live at lower densities at the edge of the city; the poor remain in high density occupancy near the city centre. The quantity of land that may be bought should increase with distance from the centre, but commuting costs will rise with distance from the centre so that the quantity of wealth available for land will decrease. Each household represents a balance between land, goods, and accessibility to the workplace.

Alonso also explained that higher-income groups, who are less constrained than lower-income groups in their choice of residential location, may prefer the accessibility to the CBD offered by the inner city to the space, quiet, and cheaper land of the suburbs, so that *gentrification may result.

The assumptions on which this theory rests range from all land being of equal quality to lack of planning constraints. This means that the theory is a long way from reality, although it does reflect some aspects of urban morphology.

alp The shoulder of land above a *glacial trough. The alp runs from the break of slope above the trough to the summer snowline, and the term is also used in Switzerland to signify summer pasture there.

alpha index, α index In *network analysis, the ratio of the actual number of circuits in a network to the maximum possible number of circuits in that network. It is given as:

$$\alpha = \frac{e - v + p}{2v - 5} \times 100$$

where e = edges, v = vertices, and p = number of graphs or subgraphs. Values range from 0%—no circuits—to 100%—a completely interconnected network.

Alps A chain of fold mountains extending from eastern France and northern Italy through Switzerland to Austria. This chain has given its name to a number of geomorphological features: an **alpine glacier** is synonymous with a valley glacier, and **alpine topography** denotes those features of glaciation, such as *cirques, *glacial troughs, *hanging valleys, and *truncated spurs, found in upland areas. The term is also used for the Southern Alps of New Zealand.

The **alpine orogeny** was a mountain-building movement in the *Tertiary responsible for *fold mountains as far apart as the Andes, Japan, and the East Indian *island arcs; areas still characterized by earthquakes and volcanic activity.

altimeter An instrument used to plot altitude. The **pressure altimeter** is based on the change of pressure with height and the **radio-altimeter** on the time that radio waves take to 'bounce' back to a recording device on an aeroplane.

altimetric frequency curve A frequency curve which shows how often various heights of land occur within a given area. Such curves may be used to identify the remnants of *erosion surfaces.

altimetry The measurement of heights.

altiplanation The process of diminishing relief through *periglacial processes; chiefly by *gelifraction, *nivation, and the accumulation of debris. Note, however, that J. Tricart (trans. E. Watson, 1970) described altiplanation as 'a theoretical process seldom realized in nature'. A similar uncertainty is attached to the features known as **altiplanation terraces**; irregular, gently inclined benches thought to be formed by *nivation at the edge of snow-filled hollows under *periglacial conditions. They may be up to 1000 m long and 5–10 m high, some of waste, and some cut into solid rock. An **altiplanation surface** is a flat hilltop occurring in a periglaciated environment.

alto- Prefix referring to *clouds between 3000 and 6000 m high, as in **alto-stratus**, a grey sheet of medium cloud.

ambient Surrounding, of the surroundings. **Ambient temperature** is the temperature surrounding a given point. The **ambient air standard** is a quality standard commonly used in the USA for air in a particular place, as defined in terms of pollutants. Industrial discharges of pollutants must not cause the standard to be breached.

amenity Pleasantness; those aspects of an area such as housing, space, and recreational and leisure activities which make it an attractive place to live in. By this definition, a 1995 survey found that Henley-on-Thames had the highest amenity of any town in the UK.

amphidromic point The point around which *tides oscillate. Thus, while there are tides along the coasts of East Anglia and the Netherlands, there is a point in the sea between the two where there is no change in the height of the water; the tidal range increasing with distance from this point. High water rotates around the amphidromic point; anticlockwise in the Northern Hemisphere and clockwise in the Southern Hemisphere.

ana- Rising. *See* ana-*front, *anabatic wind.

anabatic wind An upslope wind caused by local heating. For example, when the sides of a mountain valley are heated more intensely, and more rapidly, than the valley floor, the warmer air will rise upslope, creating an anabatic wind, known as a *valley wind.

anabranching channel A *distributary channel which departs from the main channel, sometimes running parallel to it for several kilometres before rejoining it. Brice (quoted in H. W. Shen, 1981) specifies that islands in anabranching channel patterns should be three times wider than the river width at its maximum discharge. *See also* *drainage patterns.

anaerobic Describing any organism or process which can or must exist without free oxygen from the air, such as the anaerobic bacteria which are responsible for the process of *gleying.

analogue model A model which explains a phenomenon by reference to some other occurrence; for example, H. T. Odum likened an ecosystem to an electrical circuit. Energy input was shown as batteries, food flows as electric currents, and energy dissipation by amperage and voltage chambers.

anastomosis The division of a river into two or more channels with large, stable islands between the channels. An **anastomosing channel** differs from an *anabranching channel in that these fugitive channels have distributaries of their own. See also *drainage patterns. For **anastomotic drainage,** *see* *drainage patterns.

anchor tenant In a new shopping centre, shops, such as nationally known chain stores, which will attract many customers, are encouraged and may even be granted lower rents. These are the anchor tenants which provide customers for smaller shops.

ancillary linkage A link, which may benefit both, between two different forms of land use. Thus, building societies are increasingly located in high streets. The presence of the shops attracts clients to the building societies, while the provision of cash points encourages shoppers.

andesite A fine-grained volcanic rock, taking its name from the Andes Mts. The **andesite line** is, essentially, the boundary between the *basic rocks of the oceanic *crust and islands and the *acid rocks of the continental crust in a belt surrounding the Pacific. It is the boundary between oceanic *sima and continental *sial.

anemometer A device for recording wind speed. A **vane anemometer** consists of three semi-conical cups mounted on a vertical spindle and attached to a generator. The cups are driven round by the wind; the faster they move, the greater the generator output, which may be calibrated directly onto a continuous paper trace. The use of a wind-vane keeps a **pressure-tube anemometer,** or **pitot tube anemometer** facing the wind. The wind pressure is transmitted down a tube and is then converted into a print-out.

angle of dilation In studies of *sedimentation, the change in the orientation of grains on a *shearing surface. This angle is related to the angle of internal shearing resistance; that is, the angle measured in a *shear box to give the friction angle of the Mohr–Coulomb equation, and to the angle of plane sliding friction. This latter is the angle at which loose particles will begin to slide down a surface; that is, the static angle.

angle of repose In studies of *sedimentation, the angle at which granular material comes to rest. The angle of repose of sand, for example, is between 30° and 35°.

angular momentum *See* *conservation of angular momentum.

annular drainage *See* *drainage patterns.

anomie The lack of traditional social patterns within a group; a lack of norms in a society leading to conflict and confusion. The term is associated with the work of Émile Durkheim, who associated anomie with rapid economic change.

L. Wirth claimed it to be a consequence of *urbanism, where life can be led under conditions of anonymity and where, he claimed, social relations are transitory. Individuals become alienated from their 'folk' backgrounds, no longer feeling part of a group; social norms are so muddled and weak that people, unclear or unhappy about them, tend to challenge or ignore them. This would imply, therefore, that urban populations are more disaffected and lawless than rural populations, but this theory has been contested; *see* *Wirthian.

Antarctic Denoting regions south of the Antarctic circle, 66° 32' S (often taken as $66\frac{1}{2}°$ S). The major land mass within this zone is **Antarctica**. Within the **Antarctic circle**, the sun does not rise on 21 June (winter *solstice in the Southern Hemisphere) or set on 22 December (summer solstice in that hemisphere).

Antarctic air mass These *air masses originate within the *Antarctic and have similar properties to *Arctic air masses.

Antarctic meteorology Winters are severe, with characteristic double temperature minima, i.e. two separate occasions of minimum temperature, due to the absence of *insolation for several winter months, and to the frequent exchange of air with that of lower latitudes. *Blizzards are common. Temperatures rise in late summer as the long waves of the *westerlies bring incursions of warmer air. Nevertheless, precipitation is still almost always in the form of snow, since maximum temperatures, occurring at the summer (December) *solstice rarely exceed 0 °C.

Antarctic plate A major lithospheric *plate, almost entirely bounded by *constructive plate margins.

antecedent Prior to, before, as in **antecedent** *drainage patterns. For example, the creation of the incised valley of the Wind River, Wyoming, may be due to the river being older than the dome it cuts through, so that river erosion kept pace with the uplift of the Owl Creek Range. **Antecedent moisture** is the amount of moisture already present in the soil before a specified rainstorm.

anthracite A hard, compact type of coal containing over 85% carbon, burning smokelessly and slowly, generating much heat.

anthropogenic Brought about by human agency; *see* *anthropogeomorphology.

anthropogeomorphology The study of the human effects on the physical landscape. This effect may be direct, through construction, excavation, or hydrological interference (damming, dredging, canal construction, and so on), or indirect, when human agency has unlooked-for consequences. Examples of the latter include accelerated *soil erosion, the generation of *earthquakes through the construction of reservoirs, subsidence caused by mining, or through the melting of *thermo-karst, or slope failures caused by undercutting.

anticline *See* *fold.

anticlinorium *See* *fold.

anticyclone A region of relatively high *atmospheric pressure frequently thousands of kilometres in diameter, often formed as response to *convergence in the upper *atmosphere, and also known as a high. The name comes from the circulatory flow of air within the system; **anticyclonic circulation** has a local circulation opposed to the earth's rotation, that is, clockwise in the Northern Hemisphere, and anticlockwise in the Southern Hemisphere. As air near ground level flows into an anticyclone, its absolute *vorticity decreases, and it is therefore subject to *divergence; this infers the descent of air. Anticyclones appear on weather charts as a series of concentric, widely spaced isobars of 1000 mbs and above.

Cold anticyclones, also known as continental highs, form when the interiors of continental land masses lose heat in winter through *terrestrial radiation, cooling the air above by contact to form shallow highs. Such systems are semi-permanent over Siberia and north-west Canada in winter, and are marked by subsidence which inhibits cloud formation, maximizing radiative cooling, and thus making the anticyclones self-sustaining.

Subtropical anticyclones are warm anticyclones, which form due to subsidence below the convergence associated with the westerly sub-polar *jet stream at its poleward limit—the northern limit of the *Hadley cell circulation. The descending air does not sink to ground level, but spreads over a cooler surface layer to form an *inversion. These anticyclones are at their strongest at 32° N and S. Three permanent subtropical anticyclones exist in the Northern Hemisphere; the North Atlantic, North Pacific, and North African *highs, and three in the Southern Hemisphere; the South Atlantic, South Pacific, and South Indian Ocean highs. In each hemisphere the highs are separated by low *cols, which are important for *meridional movements in the atmosphere. Subtropical anticyclones are responsible for stable atmospheric conditions, and thus, fine, hot, dry weather.

In *mid-latitudes anticyclones are often located beneath the leading edge of ridges in the upper-air westerlies, where they may be associated with *blocking weather patterns. When the warm Azores high moves north-east to the British Isles in summer, the weather is unusually fine. This is due to the compression of the air as it descends, causing *adiabatic warming. Winter anticyclones may bring cold, frosty weather, or fog. *See* *anticyclonic gloom.

anticyclonic gloom In winter, in north-west Europe, fog, or poor visibility caused by broad, persistent sheets of strato-cumulus cloud at the base of an *inversion trapping polluted air below. The inversion results from the arrival of a cold *anticyclone.

antidune A ripple on the bed of a *fluvial channel which travels upstream.

anti-natalist Concerned with limiting population growth. The People's Republic of China has pursued anti-natalist policies, notably the 'one-child' strategy, for over a decade. While anti-natalist government policies may be instrumental in lowering birth rate, state coercion may have unexpected and damaging results; reports in 1995 suggested that abortion of female children had become common in China, so that male:female sex ratios at birth had become grotesquely imbalanced (rising, in one province to 140:100), and so that female infanticide and the abandonment of girl babies is common. Research on population growth suggests that improving opportunities for women is the best contraceptive. *See* *birth rate, *demographic transition model.

antipodal bulge The tidal effect on the earth's surface at the point where the lunar attraction is weakest.

antitrades Westerly winds in the upper atmosphere above, and in contrast to, the easterly *trade winds at ground level.

anvil shape In meteorology, the flattened, frozen, upper part of a mature *cumulo-nimbus cloud, which is much wider than the base.

anyport A model of port development suggested by Bird. Initially, a primitive port grows around a natural, sheltered harbour. As the port grows, marginal quay expansion occurs and the port's quays are of greater size in relation to the town. The next step—marginal quay elaboration—sees the extension of jetties and the cutting of docks. After this, as the port continues to grow, there is dock elaboration as new docks are excavated downstream. The next stage—simple linear quayage—sees the rationalization of existing quays and docks to provide better facilities. Finally, specialized quayage develops with facilities for *containers and for *Ro-Ro traffic. As these seven stages occur, the activity of the port moves seawards.

apartheid The system of racial segregation first promulgated by the largely Afrikaner National Party of South Africa in 1948. **Petty apartheid** meant the separation of facilities such as lavatories, transport, parks, and theatres into two groups: white and non-white. On a much larger scale was the allocation of 12% of the land area into 'independent republics', or 'homelands', for the African population, which comprised 69% of the population when the policy began, in 1954. The long-term aim of this policy was to restrict Africans to these 'homelands', which were to be governed and developed separately from white South Africa, while allowing African workers strictly limited rights to live in the white areas, as and when their labour was required.

With the election of South Africa's first democratic government in 1994, the last vestiges of apartheid were officially removed, but the policy will have left its mark on the South African landscape and its society for many years to come; for example, under the Nationalist Party regime, housing was zoned for racial groups so that predominantly African, Coloured, Asian, and White areas still exist, as do the homelands, to which every African was assigned, according to each one's major tribal group.

aphelion The point furthest from the sun of any body orbiting the sun. On 4 July the earth is at aphelion, 152 000 000 km from the sun. The dates of aphelion and *perihelion advance at about 30 minutes per calendar year. *See* *Milankovitch cycles.

aphotic zone In any watery environment, the deeper zone, which is not penetrated by light.

applied climatology The systematic study of climatology for an operational purpose, such as agriculture.

applied geomorphology The study of the interactions between geomorphology and human activity. Thus, applied geomorphology covers the following:
  1. The specialized mapping of landforms, such as slope elements, which affect or may be affected by human activity, and which are not mapped by other disciplines;
  2. The interpretation of features shown on aerial photographs or by remote sensing methods;
  3. The monitoring of changes in the environment, especially when those changes bring risks to society;
  4. The assessment of the causes of these changes, notably of those which develop as hazards to man;
  5. The remedies to such hazards;
  6. The recognition of the consequences of human activity in geomorphology.

applied meteorology The application of meteorological data to specific down-to-earth problems.

appraisive image The meaning invoked by a particular place, usually with regard to its attractions. A distinction is made between an affective, emotional response to an image and an evaluative response, where an objective judgement is made between places. Put more simply, an individual may prefer the scenery of Snowdonia to that of the South Downs because the former is, subjectively, more 'awe-inspiring', or because, objectively, it displays more marked breaks of slope and higher relief.

appropriate technology A technology which uses simple techniques and large amounts of labour; that is, a labour-intensive, low technology, capital-saving method of working which, it is argued, is more suited to relatively impoverished, less developed countries suffering from high unemployment and under-employment. Thus, a contrast may be drawn between a high-technology, capital-intensive irrigation project, such as the South Chad Irrigation Scheme, and the use of hand-built, low stone walls (diguettes) in the Sahel as a means of improving agricultural water supply. However, governments of LDCs often prefer capital-intensive projects because, although they provide less immediate employment, they promise higher rates of growth.

aquaculture The use of waters, other than the sea, for agricultural production, usually the production of fish. *See also* *fish farming.

aquiclude A rock, such as London clay, which does not allow the passage of water through it; an impermeable rock. Such a rock will act as a boundary to an *aquifer.

aquifer A rock, such as chalk, which will hold water and let it through. A **confined aquifer** is one sandwiched between two *impermeable rocks (see *aquiclude), and an **unconfined aquifer** is one where the *water-table marks its upper limit. A **perched aquifer** is an unconfined aquifer upheld by a small *aquiclude. Water runs into aquifers where the rock is exposed to the surface or lies below the water-table.

Arabian plate A minor lithospheric *plate, currently colliding with the Iran plate. Its boundary with the Indo-Australian plate forms the Owen fracture zone.

arable land Originally meaning fit for cultivation, as opposed to pasture or woodland, the term is now applied to agricultural land used for growing crops.

arch In coastal geomorphology, an arch is made when two caves occurring on either side of a headland are cut until they meet. Durdle Door, Dorset, is a British example, and arches are common on the coast of the French Pays de Caux. Arches are relatively temporary features of the landscape. Roof falls cut off the seaward end of the arch, which is then left as a *stack.

archipelago Originally meaning an island-studded sea, such as the Aegean Sea, but now referring to a group of islands, such as the Bismarck Archipelago, to the east of Papua New Guinea.

Arctic Denoting regions within the Arctic Circle, i.e. north of 66° 32′ N (often taken as $66\frac{1}{2}$° N). Within these regions the sun does not set on June 21 (the summer *solstice in the Northern Hemisphere) nor rise on December 22 (the winter solstice in the Northern Hemisphere). In climatology the Arctic is defined in terms of the treeless zone of *tundra and of the regions of *permafrost in the Northern Hemisphere.

Arctic air masses are exceedingly cold, with the Arctic Ocean as their source region. Such air masses should not be confused with polar *air masses.

Arctic meteorology The Arctic regions experience an annual cycle of winter 'night' and summer 'day'. In winter, highs and lows traverse the area, but most have little effect on surface weather except for *cold lows which cause medium- and high-level clouds. These partially offset radiational cooling. Most weather results from the intensely cold ground air which is chilled by contact with land losing heat from strong *terrestrial radiation, since winter clouds are otherwise scarce. Only infrequently do depressions penetrate the *inversions so formed. Winter temperatures are close to −40 °C. Although snowfall is slight, winds cause frequent blizzards and drifting. In spring, days are longer and sunny but temperatures remain low because incoming *solar radiation is reflected back into the atmosphere from the snow surface. In summer, some depressions bring thicker cloud and light rain. The snow- and ice-melt in June and July keep air temperatures over the pack ice close to 0 °C. Skies are usually overcast over coastal areas, but by late afternoon temperatures in inland areas may rise to 15–20 °C.

Arctic sea smoke A form of *steam fog common in coastal seas around cold land masses such as Labrador and Norway. As very cold air from the land passes over the warmer sea, it is rapidly heated. Convection currents carry moisture upwards, which quickly recondenses to form fog.

area One of the basic terms of reference of spatial analysis; the other two are *point and *line. It may be defined as the extent of a surface or expanse of land, now usually measured in metric square units.

Area of Outstanding Natural Beauty, AONB An area, supervised by the relevant local authority, in which development is very carefully considered so that the beauty of the landscape is not diminished. 6% of England and Wales is covered by AONBs.

areal differentiation A recognition of the different regions of the earth's surface, the study of the way human and physical phenomena vary over the earth's surface which is aimed at interpreting the variations in the character of the settled world. As such it may be seen as *chorology, or traditional regional geography. As *systematic geography developed, using the *quantitative approach, many geographers abandoned regional geography from the mid-1960s onwards.

In the 1980s, areal differentiation was again seen as a central theme in *geography. The major topics in this field emphasized by *humanistic geographers include the social construction of space, the sense of place, and the *iconography of landscape, while other geographers, notably *Marxist geographers, are concerned with those spatial variations in the quality of life and in economic activity which are referred to as *uneven development. Another approach—that of *contextual geography—considers the interaction of place and human agency in producing geographical sameness and difference.

*Systematic themes have by no means been abandoned, but it is generally recognized that every process will be modified according to the unique nature of each separate environment.

areic Without surface drainage, that is, without streams or rivers. Areas of permeable rocks, such as limestone, often lack surface drainage.

arenaceous Sandy in texture, or applied to rocks composed of cemented, usually quartz, sand. Arenaceous rocks include quartzites and greywackes.

areography The study of the spatial distribution of plant and animal taxa.

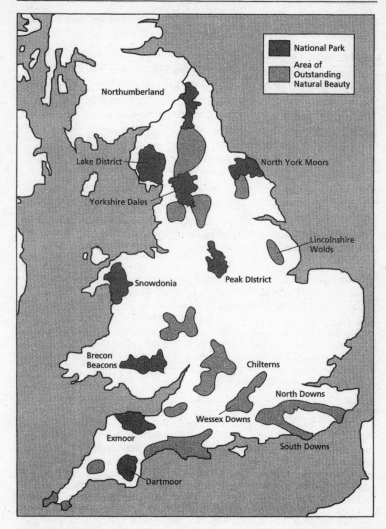

FIGURE 4: *AONB*

arête A steep knife-edge ridge between *corries or *glacial troughs in a glacially eroded, mountainous region. Striding Edge, Helvellyn, Cumberland, is a British example of an arête, but arêtes are common in any landscape of glacial erosion.

Arêtes seem to have formed by the postulated backward extension of the corries into the mountain mass. Others are moulded by *nivation and *frost wedging where mountains protrude through glaciers. *See* nunatak. Arêtes may be embellished with *gendarmes.

argillaceous Clay-like in composition and texture, referring to rocks containing *clay minerals and clay-sized particles, for example, shale.

aridisol A soil order of the *US soil classification, found in arid environments. It is a *desert soil, predominantly composed of minerals, and often high in accumulations of water-soluble salts.

aridity The degree to which a climate lacks moisture; in *meteorology, the reverse of *humidity. C. W. Thornthwaite's **aridity index** is based on the relationship between precipitation and potential evapotranspiration, and indicates the degree of water deficiency:

$$\text{aridity index} = \frac{100d}{n}$$

where d is the sum of monthly differences between precipitation and *potential evapotranspiration when the former is less than the latter, and n is the sum of monthly values of evapotranspiration for the deficient months.

arithmetic mean A numerical value representing the average worth of a set of data. It is calculated by adding together all the values of a set and dividing the total by the number of values. It is quick and easy to calculate, but the presence of an unusually high or low value distorts the mean from a truly central value. Furthermore, it is unreliable when the data set contains only a few values. The mean of *grouped data may be established by multiplying the mid-point of each class by the number of observations in each class, summing these figures, and dividing the sum by the total number of values in the data set.

arkose An *arenaceous rock composed of quartz and more than 25% feldspar.

armouring The concentration of grains on a river bed of sufficient coarseness to protect the finer material below it from erosion and thus stabilize the bed. An **armour layer** is characteristic of an equilibrium channel in a heterogeneous sediment, and, although migratory during floods, will re-form when normal flow is re-established. The layer is generally well sorted and one or two grains in thickness.

arroyo A *gully which is rectangular, rather than V-shaped, in cross-section. Arroyos occur in arid environments and result from the entrenchment of stream channels caused when run-off increases, either through climatic or land use changes.

artesian basin A *syncline of permeable rocks with outcrops at the crest of the syncline. Water from rain or streams seeps into this *aquifer, moving towards areas of lower *hydraulic head. Eventually the rock becomes saturated and the water is under pressure. If a borehole is sunk at depth to tap the water, an artesian well forms from which the water will initially flow upwards without pumping. The term comes from the basin of Artois, in north-west France, and the most famous British example is the London basin where the *aquifers are chalk and Lower Eocene sandstones and the *aquiclude is London Clay.

artificial rain *See* *cloud seeding.

ash *See* *pyroclast.

ash cone *See* *scoria cone.

ash fall During and after a volcanic eruption, volcanic ash may be carried hundreds of kilometres, depending on prevailing wind conditions. After the 1980 eruption of Mt. St Helens, USA, ashfall covered an area of some 20 km^2. Ash may sterilize soils, bury crops, and cause roofs to collapse under its weight.

ash flow *See* *pyroclastic flow.

Asiatic mode of production A *mode of production characterized by an absence of three elements: of private property, of *urbanization, and of a bourgeoisie. Karl Marx (Hindess and Hirst, 1975) believed that all these were absent from Asian societies, which were characterized by 'communal appropriation', and that these absences explained why *capitalism arose in Western Europe, and not in Asia. These ideas have been strongly criticized as being not only Eurocentric but ill-informed, since they represent an image of Asia as being 'unchanging'; an image which is backed up neither by its archaeology nor its history.

aspect The direction in which a valley side or slope faces. In deeply cut east–west orientated valleys, the slopes facing the equator receive more sun and are more attractive to settlement than the shaded sides of the valley. *See* *adret and *ubac.
    Aspect may be an important factor in the formation of landforms, since slopes facing away from the equator may be 6 °C colder than their opposites; C. B. Beaty (*Jour. Geol.*, 1962) estimated that gradational processes are 2–3 times as active on northward-facing slopes in the Northern Hemisphere. *See* *asymmetry. Thus, in the Jotunheim of Norway, for example, 70% of corries lie on the north side of the massif.

asphalt A naturally occurring, semi-solid, bituminous rock found, for example, in Trinidad. It may also be manufactured from crude oil.

assart In medieval Britain, the taking into cultivation by individual farmers of previously waste, often forest, land.

assimilation Also known as *acculturation, or integration, this is the integration of an immigrant, outsider, or subordinate group into the dominant, host community. Initially the migrant group is segregated from the host culture but then there is often a blurring of cultural lines, and the concept of assimilation does imply that the minority group eventually takes on the values of the host, or *charter group. This contrasts with the view that new groups can affect the values of the dominant group, or live alongside it in a *multicultural society.

The rate of assimilation depends on the race, religion, customs, occupations, and cultures of the migrants and the dominant group. **Behavioural assimilation** is the absorption of the incoming group into the host community, as the newcomers absorb the culture and history of the *charter group, while **structural assimilation** is concerned with the way incomers are distributed throughout society, in occupational and social groupings. Urban geographers note that the degree of *residential segregation is inversely correlated with the level of assimilation of urban immigrants.

assisted area A part of the UK where government intervention is thought to be necessary to boost economic development or at least to halt its decline. *See* *development area.

associated number Also known as the König number, the associated number of a *node is the number of *edges from that node to the furthest node from it. This is a *topological measure of distance, in edges rather than in kilometres. A low associated number indicates a high degree of *connectivity. The measurement may be used to measure the *accessibility of a transport network.

association, plant association A *plant community unit. The term has been variously used, ranging from a large-scale area of *climax vegetation to a plant community. One widely used definition of an association is a floral assemblage with a characteristic dominant and persistent species, although characteristic combinations of species may be used.

association coefficient Synonym for *correlation coefficient.

asthenosphere That zone of the earth's *mantle which lies beneath the relatively rigid *lithosphere, between 50 and 300 km below the surface. The asthenosphere is composed of hot, semi-molten, and therefore deformable, rock, within which convection currents occur. Descending convective limbs can penetrate to depths of 700 km, and rising limbs are located under spreading centres (mid-oceanic ridges).

The asthenosphere is approximately commensurate with that zone of the mantle which transmits seismic waves at low velocity.

asylum migration The international movement of *refugees and persons who, while having suffered generalized repression, violence, and poverty, do not qualify as *refugees under the strict requirements of the 1976 UN protocol. Currently, the only option for non-EU nationals unable to use the family reunion scheme, but wishing to migrate to the EU is to claim asylum. The number of people claiming asylum in Western Europe grew from under 100 000 a year in the early 1980s to around half a million a year in the early 1990s. The United Nations High Commissioner for refugees estimates that half the claims for political asylum in Europe are fraudulent and the pressure of increasing numbers of false claims has forced receiving governments into a more hardhearted attitude; only 4% of the claims made in Germany in 1990 were accepted as genuine, for example.

asymmetry Having a structure which cannot be divided into two balancing parts; lacking symmetry. An **asymmetrical fold** is one in which one side of the fold dips more steeply than the other; an **asymmetrical valley** has one side steeper than the other. Such valleys are common in present or past *periglacial environments, such as the south German Alpine foreland, where aspect plays a major part in the *frost-based processes which are responsible for modelling the valley sides, since aspect can be critical in determining where frost forms. In the Kenai peninsula, south-facing slopes are consistently steeper than those which face north, where winter snows last until June.

Other asymmetrical valleys reflect the underlying geological structure; the valley of the lower Conway, Clwyd, is an example.

Atlantic Period In north-west Europe, the period from around 7500 to 5000 BP, of *oceanic climate, when temperatures were warmer than at present.

Atlantic-type coast A coastline where the trend of ridges and valleys runs transverse to the coast. If the coastal lowlands are inundated by the sea, a *ria or *fiord coastline may result. The coast of south-west Eire is an example. *Compare with* *Pacific-type coast.

atmometer A device for measuring *evaporation rates.

atmosphere 1. The layer of air surrounding the Earth, with an average composition, by volume, of 79% nitrogen, 20% oxygen, 0.03% carbon dioxide, and traces of *rare gases. This surprisingly uniform composition is achieved by *convection in the *turbosphere and by *diffusion above it, especially above 100 km, where diffusion is rapid in the thin atmosphere, and stirring is weak. Also present are atmospheric moisture, ammonia, ozone, and salts and solid particles. The atmosphere is commonly divided into the *troposphere, the *stratosphere, and the *ionosphere.

Since the troposphere contains the majority of the atmospheric mass, and virtually all of the atmospheric water vapour, most *weather events occur within it.

2. A unit of air pressure; one atmosphere is equal to the pressure exerted by the weight of a column of 760 mm of mercury at 0 °C, under standard gravity, at sea level.

atmospheric boundary layer *See* *planetary boundary layer.

atmospheric cells Air may move, with a circular motion, northwards or southwards in a vertical cell, such as the *Hadley cell which extends roughly from the equator to 30° N. This cell was thought to be the result of *convection, and is hence known as a thermally direct cell, but its origin may be more complex.

The Ferris cell is an indirect cell, driven by the Hadley and Polar cells.

Atmospheric cells are major components in the transfer of heat and momentum in the atmosphere from the equator to the poles. In the 1950s, the existence of horizontal cells, also fulfilling this role, was established.

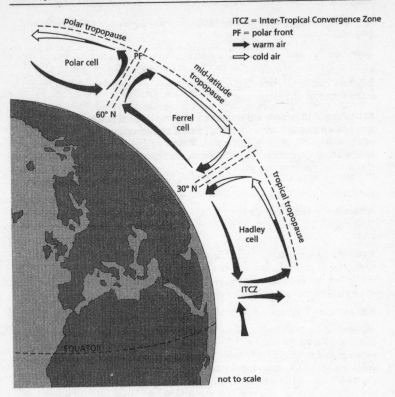

FIGURE 5: *Atmospheric cell*

atmospheric heat engine The system of energy which drives and controls the nature of the pressure, winds, and climatic belts of the earth's surface. N. Shaw described the weather as 'a series of incidents in the working of a . . . natural engine'. In the atmospheric heat engine, the heat source is *solar radiation, and the heat sink is *terrestrial radiation. Mechanical energy is expended by the 'engine' in the form of atmospheric processes. *See* *general circulation of the atmosphere.

atmospheric moisture Water, in liquid or gaseous form, present in the atmosphere.

atmospheric pressure The pressure exerted by the *atmosphere as a result of gravitational attraction exerted on the column of air lying above a particular point. Atmospheric pressure, measured in *millibars, decreases logarithmically with height.

atoll A *coral reef, ring or horseshoe-shaped, enclosing a tropical *lagoon.

The largest is Kwajalein, in the Marshall Islands, which is 120 km across. Most of the world's atolls are found in the Indian and Pacific Oceans, and all are sensitive to fluctuations in relative sea level. A fall in sea level leaves atolls high and dry some metres above the water line, and unable to grow; and a rise can shade out the coral, since the supply of sunlight usually becomes inadequate at a depth of about 45 m. An **atollon** is a small atoll on the margin of a larger one.

atomism A philosophical opinion which reduces knowledge to its smallest elements, such as human beings, and thus does not recognize larger configurations, such as social structures and social institutions. This view would run counter to the concept of social geography.

atomistic economy *See* *segmented economy.

Atterberg limits Limits used to classify soils. The **liquid limit** is the minimum moisture content at which the soil can flow under its own weight; the **plastic limit** is the minimum moisture content at which the soil can be rolled into a 3 mm diameter thread without breaking; the **shrinkage limit** is the moisture content at which further water loss will not cause further shrinkage.

attrition In geomorphology, the wearing away or fragmentation of particles of debris by contact with other such particles, as with river pebbles.

aureole *See* *metamorphic aureole.

Aurora Borealis Coloured and white flashing lights in the atmosphere north of the Arctic Circle. The lights are the result of the ionization of atmospheric molecules, at low temperatures, by solar and cosmic radiation. This phenomenon also occurs in the *Antarctic, where it is termed Aurora Australis.

autecology The ecology of particular species and individual organisms. It refers particularly to the relationships between species and their environments and within species, especially to the way organisms act within communities.

authority constraint In *time-space geography, a limit to an individual's actions because some activities are available only at certain authorized times. Thus, many parents are able to work during the limited opening hours of a day nursery, for example.

autochthonous Referring to features and processes occurring within, rather than outside, an environment. An **autochthonous rock** has been formed *in situ*; coal is an example.

autogenic succession In ecology, a change in the environment brought about by living organisms, particularly by plant life. The environmental change then brings about changes in the plant *succession. Thus, in a psammosere succession, the *pioneer species may provide enough organic material to improve the water-holding capacity of the soil, and thus pave the way for a less xerophytic species, which may then out-compete the pioneer. This compares with an *allogenic succession, although the two types are not always clearly distinguishable.

autonomous activity *See* *basic activity.

autotrophe An organism which uses light energy to synthesize sugars and proteins from inorganic substances. Green plants are by far the most common autotrophes.

avalanche A rapidly descending mass, usually of snow, down a mountainside. **Powder avalanches** consist of a moving amorphous mass of snow. **Slab avalanches** occur when a large block of snow moves down a slope and can cut a swathe through the soil and sometimes erode the bedrock if the snow is wet.
    The impact of such avalanches on humans is growing in most developed countries because of the increasing recreational use of alpine areas; avalanche-related deaths rose in the USA, for example, from 12 in 1961/2 to 24 in 1981/2.
    Avalanches of other substances are forms of *mass movement, and are distinguished by the type of material involved, e.g. **debris avalanche, rock avalanche**. The latter occur when *jointing in rock persists until the rock loses internal cohesion; until some sections are, effectively, masonry blocks, held together only by the friction between them. If this frictional force is lessened through water seepage or weathering, or if lateral support is removed, failure will occur, sometimes on a massive scale. *See also* landslide.
    An avalanche may also be triggered off by its own weight, by undercutting at the foot of the slope, by the pressure exerted by water in the pores of snow or debris, or by earthquakes. Preventative measures include the planting of trees, the erection of fences and splitter wedges, and the close monitoring of avalanche-prone slopes so that human use is banned at times of high risk.

avalanche wind The rush of air formed in front of an *avalanche. Its most destructive form, the avalanche blast, occurs when an avalanche stops abruptly.

aven A vertical upward opening linking a cave with the surface, or with another cave or passage.

average A measure of *central tendency; the most representative value for a group of numbers. The term is usually synonymous with the *arithmetic mean.

avulsion The slicing off of a meander, so that the river course becomes straighter, and an *ox-bow lake is left behind.

awareness space Any locations known of by an individual before a decision about such places is made. For example, an industrialist will choose to locate in a site of which he has previously been aware rather than in somewhere he has no knowledge of.

azimuth Most commonly, the length in degrees of the arc of the horizon between a given point and true north, measured clockwise; a horizontal direction measured in degrees.

azonal Referring to a soil without *soil horizons, such as a young soil developing on a bare rock surface. *Alluvium and *sand dunes are examples of azonal soils.

B

background level The naturally occurring level of pollution or radiation in the environment, against which the human input can be measured.

backhaul rate A cheap transport rate offered to a customer using a service which would otherwise be unused. For example, empty iron-ore wagons returning to the point of supply may be used to haul other goods like coal at a discount rate since otherwise they would return empty or containing only ballast.

backing Of winds in the Northern Hemisphere, a change in direction in an anti-clockwise movement, e.g. from westerly to southerly. The converse applies in the Southern Hemisphere.

backshore That part of a beach found between the ordinary limit of high tides and the point reached by the very highest tides. On cliffed coastlines, the backshore is the section of cliff foot and *shore platform affected only by storm tides; on low, shingle coasts it can take the form of a *berm; on sandy coasts, *foredunes may develop in this zone.

backswamp On a *flood plain, a marshy area where floodwater may be confined between the river *levées and higher ground. When artificially drained, backswamp areas yield fine-grained soils, often rich in organic matter.

backwall The curved, steep head wall of a landslip scar or a *cirque. In resistant, glaciated rock, corrie backwalls may be imposing; the backwall behind Llyn Llydaw, Snowdonia, is 300 m high.

backwash The return flow of water downslope to the sea after the *swash of a wave has moved upshore. This flow is an important factor in determining the gradient of the beach; because coarse beach matter reduces the volume of the backwash through percolation, a steeper slope is needed to produce the more rapid flow required if the balance between swash and backwash is to be maintained. Steep waves and long waves are linked with stronger backwash, and hence flatter beach gradients.

When the path taken by the retreating backwash differs from the path taken by the incoming swash, the movement of sediment known as *longshore drift will occur.

backwash effect Gunnar Myrdal (1957) argued that economic growth in one area adversely affects the prosperity of another. Wealth and labour moves from poorer, peripheral areas to more central regions of economic growth and the industrial production of wealthy regions may well undercut the industrial output of the poorer regions. This draining of wealth and labour together with industrial decline is the backwash, or polarization effect, and is a feature of *core–periphery relationships. *See* *spread effect.

backwearing The erosion of a slope whereby the slope maintains a constant angle as it retreats; that is, *parallel slope retreat. Compare with *downwearing.

bacteria Single-cell organisms. Their importance in geography lies in their role in the formation and development of soils. *See* *gley soils and *polysaccharide gums.

badlands Usually arid lands, generally bare of vegetation, which have been cut into to form a maze of ravines and sharp-crested hills. This dissection is aided by the lack of vegetation, high *run-off and heavy sedimentation which increases *abrasion, and *drainage densities may develop which are ten times greater than those under humid climates. Non-fluvial processes, such as *mass wasting, *piping, and tunnel erosion are also important. Badland topography develops best in weakly-consolidated sediments. *See* *donga. The term comes from the Bad Lands of South Dakota but is now applied generally.

bajada (bahada) A series of *alluvial fans which have coalesced along the foot of the mountains to form a gently sloping plain of unconsolidated sediments. One strip of coalescing fans in the San Sanquin valley of California is some 20 km wide, with gradients from 1:35 to 1:530.

balance of payments A comparison between the payments made by one country to other nations of the world and the revenue it receives from them. If receipts exceed outgoings, the balance is positive. The **capital account** records payments made in settlement of old debts or establishment of new ones; the **current account** shows payments made on goods and services, including interest payments. The **balance of trade** is a similar record, but registers only visible exports and imports.

balanced growth A strategy of growth with an equal emphasis on agriculture and industry. Agricultural development provides the food required and releases labour from the land to engage in industry. Industrial wealth stimulates markets for agricultural growth—or such is the theory. Unbalanced growth denotes a strategy which focuses on agriculture or industry alone.

balanced neighbourhood A neighbourhood which contains groups from all levels of society. Such a neighbourhood does not usually occur spontaneously and has to be planned, usually in order to counteract the processes of social *segregation. It is hoped that a balanced neighbourhood will benefit 'lower' groups when they come into contact with 'higher' groups who will develop powers of leadership, and it is argued that the creation of a balanced neighbourhood will stabilize an area and put an end to the processes of *invasion and succession.

Planning a balanced neighbourhood has been criticized; it is argued that there will be no sense of community when the members come from disparate backgrounds, and that conflict between unequals is more likely to be the result. Furthermore, balanced neighbourhoods would be unlikely to survive in a free housing market. A more fundamental criticism would be that it is based on an outmoded view of society.

bank erosion The erosion of material from the side of a river channel, not only by fluvial processes, but also by frost heave, groundwater sapping, and slope failure. Rates of erosion vary with bank composition and moisture content, bank vegetation, and speed of flow; rates are highest on the outer bank of meander bends or where bars in the channel have diverted the *thalweg.

bank storage Water held in the river bank which may contribute to stream flow, but, in arid conditions, may cause a decrease in discharge as water percolates from the river to the banks. This decrease has to be taken into account when increasing river flows by adding water from reservoirs.

bankfull discharge The *discharge of a river which is just contained within the banks. This is the state of maximum velocity in the channel, and therefore of maximum competence for the transport of debris load.
   Bankfull discharge is assumed to be a major determinant of the size and shape of a river channel, but it is difficult to measure in the field, and a wide variety of field procedures exists for this measurement. Quoting *return periods for bankfull discharge is a tricky business because over a dozen methods are available, but the frequency of its occurrence seems to vary with climatic regimes.
   During rising flood stages, the *thalweg tends to switch from the outside curves to the centre and can straighten out at the bankfull stage, scouring the *point-bar deposits.

banner cloud A cloud developing to one side of a peak, elongated by the *prevailing wind; the Matterhorn is often adorned with a banner cloud. As air rises over mountain barriers, condensation may occur, and where the mountain is high enough for air not to rise entirely above it, banner clouds may form.

bar 1. On a gently sloping coastline, a submarine accumulation of marine sediment, which may be exposed at low tide, most often formed where steep, destructive waves break, moving landwards outside the bar and seawards inside it. Such bars can be called **break-point bars**, and are very common along low coasts, such as the German Baltic. The crest of the bar generally runs parallel to the coast, but may extend across an *estuary or a bay, when it is known as a **bay bar**. Some bay bars entirely enclose the inlet and a *lagoon may then form on the landward side. The formation of *offshore bars, which are located further out to sea, is thought to result from the breaking of larger waves, which erode the sea bed and throw up material ahead of them to form ridges.
   2. In a *glacial trough, such as the Nant Ffrancon of North Wales, a transverse rocky barrier.
   3. Within a river, a deposit of *alluvium which may form temporary islands. Deposition takes place in areas that are away from the threads of maximum velocity and *turbulence. It can begin with two particles coming to rest so that the upstream particle shields the downstream one.

This self-accentuating process creates the *traction clog which will eventually become a bar.

Alternating bars develop as patches of *alluvium, often regularly spaced, along alternate sides of a straight channel.

Braid bars form within a channel and cause the river to split up. Braid bars are roughly diamond shaped and are generally aligned along the course of the channel. **Point bars** form on the inner curves of a meandering river where the *discharge is low.

barchan *See* *sand dune.

baroclinic A term applied to sections of the *atmosphere where trends in pressure (pressure surfaces) are at an angle to trends in temperature. (The precise definition refers to the intersection of *isobars and **isopycnals** (levels, or surfaces, of equal density), but, for most purposes, *isotherms can replace isopycnals.) The number of intersecting isobars and isopycnals is a measure of **baroclinicity**. Strong baroclinicity indicates a steep horizontal temperature gradient with associated *thermal winds. The situation occurs thus: with distance from the equator, patterns of *isobars and *isotherms are broadly alike, each sloping from a maximum, to a minimum at the poles. However, the slope for isotherms is much steeper than for isobars, so that at some point the two intersect, and this happens most often in *mid-latitudes in winter.

In such areas, known as **baroclinic zones**, the isotherms at height show a sudden polewards fall (i.e. there is a sudden increase in the *meridional temperature gradient), cutting across the vertically zoned isobars. (Imagine temperatures falling very rapidly polewards while pressure remains constant in that direction, but falls with height.) In the baroclinic zones of the mid-latitudes, spontaneous generation of weather systems such as depressions and thunderstorms is common. These are **baroclinic disturbances**, characteristically, on *synoptic charts, with strong meridional pressure gradients in the constant-pressure surfaces and vertical *wind shear. When the temperature gradient along the *meridians is very steep, *atmospheric cells break down into cyclonic and anticyclonic *eddies. This failure is known as **baroclinic instability**, and is characterized by the ascent of warmer, and the descent of colder, air.

barometer The **mercury in glass barometer** reflects atmospheric pressure which pushes an exposed column of mercury up an upright glass tube with an end which has been partly evacuated, and sealed. This response may be expressed as:

$$p = g.\rho h$$

where p is the pressure, g the acceleration due to gravity, ρ is the density of mercury, and h is the height of the column of mercury. The mercury-in-glass version is slow to react to changes in pressure, difficult to transport, but accurate, although barometric corrections must be used to take into account changes in local gravity and in the temperatures of the mercury and the scale.

The **aneroid barometer** depends on the response of a partially evacuated capsule of metal to changes in atmospheric pressure; it contracts as pressure rises, and expands as pressure falls. Domestic versions tend to be inaccurate.

barometric gradient *See* *pressure gradient.

barotropic A term applied to *atmospheric conditions where trends in pressure (pressure surfaces) align with trends in temperature, as in the ideal *air mass; the reverse of *baroclinic.

barrage A structure built across a river or estuary in order to restrain or use water; for example, the barrage at La Rance, near St Malo, France. The Thames barrier was constructed for flood control, but other structures are built to store irrigation water. Some writers distinguish between barrages, built simply to restrain water, and dams, also used to generate *hydroelectric power; others seem to distinguish between the two on the height of the structures (it is not unusual to read of dams *and* barrages on the Nile, for example) but these are not hard-and-fast distinctions.

barrier In *geomorphology, a general term for any offshore depositional form, usually running parallel to the coast or across an *estuary, and above water at normal tides. In north-west Europe, barriers are thought to have originated during the *Flandrian transgression when the sea engulfed coastal plains. The clay and silt were carried offshore and the coarser sand and shingle are thought to have been pushed landward. A **barrier bar** is a longshore shingle bar, formed off a gently sloping coast made of unconsolidated sediments; a **barrier beach**, such as Chesil Bank, Dorset, is a small shingle feature protecting a coast from erosion. Many barrier beaches have formed on the shores of the Mediterranean; the Andalusian coast near Calahonda furnishes a good example.

barrier island A long, narrow offshore island, usually having beaches and *dunes on the seaward side, often with *lagoons on the landward side, such as the islands enclosing Pamlico Sound, off the coast of North Carolina. The formation of barrier islands has been variously explained as:

- the drowning of beach ridges
- the *aggradation of submarine bars
- the *progradation of *spits.

barrier reef A *coral reef, stretching along a line parallel with the coastline but separated from it by a wide, deep *lagoon, the most famous of which is the Great Barrier Reef off north-west Australia. Also classified as barrier reefs are those coral reefs encircling islands, one of the best examples being the Turk group in the Caroline Islands. The coral on a barrier reef builds up to the level of low tides. *See also* *coral reef.

barrow A communal burial mound built from the Stone Age until Saxon times. Long barrows, up to 100 m long and 20 m wide, were the earlier form, while round barrows were introduced during the Bronze Age. Both are common on the English Wiltshire Downs, for example.

basal complex, basement complex The ancient *igneous and metamorphic rocks which lie beneath Precambrian rocks and constitute the *shield area of the earth's continents.

basal concavity A gentle, concave element at the foot of a slope, characteristically of 1°–10°, although 15° slopes develop in the humid tropics. Basal concavities are characteristic of arid and semi-arid regions (see *pediment), but are found in other climatic regimes. They appear to extend headwards over time.

basal sapping Erosion at the foot of a slope. It is commonly brought about in tropical areas through chemical erosion where there is an accumulation of water, in temperate areas by scarp-foot springs, or, in glacial environments, such as the foot of corrie walls, by plucking.

basal slipping, basal sliding The advance of a glacier by comparatively rapid creep close to its bed. This form of ice movement occurs in *warm glaciers where the ice at the base is at its *pressure melting point. Basal sliding accounts for the majority of flow in warm glaciers, especially where the gradient is steep, the ice thin, or where *meltwater is present; there is a direct relation between the slip rate of the glacier and available lubricating water.

basalt A dark-coloured, very fine-grained *igneous rock derived from volcanic upwellings, and mainly composed of plagioclase feldspar, pyroxene, and magnetite. Molten basalt spreads very rapidly to form lava sheets often hundreds of kilometres across, such as the Deccan Plateau of India, and basalt flows cover about 70% of the earth's surface. Basalt may crack into hexagonal columns, as at the Giant's Causeway, Co. Antrim.

    Basalt is the principal rock of the sea bed and is associated with oceanic *plate margins. It accounts for 90% of all volcanic extrusions.

base See *alkali. The term 'basic' as applied to rocks has a completely different meaning. See *basic rocks.

base flow The usual, reliable, background level of a river, maintained generally by seepage from groundwater storage, but also by *throughflow, which means that the river can maintain the base flow during dry periods. With prolonged drought, baseflow itself will diminish, the rate of flow falling in a depletion curve.

base level The theoretically lowest level to which the course of a river can cut down. This level may be sea level, the junction between a tributary and the main river, or the level of a waterfall or lake, but streams rarely erode as far as base level. Base level may alter due to *eustatic or *isostatic change, and may be termed positive changes of base level if the land sinks relative to the sea, or negative changes of base level if it rises. Large positive changes in base level can initiate invigorated erosion (see *rejuvenation). **Marine base level** is the lowest point at which marine erosion occurs, perhaps as low as 180 m below the surface.

basement Intensely folded *metamorphic or *plutonic rocks, often *Precambrian, covered by relatively undistorted sedimentary rocks.

basic Of lava and rocks, dark, dense, and containing 50% or less of silica. Examples include *basalt and gabbro. Compare with *acid rock.

basic activity, basic workers According to *economic base theory, basic activities are those which contribute directly to the wealth of a city or region because they bring in money from outside, as opposed simply to serving the needs of the inhabitants; that is, they are export activities. They are also termed **autonomous activities, carrier industries**, and **exogenous activities**. In the context of cities, they may be termed **city forming activities/workers**, since they bring wealth into the city by providing goods and services for the *umland, thereby earning the income necessary to finance the city's needs.

basic/non-basic ratio In and around a city, the ratio of basic (city forming) to non-basic (city serving) workers. This ratio expresses the power of the city as it influences its region, but is difficult to calculate. One method is to collect data from firms about the percentage of sales within the city and outside it. A firm of 100 workers may sell 60% of its product in city markets. 60 of its workers are therefore classified as non-basic while 40 are classified as basic. It is also possible to establish the basic/non-basic ratio by the use of questionnaires for a sample population.

The ratio seems to be linked with city size; the larger the city the higher the proportion of non-basic workers. This is because the amount of trade within the city increases as the city grows. Difficulties occur in the calculation of the ratio because of problems arising from the definitions of urban and rural regions, and because many workers are involved in output which is both basic and non-basic. See *economic base theory.

basin A major relief *depression, guided by *structure, or formed by *erosion. **Basin and range** describes a landscape where ridges made of asymmetric *fault blocks alternate with lowland basins. In the USA, the basin and range country lies between the Sierra Nevada and the Wasatch Mountains.

basin cultivation A form of tropical cultivation where low earth ridges enclose small fields—basins. The ridges are to check run-off from heavy tropical rains, conserving soil moisture and limiting soil erosion.

bastide A planned, fortified strong point and centre of economic development created in the Middle Ages, mostly in France. The rectilinear street pattern contrasts strongly with the irregular, cramped layout of most medieval towns.

batholith A massive, frequently *discordant, intrusion of coarsely textured plutonic rocks, at least 100 km^2 in area and extending 20–30 km down into the layer of *magma, which may be composed of several *plutons. *Granite batholiths, such as Dartmoor, tend to form as *domes. Erosion may expose all or part of the upper surface of the batholith.

bathymetry The measuring of water depth, mainly of seas and oceans but sometimes of deep lakes.

bauxite The major ore of aluminium, usually occurring as a form of clay which results from the weathering of tropical rocks. Its main constituents are aluminous laterite and hydrous aluminium oxides. Australia, Brazil, and Jamaica are major producers.

bay A wide-mouthed recess in the line of the coast, filled with sea water and with open access to the sea.

bay bar *See* *bar.

bayhead At, or pertaining to, the top a bay, as in **bayhead barrier**, a *barrier beach protecting the head of the bay, but separated from it by a *lagoon, **baymouth barrier**, a barrier partially confining a bay, and **bayhead beach**, a shingle and sand beach at the head of a bay.

baydzharakh, baydjarakh A dome-shaped polygon, some 3–4 m high and up to 20 m wide, found in a *periglacial landscape, and formed as the *ice wedges begin to thaw.

beach An accumulation of *sediment deposited by waves and *longshore drift along the coast. The upper limit is roughly the limit of high tides; the lower of low tides. Beach material is very well sorted and the size range tends to be very limited at any particular beach; pebble beaches usually have very little sand, and sand beaches have little shingle. The size of the sediment determines the slope of a beach; shingle and pebble beaches are steeper than sandy ones.

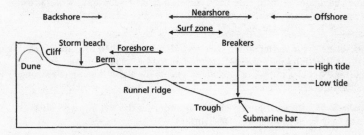

FIGURE 6: *Beach* (after King)

beach budget Most beaches are in equilibrium—the material removed by erosion is compensated for by deposition. Along British coasts during winter, waves tend to be destructive, erosion outstrips deposition, and gradients steepen, but, with summer conditions, wave period lengthens, waves become constructive, deposition dominates, and gradients slacken, so that there is an annual pattern to the beach budget.

If this budget is upset by the building of substantial breakwaters or sea walls, sand beaches may disappear as their source of supply is cut off. Since beaches contribute to the protection of the coastline, the disappearance of a beach can lead to the problem of increased marine erosion.

beach rock A calcareous sandstone crust which forms near groundwater level on tropical beaches by calcium carbonate precipitation from sea and/ or fresh water. A sufficient quantity of calcium carbonate in the sea water is necessary for cementation of the sand grains, and this may be derived from algae, coral, and shells. Beach rock is common on the islands of the Great Barrier Reef, Australia.

Beaufort scale A scale of wind strengths, devised in 1805 by Admiral Sir Francis Beaufort, and modified in 1926. The scale ranges from light winds (1–3) to breezes (4–6) and to gales and hurricanes (7–12). Wind speeds are now generally expressed in metres per second or miles per hour.

bedding plane The surface separating distinct rock *strata.

bed-floor roughness Also known as channel roughness, this is the frictional force of a river bed. A rough bed of boulders, pebbles, and potholes exerts more friction than a smooth, silky channel. *See* *Manning's roughness coefficient (n).

bedforms, sedimentary bedforms In hydrology, forms such as ripples and hollows moulded on a river bed by a flow of water. Bedforms range in size from ripples in the sand, a few centimetres apart, to 'dunes' tens of metres in length.

Bedforms appear to develop further as stream power increases. Initially, the bed is flat; the plane bed. With an increasing discharge, small ripples form and develop into dunes. There follows a transitional zone with a plane bed again. With further increases in discharge, standing waves are set up in the water, creating dunes and anti-dunes. Additionally, the pattern of bedforms has an effect upon the river flow.

bed load The material which is moved along a river bed by rolling and pushing (**traction load**), and *saltation. Bed load is usually composed of sands and pebbles but when the water level is high and the current strong, boulders may be moved.

bedrock The unweathered rock which underlies the soil and *regolith or which may be exposed at the land surface.

bedrock fracture Bedrock is broken up by the impact of large blocks of rock which are incorporated into glacier ice and are moving with the glacier.

behavioural environment The perceived environment; the impression people have of their environment, which is the basis for decision-making, as individuals organize the facts of the phenomenal world into patterns and give them values and meanings in accordance with their cultural contexts. Compare, for example, one young person's view of Florida— Disney land—with another's—'full of old people'—and with its image for the elderly and rich. How you perceive Florida will affect whether you decide to go there.

behavioural geography This view of geography counters the simplistic

views of geographical *determinism and *neoclassical economics and
suggests that, far from being an *economic man, an individual is a
complex being whose perception of the environment may not correspond
with objective reality. A distinction is made between the objectively
observed environment—things as they are—and of the perceived
environment—things as they are seen by the individual. *See* *behavioural
environment. Individuals react to their perceptions, rather than to the
*phenomenal environment. Furthermore, their decisions may not be
rational, or optimizing, but may depend greatly on chance. *See* satisficer.

Behavioural geography is concerned with understanding the flow of
events which produce, reproduce, or transform a system; an analysis of
processes rather than outcomes. It is concerned with the selectively
abstracted structures (*mental maps) which are used as part of the
decision-making process, whether by individuals or by corporations. The
value of such decisions depends upon the perceptions of the decision-
maker and his or her ability to respond to that perception. *See*
*behavioural matrix.

behavioural matrix A way of looking at two variables which affect the
quality of judgements made by a decision-maker; the ability of the
decision-maker and the quality of the available information. The diagram
shows a grid of cells, perhaps ten by ten. As you move *across* the grid, from
left to right, the ability of the decision-maker to use incoming information
improves. As you move *down* the grid, from top to bottom, the quality of
incoming information rises. This means that the poorest decisions will be
represented by a position top left, where ability and information are both
poor, and the best at bottom-right, where both variables are at their best.
All the intermediate cells represent possible combinations of the two
variables.

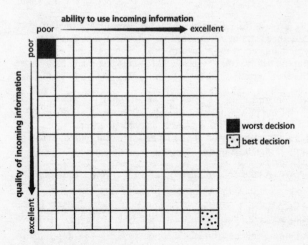

FIGURE 7: *Behavioural matrix*

As firms grow, they should receive more information and be better able to use that information. This results in a diagonal movement in the quality of choices made towards the lower right of the matrix which symbolizes perfect decision-making. A firm with high ability and a high level of information would tend to seek the *optimum location, while a *satisficer, located near the top left of the matrix will have a more random location pattern.

Criticisms of the matrix include the observation that information and the ability to use it are not independent of each other, and that only crude and arbitrary measures can be used to quantify the levels of the two variables.

behavioural model A *model which takes into account the vagaries of human nature rather than depending on the concept of *economic man.

behaviourism The view that the actions of an individual occur as responses to stimuli. Through constant repetition, the individual learns to make the same, 'correct' response to a given stimulus; the 'classic' example comes from the experimental work of I. P. Pavlov (trans. and ed. G. V. Anrep, 1927), who rang a bell before he fed his dogs. He was thus able to condition dogs into salivating when they heard the sound of a bell, even when the food was no longer provided.

Thus, some psychologists claimed, it should be possible to predict the learned behaviour that the individual would act out for each stimulus. This is the stimulus–response model.

Behaviourism has been widely rejected by social scientists who note that it over-simplifies human behaviour and takes no account of the mental processes involved in the perception of, and response to, a stimulus; it neglects all the aspects of human behaviour which cannot easily be observed.

benchmark 1. A mark, the height of which has been determined in relation to *Ordnance Data by spirit levelling. The most common is the cut bench mark which appears thus: ⌐ cut into stone or brick-work.

2. In *Geographic Information Systems, a standard test made to permit comparisons between systems.

beneficiation Concentrating the mineral content of an ore by *ore-dressing, smelting and pelletizing. Beneficiation usually takes place close to the site of an ore body prior to its transportation to a manufacturing region; it is carried out in Liberia, for example, to save transport costs on Liberian iron ore.

Benguela current A cold *ocean current off the south-western coast of Africa.

Benioff zone Named after Hugo Benioff (1954), an inclined zone of earthquakes, plunging below the earth's surface at an angle between 30° and 80°, but commonly at around 45°, and extending to a depth of 300–400 km. Benioff zones are associated with the downward movement of a lithospheric *plate at a destructive plate margin. The plate is compressed, and earthquakes seem to take place within it.

benthic Occurring at the base of bodies of water: lakes, oceans, and seas. Benthos refers to life attached to the bottom or moving in the bottom mud.

Bergeron–Findeisen theory Proposed by T. Bergeron and subsequently modified by W. Findeisen, this is a theoretical explanation of the way droplets of *precipitation form in a cloud composed both of ice crystals and liquid water drops, as minute droplets of water coalesce around ice crystals. The nub of the explanation lies in the differing saturation *vapour pressures of ice and water; *saturation vapour pressure is larger over supercooled water than over ice, so that when water droplets and ice crystals exist together in a cloud, the vapour between these particles cannot simultaneously be in equilibrium with respect to both states. If the vapour is in equilibrium with the ice crystals, it is too small for equilibrium with the water droplets, which will begin to evaporate. As a result, saturation vapour pressure is too big for equilibrium with the ice, so that vapour condenses onto the ice crystals, and the ice crystals grow at the expense of the water droplets. The crystals will have grown large enough to form *precipitation when their fall-speed is greater than the upward movement of air currents. As they descend they may melt to form rain. Without the presence of ice-crystals, water droplets can remain unfrozen at temperatures of minus 20 °C. (Such droplets are supercooled.)

  This theory does not explain precipitation from tropical *cumulus clouds, which can give rain when the cloud-top temperature is 5 °C or more. *See also* *coalescence theory.

Bergman's rule This states that, in warm-blooded animals, species living in cold climates tend to be larger than related species living in hot climates. Some writers question the validity of this 'rule'.

bergschrund A deep, tensional *crevasse formed around the head of a *cirque glacier. The crevasse forms as ice falls away downslope. Often, a sequence of bergschrunds forms.

berm A low embankment or ridge on a sand beach, constructed by *swash or breaking waves.

beta index, β A simple measure of *connectivity relating the number of *edges to the number of *nodes. It is given as:

$$\beta = \frac{\sum e}{\sum v}$$

where e = edges, v = vertices (nodes).

  The greater the value of β, the greater the connectivity. As transport networks develop and become more efficient, the value of β should rise. *See* *alpha index, *cyclomatic number, *König number.

betterment migration Migration with the object of improving the migrant's material circumstances; economic migration.

bevelled cliff A sea cliff with an upper, gentle slope, but a steep, or vertical lower slope, or *free face. Its formation may be structural, when a softer rock layer overlies more resistant strata, or may be the result of changes in sea level; a relative fall would have seen slope wasting and the slackening of the gradient, and a subsequent rise would see renewed wave attack, steepening the lower sections. Bevelled cliffs are common on the coasts of Wales and Brittany.

bid-rent theory A **bid-rent curve** is a graph of the variations in land rents payable by different users with distance from some point in the market, usually the *CBD. Since transport costs rise with distance from the market, rents generally tend to fall correspondingly, but different forms of land use (retail, service, industrial, housing, or agricultural) generate different bid-rent curves. For example, retailers will be willing to pay high rents for sites near the CBD where accessibility is of prime importance, but will be unwilling to pay much for sites more than about 500 m from the *peak land-value intersection, because the distance shoppers are willing to walk is surprisingly short. The curve for industry starts lower—manufacturers cannot afford the high rents that retailers can—but drops away less sharply because pedestrian access is not such a key point. Bid-rent theory shows that each land-user will outbid the others at certain points (as the curve for each stands above all the others on the graph). At that point, the successful, highest competing land use will predominate, and the theory posits a series of land-use rings around the CBD. As with most locational models in human geography, the usual caveats apply: the theory takes no account of relief variations, lines of communication, planning constraints, and so on. None the less, *see* *Alonso model, von Thünen model.

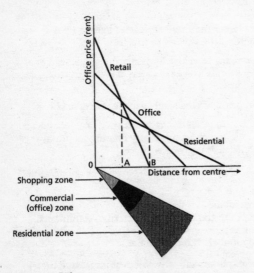

FIGURE 8: *Bid-rent*

bifurcation ratio In a drainage basin, the ratio of the number of streams of a given order to the number of streams of the next, higher order. The ratio varies with the different classifications of stream orders.

Stream ordering after Strahler

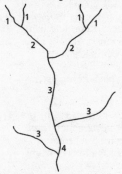

FIGURE 9: *Bifurcation ratio*

binary distribution A *city-size distribution in which a number of settlements of similar size dominate the upper end of the hierarchy. This distribution is said to be characteristic of nations with a federal political structure, such as Australia. *See also* *rank-size rule.

binomial distribution A theoretical frequency distribution which is used in sampling to test whether the characteristics of a random sample are representative of the whole: the *population. For example, if it is known that half the population is male, then the probability of sampling one male at random is 0.5, of sampling two consecutively at random is 0.5^2, i.e. 0.25, of sampling three consecutively at random is 0.125 (0.5^3), and so on. If the findings of the sample match this probability, then it is representative. In large samples, the binomial form has the same pattern as a *normal distribution.

bioclimatology The study of the relationships between climate and living things.

biodegradable That which can be decomposed by naturally occurring organisms such as bacteria; apple cores are biodegradable while polythene bags are not.

biodiversity The varied range of flora and fauna. The maintenance of biodiversity is seen to be of critical importance, both in environmental terms and as a resource for human survival.

biogeochemical cycle The cyclical movement of energy and materials within *ecosystems. This cycle is applied to specific materials, such as copper or carbon.

biogeography The study of the distribution of life forms, past and present, and the causes of such distributions. Biologists usually omit man from these studies while geographers may stress human intervention.

biohistory An approach to *human ecology which stresses the interplay between biophysical and cultural processes. Its starting point is the study of the history of life on earth, and of the basic principles of evolution, ecology, and physiology, and of the sensitivities of ecosystems and living organisms. It then considers the biology and innate sensitivities of humans, the emergence of the human aptitude for culture, and its biological significance. It is particularly concerned with the interplay between cultural processes and biophysical systems, such as ecosystems and human populations.

biological control The attempt to reduce numbers of pests by the use of predators, either from within the community, or by introduction from outside, rather than by chemicals. Although there have been some successes, notably the control of prickly pear in Australia by importing the moth *cactoblastis* from Latin America, most efforts have had at most partial success and have had to be backed up by the use of pesticides. It should be noted that exceptionally effective predators are not usually indigenous.

biological magnification The build-up of toxins from pesticides, herbicides, and domestic and industrial waste such that these toxins are more and more concentrated in living organisms with movement up the *trophic levels of food webs. For example, in the 1950s, DDT was used as an insecticide near certain lakes in the USA. The toxin ingested by the midges was absorbed by the fish which ate them, and then by the grebes which lived on the fish, causing a severe decline in the birds' ability to reproduce. The toxins are not easily broken down and hence they accumulate in the organisms at the top of food webs.

biological oxygen demand, BOD The requirement of oxygen for respiration by aquatic organisms. Deeper levels of water may have insufficient oxygen since the water is too deep for the solution of atmospheric oxygen.

biomass The total mass of all the organisms inhabiting a given area, or of a particular population or *trophic level.

biome A naturally occurring community characterized by distinctive life forms which are adapted to the broad climatic type. Major biomes are *tundra, *coniferous (boreal) forest, temperate (deciduous) forest, *tropical rain forest (selva), tropical grassland (*savanna), temperate grassland (steppe), and *hot deserts. A biome is an idealized type; local variations within a biome are sometimes more significant than variations between biomes. The present-day biomes have evolved in the last 10 000 years.

Smaller biomes are recognized, such as rocky coast biomes or coral reef biomes. In this way, the term is not synonymous with *formation.

biosphere The zone where life is found; the outer portion of the geosphere and the inner portion of the *atmosphere. This extends from 3 m below the ground to some 30 m above it. The biosphere also comprises that region of waters, some 200 m deep, where most marine and freshwater life is found. Ecologists are uncertain as to how much longer the biosphere can tolerate the present pattern of industrial productivity.

biotic Living; sustaining or having sustained life, often describing the biological components of an *ecosystem. *See* *abiotic. **Biotic factors** are factors exerted by living organisms.

biotic potential The maximum population that an area can support.

birth control Techniques to limit family size including contraception, sterilization, and abortion. Birth control is seen as a solution to problems of *overpopulation and has been encouraged by many Third World countries. The results have not been as successful as might have been hoped, partly because reaching all sections of society in a nation requires more resources than are generally available, and takes an enormously long time. Furthermore, introducing birth control involves complex social factors—children are seen in many societies as wealth, and childbearing is seen as a major function of a married woman. The Cairo population summit of 1994 concluded that the key to population control lay in improving women's education and prospects.

Some governments do not support birth control, thinking that population is a resource, and birth control is banned by many religious groups. *See* *anti-natalist, *natalist.

birth rate The number of births in a year per 1000 of total population taken at the mid-year mark. This is the crude birth rate since it is not adjusted to take account, for example, of the proportion of the population which is of childbearing age. The **crude birth rate** may be expressed as:

$B/P \times 1000$

where B = the number of births, and P = total population. A characteristic figure for a developed country might be 11/1000 per annum: 1992 figures for *EU countries ranged from 9.8/1000 (Italy) to 15.1/1000 (Ireland). For a developing country figures of around 30/1000 per annum are not uncommon.

This crude statistic does not take into account the *age structure of the population, which indicates the number of women of childbearing age, so that it is difficult to compare crude rates between two very different populations. Because of this, many demographers prefer to use a **standardized birth rate** which indicates what the crude birth rate would have been for a population if the age and sex composition of that population were the same as in a population selected as standard.

biscuit-board topography A flat or rolling upland cut into by *cirques along its edges. The Mackenzie Mountains of Canada furnish an example.

bit In computers, a basic unit of binary data.

black body In *meteorology, a body reflecting no electromagnetic radiation. **Black body radiation** is the theoretical maximum amount of radiant energy which can be emitted by a body at a given temperature.

black box The view, used in *behavioural geography, that the workings of the human mind cannot be analysed; all that can be observed is the input and output. The input is a stimulus and the output is behaviour, but the process in between is as inaccessible as the contents of a sealed black box.

black earth *See* *chernozem.

black economy The part of the job market which is not reported, for example, to the tax authorities and does not appear in official statistics. A slang term is 'moonlighting', but in many less developed countries the similar term *'informal economy' is used.

blanket bog A continuous covering of bog, mostly comprising peat. Only steep slopes and rocky outcrops are dry. Bog formation depends on high humidity and rainfall. Blanket bog is common in upland Britain, but commercial extraction of peat for sale in garden centres is threatening this ecosystem.

blind digitizing *See* *digitizing.

blind valley A steep-sided valley in an area of *karst scenery which ends in an abrupt cliff facing up the valley. This is usually the point at which any overland drainage disappears down a *streamsink.

blizzard A strong wind (specified by the US Weather Bureau as over 50 k.p.h) which whips up particles of ice and dry, powdery snow, reducing visibility to less than 200 m. The snow in a blizzard is not always falling, but is also carried up from snow on the ground.

block, volcanic *see* *pyroclast.

block field, block stream A sheet of angular rock fragments in spreads or lines in a *periglacial landscape. Some writers maintain that this debris is formed *in situ* by *freeze–thaw. Others suggest that the blocks have ridden down on the top of saturated debris during *gelifluction, or on *rock glaciers, or mudflows.

block lava Sometimes termed *aa, this lava has a thick skin broken into jagged blocks. Its chemical composition is identical to that of *pahoehoe. These two lava formations often occur in the same lava flow. Mauna Loa, Hawaii, furnishes examples.

block mountain An area of upland identified with an uplifted area bounded by *faults.

blockbusting A technique, used by estate agents in the USA, of inclining residents of a white neighbourhood to move out because they fear that the district is to be taken over by black families.

blocking A meteorological condition, when a pressure system remains stationary for, perhaps, weeks. It is linked with a blocking pattern in *Rossby waves. Blocking anticyclones develop when the Rossby pattern changes from zonal to strongly *meridional, often forming one or two high-level, closed anticyclonic circulations, so that the *jet stream splits around the *high pressure system(s). The upper air flow then guides depressions around the edge of the anticyclone(s). Unusually long periods of fine, dry weather result in summer; winter conditions are very cold, but rather dry.

Blockschollen flow A type of flow in glaciers where velocity is steady almost to the edges, but then falls off sharply. This creates a highly erosive, strong *shear force near the valley walls.

blow-out A localized area of *deflation, especially on a coastal *sand dune. Deflation may have begun through the removal of vegetation, via grazing rabbits or the trampling of tourists' feet, or through the cutting off of sand supplies as new dunes develop near the shoreline.

blowhole A crack in the top of a cliff through which air and sea water blow. The blowhole is fed from the seaward end via joints and tunnels.

bluff A steep, almost vertical, cliffed section of a river bank. A **bluff line** is a prominent slope marking the edge of a *flood plain.

bocage A landscape of small fields surrounded by low hedges. The term was first applied to the fields of Brittany and Normandy, although field enlargement has destroyed much bocage in Normandy.

bog An area of wet, spongy ground thick with partially decomposed vegetation.

bogaz Wide, deep fissures, 2–4 m wide and 1–5 m deep, running for tens of metres across *karst lands and formed as solution deepens and widens the natural joints in the rock. As such, they are larger forms of *karren, or *grikes.

Bølling A glacial *interstadial of some 200 years duration, beginning at about 12 350 years BP.

bolsa bay A wide, bottle-necked bay with a narrow opening to the sea. Bolsa bays indicate submergence of a coastline, where the inlets represent drowned gorges through mountain chains or rocky barriers. A classic example is Saint Marie Bay, Curaçao.

Bora (fall-wind) A cold winter wind blowing down from the mountains onto the eastern Adriatic coast.
    The wind develops when a cold continental *air mass crosses a mountain range and is forced to descend because of the *pressure gradient. Despite *adiabatic warming, this cold air displaces warmer air. The term is now applied to winds of similar origin in any other region.

bore The current of the incoming tide up a river producing a wall of water which moves upstream. It is linked with the narrowing and shallowing of the river mouth. In Britain, the most famous river bore occurs on the Severn, but the largest bore in the world occurs on the Amazon, where it moves upstream at a rate of 10 m/s reaching a height of 5 m.

boreal forest See *coniferous (boreal) forest.

bornhardt A dome-shaped rock outcrop more than 30 m high, and sometimes several hundred metres in width. They commonly rise above erosional plains in the tropics, but also develop in unglaciated uplands in high latitudes. Most are formed in granites and gneisses; some in sandstones and conglomerates.

The origin of these features is problematical. The dome form may have originated as an intrusion; they may have formed through differential weathering at depth, and then have been exposed by the stripping of the *regolith, either by *parallel slope retreat or by the *downwearing and removal of soil; or some may have been up-faulted, as with the most famous bornhardt, the Sugar Loaf Mountain of Rio de Janeiro. See also *inselberg, *tor.

borral A soil order of the *US soil classification. See *chernozem.

Böserup model Ester Böserup's (1965) view that increases in population size stimulate agricultural change in subsistence societies, given no increase in land area. At the earliest stage, small families subsist through forest fallow where land is used for two years or so and is then left for twenty to twenty-five years. As population rises, *bush fallowing and short fallowing are used with increasingly intense cropping and the shortening of the fallow period. Further population growth is followed by annual cropping which consists of harvesting one crop a year with a fallow of a few months only. Multi-cropping is stimulated by further population increase and is the most intensive system of agriculture. Changes in farming technology also increase yields; for example, digging sticks are replaced, first by hoes, and then by ploughs. Weeding becomes more frequent, and manuring is introduced. These changes increase labour requirements, so that, while yields per hectare rise, yields per capita remain more or less constant.

This theory runs counter to Malthus' argument that population increase is only possible after a certain level if food supplies rise. Although Böserup's thesis was developed for subsistence populations, the Green Revolution can be seen as a response to population increases.

boss A roughly circular *igneous intrusion, a few square kilometres in area and lying at a steep angle to the ground surface.

bottomset beds Fine debris deposits, carried furthest out to sea by a river, and forming the lowest layers of its delta. These will be covered by the slightly coarser **foreset beds**; the upper layers, nearest the coast, will be covered by the coarse, horizontal **topset beds**.

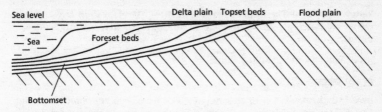

FIGURE 10: *Bottomset beds*

boulder clay A now outmoded term for glacial *till; inexact as till does not always contain boulders.

boulder field As distinct from a *block field, this is an area of boulders which is the result of *spheroidal weathering. When the weathered layers are removed by water, the corestones are left as rounded boulders.

boundary A line marking the limits of a unit of land, often a geographical region, but also of economies or societies, such as a *ghetto. Different cultural groups are divided by **ethnic boundaries**. **Physical boundaries** follow natural features such as rivers (e.g. the Rio Grande between Mexico and the USA), and **geometric boundaries** follow lines of latitude (the 49th parallel between the USA and Canada), and longitude (as in large parts of the boundary between Botswana and Namibia).

A distinction is made between **antecedent boundaries** which demarcate territories before they are settled, like the 49th parallel (above), or before they have been colonized, as in the case of many of the African boundaries established by the colonial powers at the Congress of Berlin in 1884; and **subsequent boundaries** which evolve together with the society they encompass. *See* *frontier.

boundary layer In meteorology, any layer of the *atmosphere significantly affected by its lower boundary: the earth's surface. The **laminar boundary layer** is the few millimetres above the surface; the **turbulent boundary layer**, or **surface boundary layer**, is an ill-defined layer covering the conspicuously turbulent part of the atmosphere.

bounded rationality A decision-maker has neither the time and space nor the ability to arrive at an optimal solution and many individuals may not seek to optimize at all. The idea of bounded rationality is that individuals strive to be rational having first greatly simplified the choices available. Thus, instead of choosing from every location, the decision-maker chooses between a small number. The result may be that decision-makers become *satisficers; they accept a satisfactory solution which is good enough for their purposes rather than finding the optimum answer. Early work on the theory of bounded rationality is associated with H. A. Simon (1956).

R. W. Kates (*Economic Geography*, 47) applied Simon's theory to hazards, in which context bounded rationality means that an individual responds to a hazard only after a threat is perceived, and that any ensuing action is based on a subjective assessment of a perceived range of options.

Bowen ratio, β The ratio of sensible heat to *latent heat. In arid zones, β values are much greater than unity; in humid zones they are much below unity.

brackish Of water, slightly saline.

braid bars *See* *bar.

braided channel A river channel in which have been deposited *bars and islands around which the river flows. It has been shown that, for a given *discharge, braided channels slope more steeply than meandering channels. Braiding occurs when the discharge fluctuates frequently, when the river cannot carry its full load, where the river is wide and shallow, where banks are easily eroded, and where there is a copious bedload, as is common in *periglacial environments. The position of the bars is changeable. Braiding differs from *anastomosis in that the islands are less permanent.

braunerde *See* *brown earth.

Brazil current A warm *ocean current off the east coast of Brazil.

break-of-bulk point The point at which a cargo is unloaded and broken up into smaller units prior to delivery, minimizing transport costs. This frequently happens at waterfront sites where imports are often processed to cut costs. In consequence, processing industries are common at many break-of-bulk points, such as oil refining at tanker terminals like Milford Haven. However, costs are incurred through goods being transferred from one mode of transport to another at break-of-bulk points.

break of slope A sudden change of gradient as in a river after *rejuvenation or when the river crosses a projecting band of resistant rock.

break point, breaking point The point at which the *field of influence of one settlement ends and that of another begins. This can be calculated according to *Reilly's gravity model equation:

$$d_{jk} = \frac{d_{ij}}{1 + \frac{\sqrt{P_i}}{P_g}}$$

where d_{jk} is the distance of the breaking point from town j, d_{ij} is the distance between towns i and j, P_i is the population of town i and P_j is the population of town j. The number of shops in each settlement may be used rather than population.

break point bar *See* *bar.

breccia Any rock which consists of sharp fragments of other rocks cemented together.

bright-lights district An area of the *CBD given over to entertainment and hotels.

brown earth, brown forest soil A *zonal soil associated with areas where the natural vegetation is, or was, deciduous woodland. Brown earths may have a thick litter layer and generally have an A *horizon rich in *humus and containing iron and aluminium *sesquioxides in small, crumb-like *peds. This horizon merges into a lighter B horizon which has blocky peds and is weakly developed. There is little *leaching although the soils are free-draining.

buffer state A generally neutral state which lies between two powerful and potentially belligerent neighbours. Invasion by its more powerful neighbours is often the lot of a buffer state. Poland, as a buffer between Germany and Russia, has suffered particularly badly in this respect.

buoyancy In *meteorology, **positive buoyancy** is that quality of an air parcel which allows it to rise through, and remain suspended within, the *atmosphere, due to its lighter density (possibly caused by a local increase in temperature). If some of the water in the air parcel evaporates through an *adiabatic process, the parcel will suffer a heat loss and lose buoyancy.

bush fallowing A type of *subsistence agriculture where land is cultivated for a period of time and then left for some years to recover its fertility. This was a feature of *infield–outfield cultivation in Britain, of areas of upland farming, and of the tropics, where it still survives in places.

business climate The environment of an area in relation to industrial output. This is difficult to quantify but factors affecting the business climate include local legislation, the attitude of public officials, and the availability of finance.

business cycle Economies seem to prosper and decline in cycles; some economists have detected a five-year sequence, although others argue that the fluctuations are entirely random. *See* *Kondratieff cycle.

bustee In India, a *shanty town, or spontaneous settlement.

butte A small, flat-topped, unvegetated, and very steep-sided hill of layered strata, probably the residue of a larger feature (*see* *mesa), and thought by some geomorphologists to be evidence of *parallel slope retreat. Devotees of 'westerns' will have seen buttes many times, especially those in Monument Valley, on the Colorado Plateau, USA.

Buys Ballot's law A law formulated by the eponymous Dutch meteorologist in 1857, expressing the relationship of horizontal wind direction and pressure patterns; it states that if an observer faces the direction to which the wind is blowing, the lower pressure will be to the left in the Northern Hemisphere, and to the right in the Southern Hemisphere. Wind direction can thus be predicted, given the location of the lower pressure. It is a qualitative statement of the *geostrophic wind equation.

byte A unit of computer storage data, equal to 8 *bits, or one character. A **gigabyte** is 1000 bytes; a **megabyte**, 1 000 000.

C

caatinga Light, thorny woodland, composed of drought-resistant species, found in north-eastern Brazil; either where rainfall is low or unreliable, or where the presence of pervious rocks creates a *physiological drought.

cadastre A record of the area, boundaries, location, value, and ownership of land, achieved by a **cadastral survey**. The term is also used in *Geographic Information Systems.

Cainozoic *See* *Cenozoic.

cairn A rough mound of stones piled up as a route marker, as a boundary indicator, or as a memorial.

calcrete A *duricrust made up mostly of calcium carbonate. It forms in arid climates as a result of capillary action and prolonged evaporation.

caldera A sunken crater at the centre of a volcano, formed as a result of subsidence. As the magma founders, so the centre of the volcano collapses. The Plateau of Giant Craters in Tanzania contains many impressive calderas, including Ngorongoro, which attains a diameter of 22 km.

calibration The adjustment of a *model so that it will fit special circumstances. The values used within a particular model, such as the *gravity model, need to be modified to local circumstances before it can make predictions. Thus, complex versions of the gravity model build in the variables of job opportunities, wage rates, and prices, but the amount of weight given to each of these variables will vary from time to time and from place to place. Calibration is a method of adjusting their weighting.

caliche A *chernozem soil of the American south-west containing thick carbonate deposits.

California current A cold *ocean current off the California coast, bringing frequent fog to San Francisco, for example.

calving The breaking away of a mass of ice from an iceberg, an *ice front, or a glacier. It represents the major form of *ablation from a glacial system. The calves from the Blomstrand Glacier, Spitzbergen are often 40 m or more in height.

cambering The rounding at the edge of a cap rock occurring in *periglacial landscapes where there are underlying clays. As clays thaw, they are squeezed out and the cap rock above them cracks and sags at the edges.

Cambrian The oldest *period of *Paleozoic time stretching approximately from 570 to 500 million years BP.

canal An artificial watercourse cut for inland navigation. Canals came into prominence in England in the late eighteenth century because of the poor state of the roads and the cheapness of water transport. As all-weather roads and railways developed, the canal became outmoded in Britain, but canalized rivers are an important mode of transport in larger countries, especially for non-perishable goods, mainly because of their low *line-haul costs.

Canaries current A cold *ocean current off the coast of Mauritania.

canyon An extreme type of V-shaped valley with very steep sides and no valley floor. A canyon differs from a *gorge in that the sides are stepped, reflecting alternating rock resistances. The most famous example is the Grand Canyon of the Colorado River, which is 1800 m deep with a maximum horizontal width of 25 km. **Submarine canyons** are deep troughs in the sea bed, sometimes as prolongations of river valleys on land; the Congo canyon, for example, can be traced 150 km from the land, to a depth of 200 m, and is probably of *tectonic origin. Other canyons may have formed through earth flows, turbidity currents, the rising of springs, or the slipping of sediments.

capability constraint In *time-space geography, a limit to an individual's actions because of biological needs, like food and sleep, and because of restricted facilities, like access to public transport.

capacity In hydrology, the maximum amount of debris that the stream can move as *bed load. The capacity is dependent on the *discharge and on the nature of the load; a stream may be able to carry more weight if the particles are small than if a load of the same volume were of large boulders.

capillary In a soil, the fine spaces between soil particles.

capillary action is the ascent (or descent) of a liquid in an area of small diameter, such as a soil pore, due to the combined effects of surface tension and the forces of cohesion and adhesion. *See* *salination.

capillatus *See* *cloud classification.

capital One of the four *factors of production, along with land, labour, and enterprise, capital includes all the items designed by society to further the creation of wealth. Plant, machinery, and buildings are **fixed capital** because they earn profit without circulating further, while **circulating capital**, or **floating capital** includes raw materials, fuels, components, and labour inputs which are then sold again—in the form of the product—at a profit. **Financial capital** is the money needed for production. Most economists regard the formation and *accumulation of capital as essential for *industrialization.

More sophisticated views see capital not as a 'thing', but as a social relation which can take many forms: it can be invested as money, for example, or paid out as wages, but throughout it symbolizes economic relationships between people, whether individually or in groups; it is the result of social labour achieved in the creation of goods and services.

Geographers' interest in capital generally focuses on the way in which capital brings about *uneven development, *areal differentiation, and environmental change.

capital goods, capital equipment Also known as **producer goods**, these are goods, such as machinery and equipment, which are used to create other goods.

capital intensive Using a high input of *capital, of all types, in relation to the amount of labour used. This relationship can be expressed by the capital–labour ratio. One capital intensive industry is chemicals, where nearly 70% of investment is on capital.

capitalism In *Marxist terms, an arrangement whereby one class—the capitalists, or bourgeoisie—owns the *factors of production while the workers possess only their labour, which they sell. According to Marxism, capitalism exploits the workers by undervaluing this labour. *Accumulation is a key characteristic of capitalism.

A more general usage defines capitalism as a system where the *factors of production are privately owned. Sales occur for profit in markets which are free in the sense that, subject to the constraints of the law, *entrepreneurs are able to engage in business. The implicit assumption is that individuals are rewarded in relation to their economic contribution.

capture, river capture When a river is extending its channel upstream by *headward erosion, it may come into contact with the headwaters of a river which is less vigorous. The headwaters from the minor river may be diverted into the more rapidly eroding channel. There is often a sudden change of stream direction at the point of capture; this is the **elbow of capture**. In mid-Wales, the River Rheidol has captured the headwaters of the Teifi, and the elbow of capture may be seen about 20 km to the east of Aberystwyth. *See also* *misfit stream and *wind gap.

carbon cycle Carbon is supplied to the *biosphere as carbon dioxide during volcanic eruptions. Most of this is dissolved in the sea or incorporated into calcareous sediments which then harden to form limestones and dolomites. As these rocks are folded and raised above sea level, they are subjected to solution by weak carbonic acid and form sediments once more. This is the largest and slowest of the carbon cycles. The shortest cycle involves respiration by plants and animals whereby carbon dioxide is expired, and *photosynthesis by plants which change carbon dioxide and water into organic compounds.

It has been suggested that a third carbon cycle exists in the burning of fossil fuels, causing the emission of carbon dioxide, which can be incorporated by living organisms once again. These may then be a source of fuel. It is this third cycle that seems to be out of balance; carbon dioxide is being emitted far more rapidly than can be absorbed by the oceans, or during photosynthesis. This increase in atmospheric carbon dioxide may lead to an increase in the warming of the atmosphere. *See* *greenhouse effect.

carbon dating A method of assessing the age of an archaeological specimen which is biological in origin (e.g. wood). Radioactive nuclei of carbon-14 form as cosmic rays bombard atmospheric nitrogen. Some of these radio-carbon atoms are incorporated into living matter via carbon dioxide, during photosynthesis. As the matter dies, the radio-carbon begins to decay, at a rate known as the half-life. The ratio of radio-carbon to regular carbon is measured. From this ratio, the date of the specimen may be calculated.

carbonation A form of solution where an acid, formed by the solution of atmospheric carbon dioxide in the water, dissolves minerals. The effects of carbonation are best seen in the solution of calcium-rich rocks, such as limestone, but carbonic acid will also dissolve silicates. *See also* *karst.

Carboniferous A *period of *Paleozoic time stretching approximately from 345 to 280 million years BP. This period can be subdivided into the Mississippian and the Pennsylvanian periods. During this period, massive limestones and coal measures were formed.

carnivore Any animal which eats the flesh of other animals. Within a *pyramid of numbers, **top carnivores** are usually the least numerous, largest, and most complex animals, and are at the top of the pyramid.

carrier industries *See* *basic activity.

carrying capacity The maximum potential number of inhabitants which can be supported in a given area. The concept was first advanced in *ecology, where the 'inhabitants' are plants, and was extended to livestock, but now is increasingly used in terms of the optimum number of users of facilities, whether agricultural, where agronomists are concerned with the physical ability of an area to produce livestock and crops according to the level of technology, or recreational. In both cases, the upper limit is set at the point where the environment deteriorates.

cartel A system whereby producers divide up the market between themselves, avoiding direct competition and not encroaching on each other's share of the market. This generally comprises the fixing of *quotas on investment and output, and has the effect of raising prices. Since setting up a cartel creates a monopoly situation, cartels within nations have been outlawed by many governments, but the cartel established by *OPEC in the early 1970s was instrumental in quadrupling the price of oil. This cartel later weakened, and oil prices subsequently fell.

cartogram Broadly defined as a map using statistical symbols, a more specialist usage defines a cartogram as a type of map transformation based on a scale other than a true scale. For example, a voting map of Britain may show the size of counties in relation to the numbers of voters in each electoral unit, or an economic cartogram of the world may show countries drawn in proportion to their per capita GNP. Certain 'rules' are followed, as far as possible: the shapes of the countries and regions involved are preserved, although often stylized, and they are positioned in the correct geographical locations with respect to each other. Obviously, distortions occur, but the trick is to preserve the original shapes and positions enough to make the units recognizable; computers have helped in this, but the best examples are still produced by hand.

cartography The production and study of maps and charts. Cartography includes the historical development of map-making techniques, the social conditions which give rise to cartographic methods and themes, and the aesthetics of map-making. A relatively recent concern has been the recognition of cartography as a system of information which is used to communicate something of the nature of the real world to other people; the map is a *model, to be decoded by the map reader. So that the reader is not distracted by 'noise'—anything which stands in the way of understanding—the map has to be encoded using easily understandable signs, symbols, lettering, and lines. To this end, much research has been carried out on the way in which people react to maps.

cascade A sequence of small waterfalls, such as that occurring on the White Nile below Lake Victoria, Uganda.

cascading systems *See* *systems.

case hardening The production on a rock outcrop of a resistant *weathering rind consisting mainly of iron and magnesium hydroxides, and sometimes of amorphous silica. The origin of these materials is under dispute; some writers claim they come from the interior of the rock, others that the rind is composed of wind-borne minerals. Case hardening may be related to the production of *visors over some *tafoni entrances. *See also* *desert varnish.

cash cropping The growing of a crop to produce goods for sale or for barter rather than for the subsistence of the farmer and family.

caste A position in society inherited from parents at birth and from which there is no transfer throughout life. The system is at its strongest in India where people of high caste are respected but those of the lowest caste—the untouchables—usually work in the most menial occupations.

cataract A step-like succession of waterfalls. Cataracts, such as those on the Nile, are often associated with the 'rungs' formed by the erosion of horizontally bedded rocks.

catastrophism The largely nineteenth-century belief that geological strata and other landscape elements were formed by sudden, isolated, and forceful events (such as Noah's flood), rather than by slow processes of *tectonics, *weathering, *erosion, transport, and *deposition. While catastrophism is now generally regarded as being outdated, geomorphologists are still divided as to whether sudden and dramatic events, like the eruption of Mt. St Helens in 1980, are of more geomorphological significance than slow, everyday processes.

catch crop A fast-growing plant which is intercropped between the rows of the main crop. It is often used as *green manure.

catchment area 1. In human geography, the region which is served by a city. *See* *urban field.

2. In hydrology, the region drained by a river and its tributaries; the *drainage basin.

categorical data Data which can be placed into categories which are mutually exclusive, such as age groups. **Categorical data analysis** is a statistical technique used when one or more of the variables involved in an investigation is measured at a *nominal scale, that is, where names, such as 'male' and 'female', rather than measurements, are used.

catena A sequence of soil types arising from the same parent rock, but distinct from each other because of the variations, such as drainage, *leaching, and *mass movement arising from differences in topography.

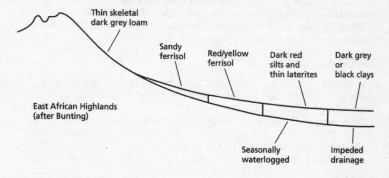

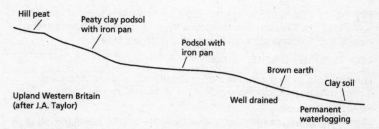

FIGURE 11: *Catena*

cation An atom, or group of atoms with a positive charge. **Cation exchange** is the process whereby a cation in solution is absorbed by a solid, replacing a different cation. Thus, in soil science, if a potassium salt is dissolved in water and applied to a soil, potassium cations are absorbed by soil particles, and sodium and calcium cations are released.

causal analysis The search for the cause or causes of particular events and objects.

causal factor One variable which causes change in another. Statistical techniques which test the strength of a postulated link between two variables, such as the volume of pedestrian flows and distance from the *CBD, include the *Student's t-test and the *chi-squared test.

causal model A *model which tries to describe causal and other relationships within a set of variables. It is basically a hypothesis about the relationship between pairs of variables, but even in the most simple model, with three variables, there will be many different versions of their relationships, and each possible model has to be investigated.

The underlying concept of a causal model is multiple causality. Thus, areas of deprivation within the city may be linked with age, social class, and ethnic origin, but these variables are also interconnected. *See* *multivariate analysis.

cave A large, natural, underground hollow, usually with a horizontal opening. The largest and best-developed caves and caverns (large caves) are found in areas of *karst scenery.

Karst caves result from solution, but *corrasion by water-borne sediments and pebbles is also important. The collapse of cave roofs causes much of the hummocky appearance of karst. **Wave-cut caves** are widely distributed, and their constantly changing outlines are related to the *jointing and *bedding of the parent rock.

cavitation The process by which bubbles in a liquid collapse close to a solid surface. Cavitation causes shock waves and usually occurs downstream of local obstructions on the bed. The process is best known to engineers because it causes serious erosion in hydraulic structures and, although much less is known about cavitation in natural channels, it seems likely that it is an important erosional *hydraulic force in streams and rivers.

cay One of a chain or row of small islands in a shallow tropical sea, so called after the Bahama Cays, developed through a combination of *coral growth and *accumulation. In Florida, the term **key** is used.

CBD *See* *central business district.

celerity In a river, the square root of the product of the acceleration due to gravity and the mean depth of the river flow. *See also* *Froude number.

cell 1. The basic unit of *spatial information in any *raster depiction of spatial *entities.
   2. *See* *atmospheric cell.

cellular clouds Roughly hexagonal patches of clouds, from 20 to 200 km across. Open cellular clouds are ring-like, and may be found over oceans in

mid-latitudes, and on the western borders of subtropical *anticyclones.
Closed cellular clouds form in stratocumulus over oceans, are much
smaller, and are found on the eastern borders of anticyclones.

cellular convection *Convection occurring in semi-regular cells, as in a
cellular flock of shower clouds.

cementation The binding together of particles by adhesive materials.
Sedimentary rocks are often bound together and hardened by cementation.

Cenozoic, Cainozoic The most recent *era of earth's history stretching
approximately from 65 million years BP to the present day.

census An investigation, usually into the size and nature of a population,
but occasionally into other things such as traffic. In order to obtain
complete coverage, most governments make it compulsory to participate
in a census. A census is taken for a particular point in time, and while
some nations require their people to note where they were at the time of a
census, others ask for the respondents' place of residence.
    The first British census was taken in 1801, and the exercise has been
repeated at ten-yearly intervals, except during the Second World War.
From 1966, a 10% census was taken at the mid-point of the ten year period,
although financial problems meant that 1996 did not see a 10% census. In
the developed world, as well as a head count, a census would inquire into
birthplace, age, sex, marital status, qualifications, occupation, family
structure, and fertility. A **census tract** is a small unit of area used in
collecting, recording, and reporting census data. In the UK, census tracts
are known as *enumeration districts.

central business district, CBD The heart of an urban area, usually
located at the meeting point of the city's transport systems, which contains
the highest percentage of shops and offices. Land values are high because
of high accessibility, therefore land use is at its most intense in order to
offset rent costs. In consequence, in many countries development is
upwards rather than sideways. Within the CBD, specialist areas, such as a
jewellery or garment-making quarter, may arise in order to benefit from
*external economies. Vertical land-use zoning is also common, so that
retail outlets may be on the ground floor, with commercial users above
them and residential users higher up.
    Methods of delimiting the CBD include mapping the intensity of land
use (see *central business height index), recording the percentage of floor
space given over to CBD functions, charting high level pedestrian
flows . . . and checking with the local town planning department for the
boundaries it has established.
    The CBD is increasingly under threat due to a combination of traffic
congestion, which has led to parking restrictions, and to the growth of out-
of-town developments, including superstores. See *Alonso model,
*centrifugal and centripetal forces, *counter-urbanization.

central business height index A measure of the intensity of land use
within the *CBD. The total floor area of all storeys of a building is

compared with the ground-floor area of the building under consideration. The higher the index, the more intensive the urban land use. The index can also be used to delimit the CBD.

central place A settlement or nodal point which, by its functions, serves an area round about it for goods and services.

central place theory A theory, advanced by W. Christaller (1933, trans. 1966) and, later, Lösch (1954), concerned with the way that settlements evolve and are spaced out. Christaller envisaged an *isotropic plain with an even distribution of purchasing power. Travel costs were the same in any direction and all parts of the plain were served by a central place, so that the spheres of influence of the central places completely filled the plain. Central goods and services were to be purchased from the nearest central place and no excess profit was to be made by any central place. Christaller contended that each central place should have a hexagonal market area since this polygon represents the most effective packing of the plain and is most nearly circular.

To ensure that goods and services are freely available, central places emerge at the centre of a hexagon containing six lower-order places. One higher-order place will serve a total of two lower-order neighbours. This may mean that two distinct lower-order places are served or that the central place will serve one-third of each of the six lower-order places surrounding it. This will bring to two the total of lower-order places served, and, with the addition of the central place itself, three places are served. This method of serving the market is known as the $k = 3$ system.

A different system of hexagons would evolve if transport costs are to be minimized. The hexagon is rotated so that the settlements are located evenly at the mid-point of the hexagon's sides. Now the central place serves a half share in the surrounding six settlements: a total of three places plus the central area. Therefore $k = 3 + 1 = 4$; the $k = 4$ system.

The most efficient pattern for the administration of settlements sees all six lower-order centres inside the hexagonal area of the central place, $k = 7$.

All these places fit into a hierarchy. Higher-order places stand out from the hexagonal pattern of lower-order centres, but are themselves packed in hexagons around an even higher-order central place. Christaller envisaged this hierarchy as going all the way up to major regional centres. It should be said that hexagonal patterns are very rarely found in real life.

Lösch, who had worked independently of Christaller, extended these ideas. He plotted the ten smallest market areas, each with a different k value. Each network surrounded a common central place. Tracings of each network were laid over each other and the tracings were positioned so as to produce the largest number of places occurring for each k value. The result was a central place with city-rich and city-poor areas spread out in wedges around the major central place. Such a pattern is found around Indianapolis.

Common sense tells us that the basic postulates of these models do not exist but they still give insight into the nature of town development and distribution.

k = 4 network, explanation

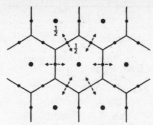

- ● higher order centre
- • lower order centre
- --→ direction and proportion of custom from lower order centres to higher order centres

(a) k = 7 network, explanation; (b) k = 7 network (an alternative orientation)

(a)

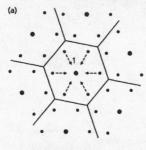

(b)

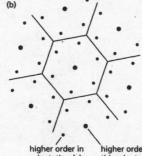

- ● higher order centre
- • lower order centre
- --→ direction of custom from lower order centre

higher order in orientation (a)   higher order in this orientation

The marketing principle, k = 3, the G-system. (Top right sector shown in full detail.)

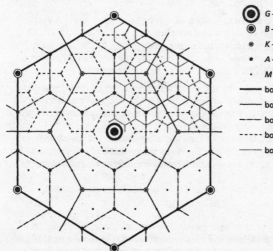

- ◉ G – place
- ◉ B – place
- ◦ K – place
- • A – place
- · M – place
- ⎯⎯ boundary of G-region
- ⎯⎯ boundary of B-region
- —·— boundary of K-region
- - - - boundary of A-region
- ········· boundary of M-region

FIGURE 12: *Central place theory*

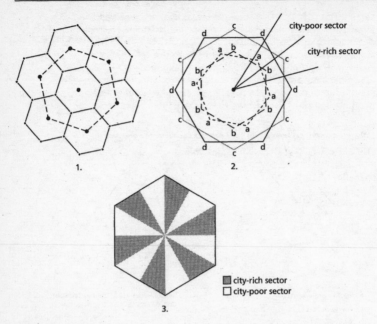

1.

2.

city-poor sector

city-rich sector

3.

■ city-rich sector
□ city-poor sector

FIGURE 12: *(continued)*

central tendency A synonym for average, most often expressed as an *arithmetic mean, a *median, or a *mode.

central vent eruption Also known as a pipe eruption, this is a volcanic eruption from a single vent or a cluster of centrally placed vents, fed by a single pipe-like supply channel from deep underground. The lava from this type of eruption accumulates around the vent to form a conical volcano, like Vesuvius, near Naples. Compare with *fissure eruption.

centrality The degree to which a town serves its surrounding area. This depends on the ease of access to the town and the range of goods and services offered. The centrality of a town is not necessarily commensurate with its population.

W. Christaller (1933, trans. 1966) used the occurrence of telephones to indicate centrality at a time when the ownership of telephones was not as universal as it is today. The equation he used to determine the **centrality index** was:

$$Z_z = \frac{T_z - E_z \cdot T_g}{E_g}$$

where T_z is the number of telephones in the central place, E_z is the population of the central place, T_g is the number of telephones in the region served, and E_g is the population of the region. Z_z is the index of centrality. The higher the index, the greater the centrality.

centrality of population A central point in the distribution of a population which has the same population numbers on every side. The **mean centre of a population distribution** is the point at which the squares of the distances of all the population from the centre is at a minimum. This is very complicated and lengthy to calculate.

The **median centre** is the point at which travel distances for all the population to converge is lowest and the **modal centre** is the area with the maximum population density.

centrally planned economy Generally speaking, this term has been used as a synonym for communist economies. The *factors of production are owned by the state. Their deployment is planned from the administrative capital which leads to the creation of an enormous bureaucratic superstructure. From there, the generalized plans will be broken into different sectors or regions. True central planning requires an enormous quantity of information which may be difficult to gather at the centre. Advocates of free market economics accuse central planning of inefficiency.

centrifugal forces In *human geography, those forces which encourage a movement of people, business, and industry away from central urban areas. These forces include traffic congestion, restricted sites, high local taxes and rents, obsolete technology, and lack of *amenity. *See* *counter-urbanization, *decentralization, *urban sprawl.

centripetal acceleration In meteorology, the force acting into the centre of a high- or low-pressure system which causes winds to blow along a curved path, parallel to the isobars. It is of greatest significance in *tropical cyclones and *tornadoes.

centripetal drainage *See* *drainage patterns.

centripetal forces Those forces which move people, business, and industry towards a centre, and are thus responsible for the growth of large central places. These forces include *accessibility, *functional linkages, *agglomeration economies, and *external economies.

centrography The study of the descriptive statistics used in measurements of *central tendency, such as *centrality of population or *population potential.

centroid In *Geographic Information Systems, the centre of gravity of an *entity, often used to reference *polygons.

chain In *geographic information systems, a line—a directed sequence of non-intersecting line segments and/or arcs—between two *nodes.

chain migration A migration process which depends on a small number of pioneers, who make the first moves to set up a new home in a new place. They send information back home, and this encourages further migration from the originating area. A British example comes from young men from the Indian subcontinent, who, after finding work and a place to stay in the UK, would send for other members of their families over the years. Chain migration is particularly apparent in international migration, where it is less easy for prospective migrants to get good information about their destinations, so that the importance of contact with family and friends is much greater. One result is that, in British cities for example, there is often a distinct segregation of migrant groups into different neighbourhoods according to the place of origin, as the original links are maintained. In 1984 Peach identified Dominicans and St Lucians in Notting Hill, Grenadians in Hammersmith and Ealing, Jamaicans south of the river, and so on.

Chain migration may also be seen as a movement in steps towards a distant objective. Groups or individuals may establish themselves in a new location for a period and then move on again with a new group occupying the space they have left. This definition accords with Burgess's (1926) view of successive waves of immigrants moving into a city. See concentric model.

chaincode In *Geographic Information Systems, a directional code for any *chain, based on 4, 8, or 16 unit-length *vectors, each of which is numbered. The beginning of the chain is referenced by x, y co-ordinates. Hence, **chaincoding**.

chalk A pure form of *limestone composed of the shells of minute marine organisms together with spherical or egg-shaped particles of calcium carbonate.

channel Any natural or man-made watercourse. The **channel capacity** of a river is its cross-sectional area in square metres, but the limits of the channel at each bank are not easy to assess. Water in the channel will experience **channel resistance** which slows or impedes the flow; friction with the bed is the major cause. See *bed-floor roughness.

Narrow, deep channels are associated with a regular discharge and coherent bank materials such as clay, while wide and shallow channels develop where discharge is irregular and the bank material incoherent, as with gravel.

Various classifications of channel types have been suggested. The typology may be based on the shape of the channel and its network, the nature of the bank and bed and the way in which these affect the channel, or the capacity of the river to carry sediment. See *hydraulic radius, *wetted perimeter. **Channel flow** is run-off within the confines of a channel—as compared to sheet flow.

channel order See *stream order.

channel roughness See *bed-floor roughness.

chaos theory The meteorologist, E. Lorenz, working in the 1960s on computer models of atmospheric processes, found that very minor differences in some initial atmospheric conditions would lead to completely different global conditions; it was Lorenz who cited the 'butterfly wing' analogy. The unexpected importance of these minuscule disparities is explained by the non-linear relationships in parts of computer models of atmospheric processes; that is the very disproportionate effect of some factors. Thus, the response of the atmosphere is chaotic; that is, irregular and unpredictable.

Chaotic behaviour is not random behaviour, since current conditions are still linked to future conditions (*see* *Laplacian determinism), but the future is only predictable on a limited scale. Chaotic systems do not repeat themselves exactly, but they often behave in a loosely recurrent fashion. This explains why short-term weather forecasts can be upwards of 80% accurate, but long-term forecasts can be wildly imprecise.

chapparal A *biome characterized by short, woody, and dense bushes having permanent, thick, hairy or leathery leaves to restrict loss of water through transpiration in summer. Seed-eating rodents and birds abound, together with small mammals. All may be preyed upon by wolves and big cats.

Chapparal is found on the west coasts of continents between 30 and 40° N and S of the equator, most notably in California. Summers are dry with average temperatures above 20 °C and winters are mild and moist.

charter group A group representing the most typical culture of a host community as a model for immigrants. This is often established by the first settlers. Thus, English-speaking, white, Protestant culture dominates the USA in spite of the influx of many different groups, and the pre-existence of Native Americans.

chase In Britain, an area of unenclosed land, such as Cannock Chase, Staffordshire, once used for hunting.

chatter mark In glaciology, a mark made upon a surface by a rock embedded in ice. Chatter marks are crescentic in shape and have been attributed to tension in a rock as ice pulls across it. The horns of the crescent point in the direction of the ice flow.

chelate A chemical substance formed from the bonding of compounds to metallic *ions, especially ions of aluminium, iron, and magnesium. The *leaching out of chelates is **cheluviation**.

chelation The formation of *chelates. By this process, some relatively insoluble materials may become soluble and be released into the soil.

chemical weathering The breaking down *in situ* of rocks. The main processes are: *carbonation, *hydrolysis, *oxidation, *solution, and attack by organic acids. Chemical weathering is at its most rapid in hot and humid climates, but present even in Arctic environments. *See* *mechanical weathering.

chernozem A *zonal soil with a deep A *horizon, rich in *humus from decomposed grass, and dark in colour.

Also known as *black earths*, chernozems develop on mid-latitude continental interiors where grassland is, or was, the natural vegetation.

Chernozem development is associated with temperate continental climates which have marked wet and dry seasons. There is sufficient moisture to permit the decay of the grass litter into humus, but not enough for *leaching to be significant. The B horizon is lighter brown, but is often absent. The lower horizons are often rich in calcium compounds. In *US soil classification chernozems fall into the category of mollisols, sub-order boroll.

chestnut soil A *zonal soil found in grasslands more arid than those under which *chernozems develop. These grasslands are sparser, so that there is less *humus.

The *xerophytic nature of much of the grassland under which chestnut soils develop also retards the development of humus. In the B *horizon, there is an accumulation of calcium carbonate. In *US soil classification, chestnut soils fall into the category of mollisols, sub-order xeroll.

chevron A V-shaped, triangular, erosional microform, characteristically developed on the shallow flanks of *cuestas in arid zones, such as the Aures Mountains of Algeria. They form as surface water, flowing downslope, becomes concentrated into a number of major channels.

child–woman ratio The ratio of the number of children below five to the number of women of child-bearing years, which may be expressed as:

$$P0-4/Pf15-44$$

where P refers to a population, the numbers refer to their ages, and f denotes women.

In the absence of universal registration of births, the child–woman ratio is a relatively good indicator of fertility.

chinook A warm, dry wind descending from the Rocky Mountains of North America. The chinook occurs sporadically between December and February, bringing dramatic rises in temperature; as much as 17 °C in 15 minutes. There are three causes which act together: the replacement of the normal, cold, high-pressure cells existing over the Great Plains by warmer air streams from the Pacific, *adiabatic heating by subsidence, (the adiabatic temperature change effected as the wind descends resembles the warming of the European *foehn, föhn), and the inhibition, or destruction, of the normal, nocturnal ground inversion. The name comes from the Inuit 'snow eater'.

chi-squared test A *non-parametric statistical technique used to find the significance of the difference between one or more frequency distributions and a hypothetical, expected distribution. An assertion is made: e.g. the pebbles in a river bed become progressively smaller from source to mouth. This assertion is then restated as a *null hypothesis: there is no difference in pebble size along the course of a river. It is this null hypothesis which must be rejected if the assertion is to be proved true. The formula is given as:

$$\chi^2 = \frac{\sum (O-E)^2}{E}$$

where O = observed (actual) frequencies; the pebble sizes as measured along the course of the river. E = expected frequencies; in this example, the size of the pebbles at the first site, since the null hypothesis expects no change.

The significance of the value of chi-squared is read from a graph, using the appropriate *degrees of freedom. If the point derived from the co-ordinates falls above the 0.1% line, for example, there is less than a 0.1% probability that the null hypothesis is true and, therefore, that there is a greater than 99.9% probability that the assertion is true.

The test may also be used to check the value of a random sample, by comparing the frequency distribution of the random sample with that of the whole population. Thus, if we wanted to use a random sample of voters to predict national voting behaviour, we would need to compare the age structure and socio-economic make-up of our sample with the national patterns.

chloro-fluorocarbons, CFCs Chemical compounds which, although essentially stable at ground level, undergo an exothermic photochemical reaction in the upper *atmosphere, releasing free chlorine radicals which break down ozone in the *ozone layer; a reaction potentially hazardous to human health. CFCs have now been banned, but many remain in storage in, for example, refrigerators.

chorography, chorology The establishment and description of geographical regions. This is the oldest practice of Western geography; Greek geographers sought to describe the parts of the earth. It has been seen as a synonym for *areal differentiation. R. Hartshorne (1939) was probably the most forceful advocate of geography as a 'chorographic science', and chorography is often seen as being in opposition to *spatial science.

choropleth A map showing the distribution of a phenomenon by graded shading to indicate the density per unit area of that phenomenon; the greater the density of shading, the greater the density in reality. The data used are of the *ratio and *interval type. Examples include maps of annual rainfall, January and July temperatures, or percentages of Labour voters.

Choropleths give a clear, but generalized picture of distribution and this may mask finer details. The choice of values for the classes may affect the visual picture given, as may the areal units for which data is available. Thus, for data running from 0 to 1000, a choropleth with class intervals of 0–250, 251–500, 501–750, and 751–1000 might look very different from one with intervals of 0–333, 334–666, and 667–1000, and a boundary of a census tract, or other areal unit can put a small 'high scoring' region into a large 'low scoring' region so that, in map terms, it disappears. *See* *modifiable areal unit.

Christaller *See* *central place theory.

c.i.f. (cost, insurance, freight) pricing A form of *uniform delivered pricing in an import or export context where prices are quoted with reference to a port.

cinder cone A cone formed by fragments of solidified lava thrown out during a volcanic explosion. Some volcanoes are made of alternate layers of cinder and *volcanic ash. The best-known example is Cinder Cone, in Lassen Volcanic National Park, USA.

circuit of capital Money *capital is needed to pay for *factors of production, such as raw materials, plant, energy, and labour. Commodities may then be produced which are more valuable than the cost of all the inputs; surplus value has thus been added. The commodities are sold—turned back into money—so that surplus value leads to reinvestment. With each circuit of capital, *accumulation takes place.

In real life, the circuit is very complex, since money capital may also be used for social spending, such as education, health, or defence, invested in research and development, used by the state to monitor or supervise economic activity, or exchanged in a capital market, but all these operations can help to reproduce capital. Furthermore, although the capital is circulating in *economic* terms, much of it may actually be fixed capital; that is, fixed in spatial terms, as in a railway or a warehouse.

If the circuit of capital is broken, perhaps because one element: money, raw materials, labour, is too expensive, economic crisis will result, and capitalists might then 'switch circuits'. For example, if manufacturing industry is in trouble, investors might move to speculation in property. This is then reflected in spatial variations in prosperity: regional, national, or global.

circulation 1. The movement of capital, labour, goods, and services throughout the economy.
2. More specifically, in population geography, short-term, repetitive movements of individuals, where there is no intention to change residence permanently. For example, many West African men move to cities after harvesting their crops. Employment opportunities are greater there, but the workers return to the land before the rains to plough, plant, and weed their crops. Other examples include the movement of students between term and vacation, movements between first and second homes, and commuting. *See* *mobility.
3. The movements of air masses, or ocean currents.

circumpolar vortex A vigorous, westerly flow of air in the upper and middle *troposphere over mid-latitudes.

cirque Also known as a **corrie** or **cwm**, this is a circular, armchair-shaped hollow cut into bedrock during glaciation. The side and back walls are steep, but the front opens out downslope. Cirques may be up to 2 km across. The formation of cirques remains unclear. It is suggested that frost-shattered material falls into the *bergschrund and contributes to the erosion of the cirque. During glaciation, ice is thickest in the centre of the cirque and is thought to undergo *rotational slipping, thus *overdeepening the cirque floor but merely riding over the low bar at the mouth.

The sides and back are subject to intense physical weathering and it is

suggested that *freeze–thaw occurs at the base of the *bergschrund, but
such a crevasse could not be deep because ice becomes plastic when it is
more than 40–60 m thick (according to the elevation). Since cirques can
have back walls up to 1000 m high, freeze–thaw would not seem to be the
major formative process. Cirques seem to grow by headward extension,
biting back into the mountain mass until only *arêtes or *pyramidal peaks
remain.

In the temperate zones of the Northern Hemisphere, cirques face
outward in directions between north-east and south-east. This is due in
part to decreased insolation on north-east-facing slopes and in part to the
build-up of drifted snow in the lee of westerly winds. Cirques are widely
distributed in glacially eroded uplands, examples including Cwm Idwal,
Snowdonia, and the cirque containing the Schwarzensee in the Zillertal
Alps. There are no cirques, however, in those parts of glaciated mountains
that were completely submerged by ice.

cirro-, cirrus Like a lock of hair; referring to high, slender, ice-crystal
clouds with a feathery appearance, sometimes pulled out into 'mare's tails'
by strong upper-air winds. When these coalesce to form a thin layer of
high cloud, they are termed cirro-stratus.

Patchy, mottled, high cirrus is **cirro-cumulus**, often called a mackerel sky.
This forms above very active warm fronts. *See also* *cloud.

city Initially, a city was a European town which was the centre of a
bishop's diocese, and had a cathedral. This rather specialized term has
been superseded, and a city is now defined as a large urban centre
functioning as a central place which can provide very specialized goods
and services. There is no world-wide, or even European agreement over
limiting figures of population size or areal extent for a city.

city region The area around a city which serves and is served by the city.
Synonymous with *umland and *urban field.

city-filling activity Activity which provides goods and services for use
within the city. A synonym for *non-basic activity.

city-forming activity An activity, such as the provision of goods and
services, which contributes to the economic well-being of the city by
attracting income from beyond the city. A synonym for *basic activity.

city-size distribution The frequency with which the settlements of a
country or region occur in certain, arbitrarily defined, population size
groups, such as under 10 000, 10 000–25 000, and so on. In general, large
settlements are less frequent than small ones, but when size is graphed
against frequency, several patterns may be recognized; the most common
being the *binary, *log-normal, and *primate patterns.

See *settlement hierarchy.

clachan A *hamlet in the Scottish Highlands; usually a formless
collection of houses with a church.

class A concept which recognizes different strata in society. Pre-industrial

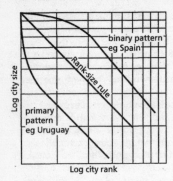

FIGURE 13: *City-size distribution*

societies saw class in terms of rank, resting on tradition and an intricate
system of rights and duties. Capitalist societies see class as defined by
socio-economic status, as in the *Registrar General's classification. Max
Weber's analysis offers a concept of class based on market relations. He
puts forward the concepts of **property classes**, which control various forms
of property, **acquisition classes**, which have marketable skills, and **social
classes**, as defined by common mobility chances. Marxists also offer an
economic analysis. The most basic was the identification of two classes:
capitalist and working class as defined by ownership of the means of
production. Feminist geographers are particularly concerned with the
intersection of class and gender: how do different classes experience
gender differences, for example?

An important recent development for geographers has been the study of
the links between space, place, and class: the role of space in the
construction of class, and the development of class-consciousness and class
practices. Very simply—why are different classes located where they are?—
and does this location reinforce their class grouping? Of key importance
here is *residential segregation and social proximity, but even architecture
can be used to reinforce class distinctions: think of the appearance of an
interwar council estate for example.

class interval The set of limits by which data are classified, as in 0–4,
5–9, 10–14, and so on. Fixing of class limits for use in *histograms or
*choropleths can change the impact of the data thus sorted. Class intervals
may be set by equal groupings at natural breaks in the data, or as part of a
logarithmic sequence. *Iteration may also be used to minimize the amount
of deviation within the classes. In an attempt to get away from the
problems which arise from setting class intervals, some cartographers have
used computer-assisted methods which generate a continuous series of
tones with no 'breaks'; unfortunately, people generally find maps with this
type of shading difficult to read.

classification Grouping phenomena into classes is a basic step in most
sciences, and geography uses an enormous range of classificatory systems.

An understanding of certain basic methodologies is useful in developing classifications for individual use. For example, geographers distinguish between **intrinsic classification** which depends on natural differences or 'breaks' in the features studied, and **extrinsic classification** which uses arbitrarily defined class limits. Similarly, there is a distinction between **monothetic classification**, which uses one criterion (such as grain size in a soil particle) and **polythetic classification** which uses a number of criteria (such as social rank, urbanization, and segregation in *social area analysis).

Attribute-based classification is based on 'present' or 'absent' evidence (a climate may or may not have a dry season), while **variable-based classification** uses a variable, such as unemployment, and forms classes on the basis of the degree to which that variable is present.

clast A rock fragment or a pebble, over 5 mm in diameter and forming part of a *sedimentary rock.

clay 1. A geological deposit, such as Oxford Clay.
2. In a soil, mineral particles less than 0.002 mm in size. When dry, clay is hard; when wet it swells and becomes pliable and sticky. **Clay colloids** are finely divided clays dispersed in water. These particles have a negative surface charge which attracts positively charged ions. These minute particles are among the most reactive constituents of a soil.

clay micelles Individual clay particles, platey in form, with a diameter of less than 2 μm, having a negative charge and therefore being able to attract cations within a soil.

clay mineral Not clay-sized particles, but a group of hydrous aluminium silicates, such as kaolinite, created by the intense weathering of rock. Clay minerals affect the physical properties of soils because they expand when wet.

clay-with-flints A deposit of clay containing flint, often in pipes or potholes, which overlies chalk. It has been suggested that this deposit comes from the weathering of chalk.

clear felling The practice of cutting down all the trees on a site. This leaves the ground unprotected against erosion and is unattractive.

cleavage 1. A division in society due to political or partisan allegiance. Examples include: Yankee versus Confederate in the American Civil War, revolutionaries versus the *Ancien Régime* in the French Revolution, or workers versus bosses.
2. The ability of a rock to split along a **cleavage plane**.

cliff A steep rock face, usually facing the sea. While an **active cliff** is still subject to the forces of marine erosion, an **abandoned cliff** is protected from wave attack by a *wave-cut platform or by a *barrier beach. As the abandoned cliff is exposed to *subaerial denudation it becomes less steep, and its upper edge more indented.

climate A summary of mean weather conditions over a time period, usually based on thirty years of records. Climates are largely determined by location with respect to land- and sea- masses, to large-scale patterns in the *general circulation of the atmosphere, *latitude, altitude, and to local geographical features.

climatic change Evidence of climatic change ranges from differing geological strata to the study of pollen grains and tree rings.

The early climatic history of the world is not well understood but it is known that during the last 55 million years the earth has been cooling, and that during the last million years there have been alternating of glacial and *interglacial episodes. *See* climatic optimum, *Little Ice Age. External causes of climatic change may include: changes in solar output; changes in the number of sunspots which seems to vary in an 11-year cycle, changes in the ellipticity of the earth's orbit, which roughly follows a 100 000-year cycle, and changes in the earth's axis of rotation, which alters the season of perihelion, and which follow an approximately 100 000-year cycle. *See* *Milankovitch cycles.

Internal causes include *vulcanicity, which can increase the density of atmospheric nuclei in the atmosphere, variations between the distribution of land and sea, *continental drift, and changes in the atmosphere–surface–ocean system. Recent research suggests that human agency is altering the climate. *See* *greenhouse effect.

climatic climax community *See* *climax community.

climatic geomorphology Also known as climatomorphology, this is the association of types of landform with different climates, as when the processes in particular climatic zones form specific landforms e.g. *periglacial landforms in *tundra climates. Many classifications have been proposed, but the thesis is a controversial one, partly because there seems not to be a consistent link between climate and dominant geomorphological process(es), and partly because climate is only one factor in the development of landforms.

climatology The study of the climates of the earth, their origin, and their role as elements of the natural environment.

climax community An *ecosystem which experiences a turnover of species, but with an overall steady state. This steady-state ecosystem is characterized by a complex, highly integrated community structure with high species diversity, but relatively low, individual population densities not subject to serious fluctuation, and can be seen as the final stage of a *succession. An example is a mature oak woodland.

A **climatic climax community** may be seen as a climax community which has developed in response to the prevailing climatic conditions; tropical rain forest is an excellent example.

climograph A graph of monthly average temperature plotted against average humidity. The monthly points are joined by a line. The shape and location of line thus drawn indicates the nature of climate in terms of heat and humidity.

clinometer An instrument used to establish angles, usually on a hillslope and more rarely along bedding planes. While commercially produced clinometers tend to be expensive, students can make home-made clinometers which perform to a reasonable degree of accuracy, at low cost.

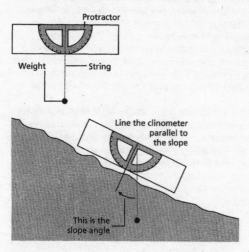

FIGURE 14: *Clinometer*

clint The raised portion of a *limestone pavement lying between the *grikes.

clisere The series of *climax communities which result from a major climatic change, as, for example, when conditions change from glacial to interglacial.

closed system A system marked by clear boundaries which admit of no movement of energy across them. The *entropy of any closed system never decreases.

cloud A visible, dense mass of suspended water droplets and/or ice crystals suspended in the air. Clouds generally form when air is forced to rise: at a *front, over mountains, or because of *convection. Clouds mirror the atmospheric processes which cause them; the approach and passage of a warm *front, for example, often follows the sequence: cirrus, cirro-stratus, alto-stratus, nimbo-stratus. At active *ana-fronts all these clouds may take on a more cumulus form. Atmospheric *convection currents are generally indicated by the presence of cumulus or even cumulo-nimbus clouds. A cumulus cloud will often form over a heated surface and then shift with the wind, so that further cumulus is formed over the same 'hot spot'. If this process continues a line of cloud, a *cloud street, is formed.

*Turbulence, generated by moderate winds, is a common cause of stratus cloud, which is often trapped beneath an *inversion. Turbulence also gives rise to a nearly continuous sheet of strato-cumulus cloud. *See also* *cloud classification.

cloud classification Clouds may be classified by form and by height.

1. Low-level clouds (0–2 km above mean sea level) stratus, S: extensive, shallow cloud sheet, often yielding *drizzle or light rain; strato-cumulus, Sc: shallow cloud sheet, broken into roughly recurring *cumuliform masses, often yielding drizzle or snow; cumulus, Cu: separate, hill-shaped clouds, (hence cumuliform) with flat, and often level, bases which are often at the same height; cumulo-nimbus, Cb: large, high *cumulus, with dark bases, often producing showers. The cloud tops are often formed of ice crystals.

2. medium-level clouds (2–4 km above mean sea level) alto-cumulus, Ac: shallow cloud sheet broken into roughly regular, rounded clouds; alto-stratus, As: featureless, thin, translucent cloud sheet; nimbus, Ns: extensive, very dark cloud sheet, usually yielding precipitation.

3. high clouds (tropical regions 6–18 km high, temperate regions 5–14 km high, polar regions 3–8 km high) cirrus, Ci: separate, white, feather-like clouds; cirro-cumulus, Cc: shallow, more or less regular patches or ripples of cloud; cirro-stratus, Cs: shallow sheet of largely translucent cloud.

To these genera, suffixes may be added:

capillatus: like a feather, or thread

congestus: growing rapidly, in cauliflower form

fractus: broken or ragged, especially of stratus

humilis: shallow

lenticularis: like a lens, especially of alto-cumulus, cirro-cumulus, and strato-cumulus

radiatus: banded

See also: *banner cloud, *lenticular cloud, *mother cloud, *pileus, *virga.

cloud droplet *See* *precipitation.

cloud seeding Any technique of inducing rain from a cloud, usually by dropping crystals of dry ice (frozen CO_2) or silver nitrate on to it. These act as *condensation nuclei. Most methods of limiting the development of *hail rely on cloud seeding using ice nucleants, or with silver oxide. In theory, hail damage can be reduced by 25% through cloud seeding, but there has been little consistent success in practice.

cloud streets Parallel lines of small cumulus. Over land, these are irregular, running downwind from a sunny slope. See *cloud. Over warm seas, they are very regular, and may arise from some atmospheric process fostering convection along parallel lines; the mechanism is not fully understood.

Club of Rome A society, formed in 1968, of one hundred members ranging from academics such as scientists, social scientists, economists, and teachers, to managers and civil servants, with the aim of understanding the workings of the world as a finite system and suggesting alternative options for meeting critical needs. Concerns have included inequality, on global and regional scales, pollution, unemployment, and inflation. Its publication *Limits to Growth* (1972) argued the necessity of slowing down population growth in order to ward off *Malthusian checks.

cluster analysis A type of *multivariate analysis which aims to group a set of variables or individuals into classes, so that the objects in each class are as like each other as possible and as unlike the other classes as possible, as defined by a designated list of characteristics and indicators. In social geography, the technique can be used to create classifications of, for example, urban areas by type. In general, the classification process begins by drawing up a table of *correlation coefficients of dis/similarity between each pair of objects. From here, the objects can be combined into larger and larger groups, or broken down into smaller and smaller ones.

clustered The layout of features, especially settlements, close together in a group. *See* *nucleated settlement.

coalescence theory *Bergeron–Findeisen's theory cannot explain the formation of all tropical rainfall since ice crystals are often absent in tropical clouds. Langmuir's coalescence theory suggests that the small droplets in clouds grow larger by coalescence until they are sufficiently heavy to fall. As they fall, they collide with other droplets, growing still bigger. Coalescence may be due to collision, but not every collision results in coalescence. The deeper the cloud, the bigger the drops grow; up to about 5 mm in diameter.

coastal dune A ridge or hill which forms when marine deposits of sand are blown to the back of the beach. The rate of formation and the extent of these dunes is dependent upon the supply of sand to the beach.

coastal plain A lowland area adjacent to the sea, which may have formed from marine sediments which have subsequently been revealed by a fall in sea level, as in the south-eastern coast of the USA.

cobble A stone of 60–200 mm in diameter, rounded, or partially rounded, by wave action or running water.

cockpit karst Also known as **kegelkarst**, this is a landscape of star-shaped hollows surrounded by steep, rounded hills, and found in tropical karst country. The cockpits, now floored with alluvium, are the hollows, or *dolines, formed by the solution of limestone. They can be 100 m deep and usually contain a *streamsink. The classic area of cockpit karst is found in Jamaica, but it also occurs, for example, in Dalmatia. *See also* *tower karst.

coefficient of concentration *See* *Gini coefficient.

coefficient of dispersion A statistical measurement which indicates the nucleation or dispersion of settlement. For a rural unit,

$$C = \frac{p \times n}{P}$$

where C is the coefficient of dispersion, p equals total population of the rural unit, P is the population outside the main village, and n is the number of settlements. The higher the coefficient, the less the settlement is dispersed.

coefficient of localization Also known as the **index of concentration**, this measures the degree of concentration of a given phenomenon, such as industry, over a set of regions. The coefficient, L, is the sum either of the positive or of the negative deviations of the regional percentage of workers in the given industry from the corresponding regional percentage of all workers in industry. A value of 0 would indicate that employment in the given industry is distributed very evenly over the regions. A value of 1 indicates extreme concentration of industry in one region only. This statistic alone is not very helpful but it may be used to compare two different regions or two different industries.

coefficient of variation, V The *standard deviation of a data set expressed as a percentage of the *arithmetic mean, this is a measurement of the amount of variation in a data set. The lower the value of V, the more the overall data approximate to the mean. It is used in comparing two apparently similar data sets.

cognition Those processes involved in the gathering, organization, and use of knowledge. Cognition is, simply, thinking, and is often compared with feeling, known as affect, and trying, known as volition. Much of *behavioural geography is concerned with cognition; the way in which people perceive and respond to outside stimuli.

cognitive dissonance A mismatch between what is perceived and what is (between cognition and reality) so that an individual may seem to act irrationally. Perhaps the most frequently quoted example is of cigarette smoking—where the smoker is well aware of the serious health risks involved but chooses to disregard them. It applies in geography to people who choose to settle in areas, like San Francisco or Los Angeles, at high risk from natural hazards.

cognitive mapping The acquisition, coding, storage, manipulation, and recall of spatial information within the mind. Cognitive mapping simplifies the complexity of the landscape and the *mental map derived thereby is held to influence behaviour.

cohesion Adhesion; the force by which materials, usually clay minerals in soils are held together. Cohesion provides a measure of the strength of a material. It is the result of chemical and electrostatic forces, and can be measured by the Mohr–Coulomb equation.

cohort A group of people who experience a significant event, such as birth or leaving school, during the same period of time, usually a year but also in five-year groups; an example would be the 'baby booomers' or 'ageing hippies'—the choice of name varies according to who is describing this particular cohort! More specifically, all children born in the UK in 1980 would form the birth cohort of 1980. **Cohort analysis** traces the subsequent vital history of cohorts; the most common type of cohort analysis uses age-groups, also known as **birth cohorts**, often in five- or ten-year age bands, to study mortality rates. The major problem of cohort analysis is to distinguish between the effects on the cohort of getting older (age effects), of common experiences like National Health orange juice (cohort effects), and particular historical events, like a war (period effects).

Cohort fertility is the total of live births born to a particular birth or marriage group.

col 1. In the landscape, a pass between two peaks or ridges. Landscape cols may have been formed by: the *headward erosion of a *cirque, *river capture, the beheading of *dip-slope valleys by *scarp retreat, or the localized *differential erosion of a ridge.
   2. In meteorology, a narrow belt of relatively low pressure, but not a depression, between two *anticyclones. Here, isobars are few and therefore winds are slack.

cold glacier Sometimes known as a **cold-based glacier**, or **polar glacier**, this is a glacier with its base well below 0 °C, unlubricated by *meltwater, and therefore frozen to the bedrock. *Accumulation is slow; snow may take 150 years to turn to ice. *See* *firn. Cold glaciers move very slowly—rates of 1–2 m per year are not uncommon—therefore they cause very little erosion through *abrasion, although *plucking may be reinforced.

cold low Also known as a **cold pool**, or a **polar low**, these terms refer to areas of low pressure and temperature in the middle *troposphere. Such lows do not appear on surface *synoptic charts, but are important in *Arctic and *Antarctic meteorology.
   A second type of cold low is the **cut-off low**, formed in mid-latitudes by the cutting-off of polar air from the main body of cold air, nearer the poles. This occurs during periods of low index circulation (*see* *Rossby waves). Such lows are associated with unsettled weather and, in summer, thunderstorms.

cold stress *See* *wind chill.

collective farm A type of farm organization identified with socialist regimes, such as North Korea and the Republic of China. The farm, made by the merging of farms which used to be owned by individuals, is owned by the state, but permanently leased to the members of the collective. The workers are shareholders, rather than state employees. The collective is responsible for paying employees and for inputs such as machinery or fertilizers, and is, in theory, self-governing, although the *centrally planned state usually sets production targets. With the collapse of communism in Eastern Europe and the former Soviet Union, this is a form of organization which is becoming much less common; and even in China rural collectivism is being replaced by private forms of agricultural organization.

collectivism A school of thought which maintains that the *factors of production and the means of distribution should be owned by all and not by individuals who might pursue their self interest at a cost to the state. It advocates public control, which is not necessarily brought about by state ownership.

collision margin In plate tectonics, the boundary between two continental plates. Such margins are a hybrid type because the two slabs of continental crust that eventually collide were initially separated by oceanic crust with passive or destructive margins. The oceanic crust is consumed leaving the two continental crusts adjoining. At such a margin, the lower plate is not consumed but, together with the upper plate, forms a double layer of crust, as in the Himalayas.

collision theory This states that raindrops grow by colliding and coalescing with each other, especially in tropical, maritime *air masses. *See also* *Bergeron–Findeisen theory and *coalescence theory.

colluvium The mixture of soil and unconsolidated rock fragments deposited on, or at the foot of, a slope.

colonialism The acquisition and colonization by a nation of other territories and their peoples. In this respect, colonialism is as old as society. The term took on a more specific meaning in the late nineteenth century when colonists saw it as the extension of 'civilization' from Europe to the 'inferior' peoples of 'backward' societies. It may also be seen as a search for raw materials, new markets, and new fields of investment. Sometimes, but not always, colonialism was accompanied by **colonization**; that is, the physical settling of people from the imperial country. Typical aspects of colonialism include: racial and cultural inequality between ruling and subject people, political and legal domination by the imperial power, and exploitation of the subject people. Many commentators see colonialism as a key cause of uneven development.

Although independence from former colonization has been achieved almost everywhere, most people accept that it has been replaced by *neo-colonialism.

colony In biogeography, a group of closely associated, similar organisms, as in a coral colony.

COMECON The Council for Mutual Economic Assistance, a trading bloc which was formed in 1949 to encourage the trade and economies of the then communist nations of East Europe and the USSR. The members were the USSR, Poland, Czechoslovakia, Hungary, Romania, Bulgaria, East Germany, Mongolia, and Cuba, with Yugoslavia as an associate. It was dissolved in 1991.

command economy A type of communist economic system, sometimes called a **centrally planned economy**, where the state controls macro-economic policy and entrepreneurial activity, but allows some freedom for economic decisions about employment and consumption at the household level. In other words, there is state control of the *factors of production

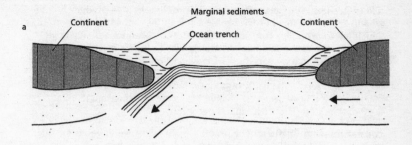

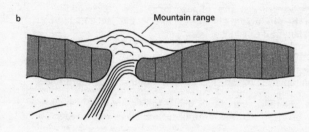

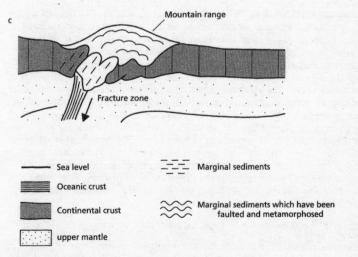

	Sea level			Marginal sediments
	Oceanic crust			
	Continental crust			Marginal sediments which have been faulted and metamorphosed
	upper mantle			

FIGURE 15: *Collision margin* (after Dervey and Bird, 1970)

and centralized, state planning—what to produce, how to produce it, and who to produce it for—but with some freedom for individual decisions, like which job to take. This contrasts with both war communism, where all spheres of decision-making are centralized (as in Russia between 1918 and 1921) and with market socialism where decisions are decentralized but state ownership remains. Command economies have been criticized because they tend to be badly organized, lack quality control or worker incentives, and have been responsible for severe environmental degradation. Compare with *market economy.

commercial agriculture The production of agricultural goods for sale. Compare with *subsistence agriculture, and *see* *cash cropping, *horticulture, *plantation agriculture.

commercial geography A form of geography, now superseded, concerned with the production and supply of raw materials, including agricultural output, and finished goods.

commercial ribbon A form of *ribbon development where service and retail outlets develop along the major routes radiating outward from the city. Such development is confined to the lots adjoining the main roads.

comminution The reduction of rock debris to fine powder, usually as a result of *abrasion and *attrition. Comminution has caused a reduction of the shingle on beaches, which may increase problems of coastal erosion. *See also* *rock flour.

commodities Very loosely, goods and services.

Common Agricultural Policy, CAP The EEC, now the EU, devised a policy designed to make its member states self-sufficient in foodstuffs, to secure farmers' living standards, to increase productivity, and to ensure reasonable prices for consumers. This has been achieved by the setting of *intervention prices; when the market price for a commodity falls to this level the EU will buy the entire production at the intervention price. It is this policy which has led to the accumulation of the notorious 'wine lakes' and 'butter mountains', and to high food prices within the Community. In an attempt to rectify these and other problems, a 1988 Green Paper, noting the diminishing importance of agriculture, stressed the necessity of shifting emphasis from *agricultural* policy to *rural* policy, so that rural industrialization, forestry, and tourism would also fall within its remit, recognized regional variations in the rural problem throughout the Community, and recommended the dismantling of price support systems to curb over-production. The EU has also introduced *set-aside policies, and the CAP now has an explicit environmental content.

common field In medieval times, an open field with common rights of cultivation and grazing.

common land Land which is privately owned, perhaps by an individual or a local authority, but over which others have legal rights.

common market An economic association of states into a single trading market having little or no restriction of movement of individuals, capital, goods, and services within it and with a united trading policy towards non-member states. The *EEC was an example.

Common Regional Policy In December 1974, the then nine members of the European Community agreed to set up the European Regional Development Fund (ERDF) to provide aid for problem areas. National definitions of problem areas were used, projects first being vetted by national authorities and then forwarded to Brussels for evaluation and decision. In 1981 the policy was revised, and 20% of the total fund was reserved for areas with very low per capita GDP and very high long-term unemployment. By 1987, the ERDF accounted for 9.1% of the total EC budget. Other EU funds for tackling social and economic problems include the European Investment Bank (EIB) and the European Agricultural Fund.

communism Historically, the principle of communal ownership of all property; basic economic resources are held in common. Modern communism is grounded in the ideas of Karl Marx. He hoped to see a society with no socio-economic difference between, for example, manual and intellectual labour, or urban and rural life. Social relations would be regulated by the maxim, 'from each according to his ability and to each according to his needs'. Centrally planned economies have been developed in accordance with this ideology and there have been many forms of communism, all supposedly seeking the classless society.

community 1. In ecology, a naturally occurring, non-random, collection of plant and animal life within a specified environment. The community is named after the physical environment, such as a freshwater lake community, or after the dominant species, such as an oak woodland community.
   2. In human geography, an interacting group of people living in the same territory: town, village, suburb, or *neighbourhood. This term is used in different ways: some see communities as having shared modes of thought and expression, and may use the term in a non-territorial sense, as in 'the gay community'. Social geographers see communities as combining human alliances with local social systems in specifically defined locations.
   Some social scientists have argued that rural communities are *gemeinschaft communities, contrasting with urban, *gesellschaft communities, but later studies claimed that, while urban districts may not have all the positive features of rural societies, they do have beneficial features.

community charge A form of local taxation whereby a fixed charge is paid per adult. The cost of the fixed charge may vary from one local authority to another. During the late 1980s in the UK, this form of taxation, which was known as the Poll Tax, replaced local authority rates. It has been replaced, in its turn, by council tax, a tax based on property rather than individuals.

commuting The movement from suburban or rural locations to the place of work and back. Commuting developed as transport systems improved; initially it was the rich who could afford to commute by train, so that exclusive suburbs developed, but the tram, and later the bus and private car, put commuting within the means of most workers, with a corresponding increase in the size of the city, and the problems of rush-hour congestion; the predicament of Parisian commuters in the public sector strikes of late 1995 indicated the crucial importance of transport systems for a majority of commuters. There is some correlation between city size and commuting distance—some workers commute to London from Leeds, for example, while the catchment area of commuters to Oxford is much smaller. The *gravity model may be used to predict commuter flows.

Commuting is usually on a daily basis but can occur weekly. Most commuting is **in-commuting** involving movement into the city to work, but **reverse commuting** also occurs where residents in the inner city travel daily to workplaces in the suburbs. **Lateral commuting** involves the journey from one residential location to another as the suburbanization of industry develops. *See* *Lowry model.

comparative advantage The advantage of some nations or regions to produce goods better and more cheaply than less favoured nations or regions. This comparative advantage leads to trade, as nations exchange those goods which they can produce more easily for goods not readily produced at home. The advantage is usually seen in resources of raw materials and labour, but very often the competitive performance of producers is based on better marketing, delivery, reliability, and quality control.

This is an important concept in understanding regional specialization, through which all regions actually benefit from exchanging the products they make best, even if each is capable of supplying all its own needs, but the system will only work properly if *free trade is permitted. The danger of regional specialization, of course, is over-dependence.

comparative cost analysis An analysis of industrial costs in order to establish a least-cost location. It is usually carried out when there is a small number of possible locations, and where there is a relatively small number of inputs; it is particularly suited to the early stages of metal manufacturing. However, it is not as easy to calculate total costs when a large number of inputs are involved. It is very difficult to express *agglomeration economies and *external economies in financial terms.

compatibility In land use, being able to accommodate more than one form of activity. In **complete compatibility**, two uses may exist side by side at the same time like walking and hill farming. **Partial compatibility** means that different activities, like water-skiing and fishing, can only take place at different times in the same place. The concept is usually applied to recreational land use.

competence In hydrology, the largest size of particle that a river can carry. Just as the *discharge of a river varies with climate, *bed-floor roughness, and so on, so the competence of the river will vary with water depth and water surface slope. The very rough-and-ready sixth power law suggests that a doubling, for example, of river velocity would increase competence by 2^6, i.e. by a factor of 64. *See also* *Hjulström curve.

competition Competition occurs when a necessary resource is sought by a number of organisms. **Intraspecific competition** occurs within a single species; **interspecific competition** occurs between different species.

complementarity An expression of mutual dependency based on an ability to produce goods in one area which are needed in another. Initially, complementarity was seen as operating in two very different regions, so that, for example, a tropical region might supply the fruit which cannot be grown in a temperate region, but it may also be seen to occur between similar environments with different regional specialities. *See* *comparative advantage. E. L. Ullman (1954) believed complementarity to be one of the three fundamental principles underlying *spatial interaction. The other two are *transferability and *intervening opportunity.

complete chain In *Geographic Information Systems, a *chain with left and right area identifiers and *node identifiers.

complex object In *Geographic Information Systems, an object which is a composite of various single, primitive objects of the same class.

components of change A classification system used in studying changes in economic activity, most often in manufacturing. Three types of change may be recognized: *in situ*, as employment grows or declines; by birth and death, as new plants open and old ones close; and through migration of plants into and out of an area.

compressional, compressive flow In glaciology, valley floors exhibit ridges and hollows, probably because the glacier was moving in different ways along its length. Where a glacier is slowing up, the flow is compressive. The planes of weakness curve upwards and outwards. In zones of compression, the ice may bring up material from the valley floor, thus eroding the valley; compressive flow accentuates pre-existing hollows.

compressive stress *See* *stress.

compulsory purchase order A directive to enforce the purchase of land by government or local authorities from a private landowner. Orders may be issued for redevelopment or for building new roads.

computer-assisted cartography The use of computer hardware and software has made *digital mapping possible, often in combination with *Geographic Information Systems. Computer-assisted methods have a number of advantages over manual cartography, such as speed, and ease of production (cartographers do not have to be as skilled as once they were), the capacity for keeping maps up to date, and the possibility of rapidly trying out a number of cartographic options.

computer graphics Images constructed under computer control, displayed on a cathode ray tube.

concave slope A slope which declines in steepness with movement downslope, also known as a *waning slope. Most often, the concave element of a slope occurs at the base, where it is known as a *basal concavity.

concealed unemployment A situation whereby individuals—primarily housewives—know that there are no jobs available and do not register as unemployed, even though they would prefer to have a job. The official figures thus underrepresent the 'true' situation.

concentration and centralization The tendency of economic activity to congregate in a restricted number of central places. This convergence of economic activity is encouraged by *functional linkages, and *external and *agglomeration economies. The centralized core tends to develop at the expense of the periphery, attracting cheap raw materials and the best brains.

Concentration and centralization occur on a number of scales; within a nation they result in regional inequality, within a continent or trading block they result in variations in national prosperity, and globally they help to bring about the contrasts we see between *North and South.

The driving force behind concentration and centralization is said to be the tendency for economic activity to be organized into larger and larger hierarchical structures; a pattern which began in the nineteenth century but which has extended into the twentieth century. Industry is in the control of fewer and fewer capitalists and the growth of *multinational corporations has seen economic activity all over the world increasingly directed from a few head offices in Europe, North America, and South-East Asia. This pattern has not been restricted to capitalist enterprises, however; in the former *command economies of Eastern Europe and the Soviet Union control was also concentrated in a small number of locations.

concentration theory The concept of a hierarchy of industrial developments whereby each level is linked to the next. Any two units, at whatever level, are linked to one unit above. Thus, two industrial plants merge into one industrial district while two industrial districts form one manufacturing town.

concentric zone theory The theory, proposed by E. W. Burgess (1926), that urban land use may be classified as a series of concentric zones.

Zone I, the *CBD, lies at the centre of the city. Zone II is in transition. It is the crowded, multi-occupied zone of the city first invaded by migrants. Within this Zone are the *ghetto areas (these are not necessarily slums). In Zone III are the working men's houses, the area of second generation immigrants, one step up from Zone II. Zones IV and V are residential; Zone IV for the better-off and Zone V for the commuters. All these zones are held to have evolved separately and without planning. They result from the competition of different socio-economic groups for land. This competition results in variations in the cost of land and, therefore, causes segregation within a city. The model assumes uniformly flat, and available, land, and ignores the importance of transport routes, but relies on the theory that city growth results from distinct waves of in-migrants, that is to *invasion and succession. In this last respect it is therefore more applicable to cities in the USA than to European cities. *See also* *sector theory, *multiple nuclei model, *Mann's model.

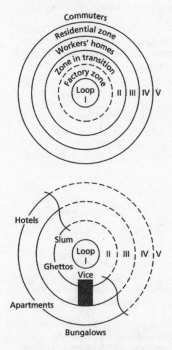

FIGURE 16: *Concentric zone theory*

concordant Complying with. In geomorphology, relief and drainage may be concordant with geological structure. A **concordant coast** runs parallel to the ridge and valley grain of the country, and is also known as a *Pacific coast—after the coast of California—or a *Dalmatian coast, after the Adriatic coast from Split to Dubrovnik.

condensation The change of a vapour or gas into liquid form. This change of state is accompanied by the release of *latent heat, which alters the *adiabatic temperature change in rising air. Condensation in meteorology can be caused by: the cooling of a constant volume of air to *dew point, the expansion of a parcel of air without heat input, the evaporation of extra moisture into the air, the fall in the moisture-holding capacity of the air due to changes in both volume and temperature, and contact with a colder material or *air mass.

The likelihood of water vapour condensing will depend on the *saturation vapour pressures of water and ice at any given temperature, and/or the presence of *condensation nuclei, since water vapour can be cooled to well below 0 °C before condensation occurs. *See* *condensation nuclei, *Bergeron–Findeisen theory.

condensation level The point at which rising air will cool to dew point, condense, and form clouds. *See* *tephigram.

condensation nuclei Microscopic *atmospheric particles, which attract water droplets, and which may then *coalesce to form a raindrop. They may be specks of dust or clay, or particles formed from industrial processes (e.g. sulphur dioxide, sulphur trioxide), known to American meteorologists as **combustion nuclei**. Condensation nuclei are vital in cloud formation, since condensation only occurs spontaneously when *relative humidity exceeds 400%. However, in accordance with *Raoult's law, soluble condensation nuclei will lower *saturation vapour pressure enough for condensation to occur. In the *Bergeron–Findeisen theory, ice crystals act as condensation nuclei, initiating condensation in unsaturated air. *See also* *cloud seeding, *hygroscopic nuclei.

conditional instability *See* *instability.

conduit A, possibly natural, channel or pipe which conveys liquids.

cone In geomorphology, a relief feature, circular or semi-circular in plan and rising to a point in the centre.

cone volcano A volcanic peak with a roughly circular base tapering to a point. Cones may be built solely of *lava or of *scoria, or of an interbedded combination of the two. Lava flows and layers of *pyroclasts form composite cones. Parasitic cones form round smaller vents on the flanks of the volcano.

confidence limits The percentage of times a given outcome in a statistical analysis could be expected to have occurred by chance. Thus, at the 99% **level of confidence**, the result will be expected to occur by chance only 1 time in 100. It is customary to check the level of confidence in the result of any investigation using a statistical analysis, generally by consulting a published table and referring to the appropriate *degrees of freedom.

conflict theory The view that more is achieved as a result of conflicting interests than by co-operation.

congelifluction, congelifluxion A form of *mass movement which occurs in *periglacial environments. Water builds up near the surface because the ground below is frozen. This water acts as a lubricant for matter sliding downslope.

congelifraction *Freeze–thaw.

congeliturbation In a *periglacial landscape, the heaving, thrusting, and cracking of the ground by *frost action. The formation of *patterned ground and the initiation of *solifluction are common results of congeliturbation.

congestion The restriction of the use of a facility by over-use. The term is generally used to indicate the slowing of urban traffic because too many vehicles are competing for too little space, but it can be applied to any excessive demand for any facility, when use exceeds *carrying capacity. Congestion on a routeway depends on the carrying capacity of the route, the volume of traffic, and the varying proportions of the total freight and passenger traffic carried by competing means of transport (the *modal split). The effects of congestion involve long, frustrating, and often costly, delays, road accidents, air pollution, and noise, all of which create an *externality to the urban economy. They are, however, difficult to quantify in terms of cost and the individual may have little control over them.

congestus See *cloud classification

conglomerate 1. A *sedimentary rock composed of rounded, water-borne pebbles which have been naturally cemented together. The *pudding-stones* of the English Chilterns are an example, being formed of flint pebbles cemented with a silica compound.
     2. A grouping of industries producing a number of unrelated products.

coniferous (boreal) forest This occurs naturally between 55° and 66° N where winters are long and very cold, summers are short and warm, and precipitation, around 600 mm per annum, falls mainly as snow. Pure *stands are common and species are relatively few. Trees are evergreen and leaves are needle-shaped, restricting surface area and preventing loss of water by transpiration. Undergrowth is sparse. Animal species are dominated by insects, and seed-eating rodents such as mice and squirrels abound. Larger animals include deer, bear, wolves, foxes, and medium-sized cats.

connate water Also known as **fossil water**, this is water laid down in sedimentary rocks and sealed off by overlying beds.

connectivity In *network analysis, the degree to which the *nodes of a network are directly connected with each other. The higher the ratio of the *edges to the nodes in a network, the greater the connectivity; and connectivity in a network is said to increase as economic development proceeds. See *alpha index, *beta index, *cyclomatic number.

consequent stream A river that develops on a newly formed surface, such as a recently uplifted coastal plain or the limb of a newly formed *anticline, and follows its slope.

conservation Conservation had its origin in the USA where attention was drawn in the 1950s to the damage done by mining in the Appalachians. It may be defined as the protection of natural or man-made resources and landscapes for later use. A distinction is made between conservation and preservation; a **conservationist** recognizes that man will use some of the fish in a lake but a preserver would ban fishing in the lake entirely.

Conservation protects resources for future use by banning reckless exploitation. It promotes an end to wasteful use of non-renewable resources, more efficient extraction methods, and recycling. A major theme is the conservation of soil, perhaps the most abused of the natural resources. Conservation is both rational, since it extends resources for the use of future generations, and morally sound. It is argued that the human race has no right to bring about the extinction of species and environments which have value on aesthetic, scientific, and recreational grounds.

Species, habitats, and man-made landscapes may be conserved: species as protected species, such as elephants in Tanzania, or the bee orchid in the UK; habitats as *SSSIs; and man-made landscapes, as with those 'safeguarded sectors'—whole neighbourhoods, or selected parts of them— created by the French 'Malraux Law' of 1962. By this French legislation, conservation has been achieved in Avignon, Chartres, and the Marais quarter of Paris, for example. Legislation in 1967 enabled British local authorities to create **conservation areas**, with further powers added in the early 1970s, and these powers have been used extensively in Bath, Chester, York, and Edinburgh.

Land use conservation considers the conflicts between human land use and the protection of the natural environment.

conservation of angular momentum is a property of all moving bodies, and is the product of a body's mass and velocity. The momentum of a body taking a curved path is angular momentum, and is the product of the body's mass, linear velocity, and the radius of the curve of its path.

Unless a force acts to change a body's angular momentum, it will remain constant. This concept is important in understanding jet streams. Imagine a body of air at the equator, moving at the same speed as the earth, 465 m s^{-1}. The radius of the earth decreases polewards. Therefore, in order to maintain its angular momentum, any body of air moving away from the equator must increase its velocity. Meanwhile, the earth below it is moving more slowly than the equatorial speed. At 30°, for example, the air is now moving very much faster than the earth below. This very fast-moving stream of air is a *jet stream.

conservative margin In *plate tectonics, a plate margin where the movement of the plates is parallel to the margin. The San Andreas Fault in California is a conservative boundary with the Pacific Plate to the west of the fault moving northwards in relation to the south-moving North American Plate on the continental side.

consolidation The reform and reorganization of land ownership in order to solve problems of *fragmentation. In the late 1940s, much Spanish farmland was impossibly fragmented, especially in Galicia, where the average number of plots per farm was 32. Accordingly, a programme of land consolidation was begun in 1952 with the aim of creating for each farmer a single plot of land equal in area to his or her previous, scattered holdings, and with reasonable access to each farmer's house. Most progress has been made in central and northern Spain. A similar scheme is the French *remembrement.

Consolidation makes *co-operatives, mechanization, and irrigation more viable, but increased production depends on the type of land held. A major problem is later fragmentation.

conspicuous consumption Consumption marked by a disregard for waste and a desire to be seen as wealthy. For example, new goods may be bought to replace serviceable but dated goods.

constant slope The straight, sloping element of a hillslope, located either on the middle part of a slope profile (between the convex element above, and the concave element below) or at the base of the *free face, and with a slope angle determined by the nature of the debris upon it. The concept fits with the theory of *parallel slope retreat.

constructive margin In *plate tectonics, a boundary where two plates are moving apart from each other, as at an *oceanic ridge, and where *magma flows upwards and outwards, as the plates move apart. These are also known as **accreting margins** or **diverging margins**.

constructive wave A low-frequency (6–8 per minute) spilling wave, with a long wavelength and a low crest, which runs gently up the slope of the beach. Such waves are thought to deposit material, because the strength of the *swash greatly exceeds that of the *backwash, which is reduced by *percolation. Spilling waves usually occur on gently sloping beaches, and are responsible for the formation of beach ridges and *berms.

consumer 1. Those organisms in all the *trophic levels, with the exception of the *producers. These include *herbivores, *carnivores, *omnivores, and *parasites. Primary consumers subsist on plant material alone. Secondary consumers feed on primary consumers, and so on.
   2. One who uses goods and services. Certain assumptions are made in economics about the consumer. He or she will use goods commensurate with their price such that a fall in the price of commodity will lead to increased consumption. So will a rise in income, and the reverse is held to be true. **Consumer goods** are bought by domestic consumers, and may be classed as *non-durable*: food and drink, and *consumer durables*: furniture, 'white goods', carpets, and so on.

contact field The pattern of contacts existing between an individual and those who surround her or him, for the *diffusion of innovations which may range from information to epidemics. Near the 'sender' the probability of contact will be strong but this will weaken progressively with distance.

container A metal box of standard size, used for the transport of cargo by road, rail, or water. Containers may be moved easily and quickly from one mode of transport to another—the whole container may be attached to a lorry or train, or swung onto a ship—and they are packed by the dispatcher so that a minimum of handling is required. **Containerization** took off in the mid-1960s, and has revolutionized transport systems in the developed world, making them more rapid but less labour-intensive, and certain ports, such as Felixstowe in Suffolk (UK), have specialized and grown rapidly as a result.

contextual effect An idea, used in *electoral geography, to explain why voting patterns vary from place to place, often diverging strongly from 'national trends'. (You may be familiar with TV programmes on election nights which proclaim 'If these trends are carried out nationwide . . .'—and then they are not!)

Predictions of the way people will vote are often founded on existing *cleavages in society, generally based on *class, and the assumption is that voters will support parties on the basis of their own self-interest. The concept of the contextual effect suggests, however, that people vote in the *context* of their local community, so that they may be influenced by the views of others in the community; 'converted', in other words, by their neighbours. In some cases, a local political culture can develop, possibly based on a particular local issue, and this may result in a pattern of voting which is quite different from national trends.

contextual theory A theory which argues that the contexts in which human activity takes place—the time, the space, and the place in the sequence of events—are crucial to the nature of that activity. T. Hägerstrand (*Reg. Stud.*, 1984), who gave birth to the idea, explained that processes are constrained and shaped by the terrestrial space and time in which they take place: every action is situated in, and shaped by, a particular space and time.

This way of thinking strengthens the re-emergence of *regional geography as a discipline, but *space is not just the context for human activity; it is also a consequence of it.

contiguous zone A zone of the sea beyond the *territorial seas of a nation, over which it claims exclusive rights. Under the terms of the UN Convention on the Territorial Sea and the Contiguous Zone (1958), the contiguous zone extends between 12 and 24 nautical miles from the coastline, but it seems that, in international law, this definition has very little force.

continent One of the main continuous bodies of land on the earth's surface. Commonly, seven continents are recognized: Africa, Antarctica, Asia, Australia, Europe, North America, and South America. Geologically, the boundaries of a continent lie offshore at the gentle slope which limits the continental shelf. In this sense, the continents include their neighbouring islands; thus Britain is part of Europe.

A **subcontinent** is a large land mass forming part of a continent, e.g. the Indian subcontinent. Large islands such as Greenland are also classed as subcontinents.

continental climate A climatic type associated with the interior of large land masses in mid-latitudes. Without the moderating influence of the sea, summer and winter temperatures are extreme. Precipitation is low, as the region is distant from moisture-bearing winds. *See* *continentality.

continental crust That part of the outer, rigid surface of the earth which forms the continents. Continental crust is thicker than *oceanic crust, but is less dense.

continental drift The theory that continents which are now separate were united in a supercontinent. The idea was inspired by the apparent jigsaw fit between the Americas and Africa.
   In 1916, the German meteorologist, A. Wegener, suggested that an original supercontinent, which he called Pangaea, split into two large continents, Laurasia to the north and Gondwana to the south. These two split again to form the continents as we know them. The intervening basins between the continents are occupied by oceans. Wegener's evidence for this theory included the presence of the same geological structures and deposits on each side of the Atlantic. Further evidence is provided by fossils of a small reptile found both in Africa and Latin America. Yet more evidence comes from a reconstruction of an ice cap radiating from South Africa which has left its mark across the southern continents.
   Wegener's ideas were ridiculed, since he was not able to suggest a means of moving continents, but he was vindicated by the development of the theory of *plate tectonics.

continental high *See* *anticyclone.

continental shelf The gently sloping submarine fringe of a continent. This is ended by a steep continental slope which occurs at around 150 m below sea level. The UN Convention on the Continental Shelf of 1958 granted states the right to mineral exploitation up to a depth of 200 m in their coastal waters, together with permission to authorize the construction of drilling rigs and the like, although such structures were not to be considered as islands. All this will be changed when the 1982 UN Convention on the Law of the Sea comes into force; the continental shelf will have a legal limit of up to 200 nautical miles from the coastline, and states will have exclusive rights to all natural resources within that limit.

continentality In *climatology, the extent to which any place on the earth's surface is influenced by a land mass, usually in terms of climate. A **continental climate** is characterized by a high annual temperature range. Since this increases with latitude, some writers measure continentality by dividing the temperature range by the sine of the latitude.

continuous media Fixed pipelines and transmission cables used for the transport of energy over permanent routes. Such networks are only profitable when they are used to the full since they are very expensive to construct.

continuous variable A variable, such as the distance between two towns, where any value may be recorded, including fractions. There are no clear cut or sharp breaks between possible values.

contour A line on a map joining places of equal heights, and sometimes equal depths, above and below sea level. The **contour interval** is the vertical change between consecutive contours.

contour ploughing A method of ploughing parallel to the contours rather than up or down a slope. It is used to check soil erosion and the formation of *gullies.

contraception Any form of birth control which prevents fertilization of the ovum.

control variable An attribute used for sorting data into categories, usually for subsequent analysis or for sampling. *See* *classification.

conurbation A group of towns forming a continuous built-up area as a result of *urban sprawl. Some geographers distinguish between a **uninuclear conurbation**, which has developed around a single, great city, such as London, and a **polynuclear conurbation**, such as the West Midlands conurbation of Walsall, West Bromwich, Wednesbury, and Wolverhampton. The term **metropolitan area** is increasingly used as an alternative to conurbation.

convection The process whereby heat is transferred from one part of a liquid or gas to another, by movement of the fluid itself. (Because, in *meteorology, the most striking weather events are due to upward convection, it is possible to forget that all convection currents also have a downward component.) Convection carries excess heat from the earth's surface and distributes it through the *troposphere.

In the *atmosphere, warmer, lighter air moves upward and is replaced by colder, heavier air. This is **free convection**, or **thermal convection**, propelled by buoyancy. The upward movement of an air parcel over mountains, at fronts, or because of *turbulence is known as **forced convection**, or **mechanical convection**. *See also* *cumuliform convection, *slope convection.

convection rain When upward *convection occurs in a parcel of moist air, the rising air will cool. Further cooling will cause condensation of the water vapour in the air, and rain may result. If the air is very moist, the cooling results in condensation, and hence the release of *latent heat. This causes the rising air to accelerate, and very tall *cumulo-nimbus clouds form.

convective condensation level The point of saturation for an air mass. It can be located on a *tephigram at the intersection of the *environmental lapse rate curve and the *saturated mixing ratio line. The saturated mixing ratio line corresponds to the average mixing ratio in the layer between 1000 and 5000 m.

convective instability *See* *instability.

convenience distance The ease, or otherwise, of travel. A town 50 km away may be well served by transport and thus 'nearer' than one 20 km away which is badly served.

convenience goods *Low-order goods like milk, bread, and occasional groceries which are frequently bought locally, with little consideration of the price charged since purchases are usually on a small scale and convenience is rated more highly than economy. More simply, convenience goods are the type bought at the corner shop which is 'open all hours', and are not to be confused with **convenience foods**, which are ready-prepared meals.

convergence 1. In *plate tectonics, the coming together of plates.
  2. In meteorology, air streams flowing to meet each other. Convergence in the lower air is usually associated with an increase in the height of the atmosphere, with air ascending, and often causes weather events. In the upper troposphere, it causes air to subside, creating *anticyclonic conditions at ground level. *See* *Rossby waves, *Inter-tropical Convergence Zone.

converging margin *See* *destructive margin.

convex slope A slope element, and sometimes an entire slope, which gets progressively steeper downhill. Convexity in a slope may be determined by structure, as on *exfoliation domes, and is associated with limestones, and with humid environments. The formation of convex slopes is said to result from weathering and the transport of debris; as the amount of debris increases downslope, the slope angle must steepen to permit the removal of the debris. This theory does not explain the imperative for debris removal, however.

conveyor belt An expansive upwards and polewards flow of air from the lower *troposphere, associated with a *mid-latitude depression. Air ascends ahead of the warm sector in the cold conveyor belt, at first parallel to the front, but then curving *anticyclonically, producing cloud and *precipitation as it does so. The warm conveyor belt rises at a typical speed of 20 cm s^{-1}, ahead of the cold front in the warm sector, turning eastwards in the upper troposphere. The moist, rising air quickly becomes *saturated, and stratiform cloud develops.

coombe, combe A *dry valley, forming a deep, rounded basin on a scarp slope, formed in a *periglacial environment. Its origin may be due to a combination of fluvial processes, such as spring sapping, and periglacial processes, such as *solifluction.

coombe rock An alternative name for a *solifluction gravel.

co-operative In *agricultural geography, an association of farmers developed in order to reduce costs and increase efficiency. Purchases of equipment, seeds, fertilizers, and fodder can be made in bulk, thus lowering costs. Conversion of crops into marketable goods can be made on a wider scale.

co-ordinate A line in a reference grid, used to locate a point. The *ordinate (y-axis) and the *abscissa (x-axis) are drawn at right angles to each other, intersecting at the point of origin of the system. The location of a point is referenced in terms of its distance from the origin along each axis. **Absolute co-ordinates** are a co-ordinate pair, or triplet, located directly from the origin of the co-ordinate system it lies in; **relative co-ordinates** are measured from another point in that system.

co-ordinate data A synonym for *image data.

copyhold A right to farm land given if the tenant was able to produce a copy of the relevant entry on the *court roll.

coral reef An offshore ridge, mainly of calcium carbonate, formed by the secretions of small marine animals. Corals flourish in shallow waters over 21 °C and need abundant sunlight, so the water must be mud free, and shallow. **Fringing reefs** lie close to the shore, while **barrier reefs** are found further from the shore, in deeper water.

A **coral atoll** is a horseshoe-shaped ring of coral which almost encircles a calm lagoon. Many coral reefs are hundreds of metres deep and yet corals will not grow at depths of more than 30–40 m. It has been suggested that deep reefs have formed during a long period of subsidence. Thus, coral forms in shallow waters and then sinks. New coral will then form on the top.

core 1. The central part of the earth. The inner core has the properties of a solid and the outer core those of a liquid. The core is dense, very hot, and probably composed largely of iron and nickel.
  2. The centre of the *core–periphery model.

core area The heartland of a nation, usually more advanced than the rest of the nation, with an intense feeling of native culture and nationality. Some writers have suggested that the core area acts as the crucible from which a state grows, and while this is true of some European states, many African states, for example, were created quite arbitrarily, and have later forged a national identity. It has also been suggested that if the core fails, the state will languish.

core region In a nation, a centre of power where innovation, technology, and employment are at a high level. Core regions may often flourish at the expense of peripheral regions. *See* *core–periphery model. To solve problems of underdevelopment, some nations have concentrated resources at planned cores like Brasilia or Tema.

core–frame concept A model of the central area of the city which recognizes a core of intensive land use indicated by high-rise buildings. Shops and offices abound and the core is the central point of the transport systems. The core is, therefore, essentially the CBD. Beyond the core lies the frame—also known as the *zone of transition—where land use is less intensive. Here are found warehousing, wholesaling, garaging and servicing of cars, and medical facilities.

core–periphery model J. Friedmann (1966) maintained that the world can be divided into four types of region. *Core regions are centres, usually metropolitan, with a high potential for innovation and growth, such as

São Paulo in Brazil. Beyond the cores are the upward transition regions—areas of growth spread over small centres rather than at a core. Development corridors are upward transition zones which link two core cities such as Belo Horizonte and Rio de Janeiro.

The resource-frontier regions are peripheral zones of new settlement as in the Amazon Basin. The downward transition regions are areas which are now declining because of exhaustion of resources or because of industrial change. Many 'problem' regions of Europe are of this type.

This concept may be extended to continents. The capital-rich countries of Germany and France attract labour from peripheral countries like Spain, Greece, Turkey, and Algeria. Higher wages and prices are found at the core while the lack of employment in the periphery keeps wages low there. The result may well be a balance of payments crisis at the periphery or the necessity of increased exports from the periphery to pay for imports. In either case, development of the periphery is retarded.

The model has been criticized on a number of counts. Most notably, it has been argued that uneven development is not the inevitable consequence of development, but of the particular *mode of production used to bring about that development, and that Friedman's model represents the effects of the capitalist mode.

corestone *See* *boulder field.

Coriolis force An apparent, rather than real, force which causes the deflection of moving objects, especially of airstreams, through the rotation of the earth on its axis. It shows up, for example, in the movement of an air stream, relative to the rotating earth beneath it. It is equal to $-2\Omega \times V$, where Ω is the angular velocity of the earth, and V is the (relative) velocity of the air stream. This apparent force has its greatest deflective effect at the poles, and its least at the equator, this deflection reducing efficiency of an *atmospheric cell to transport heat polewards.

The **Coriolis parameter** is equal to the component of the earth's *vorticity about the local vertical and, at latitude ϕ, is $2\Omega \sin \phi$. For a horizontally-moving air parcel, the magnitude of the horizontal Coriolis force on the parcel is the product of its velocity and the Coriolis parameter.

corona A set of rings surrounding a luminous body, such as the sun or moon. It is the result of diffraction by water droplets.

corrasion The erosive action of particles carried by ice, water, or wind (although the term is most often used in a fluvial context). Corrasion is another term for *abrasion.

correlation The link or relationship existing between two or more variables. Where there is a **positive correlation** between two variables, an increase or decrease in one is matched by a similar change in the other. Conversely, a **negative correlation** sees one variable increase while the other declines. Several statistical methods are used to determine the strength of the correlation, that is, the *correlation coefficient.

correlation coefficient A measurement of the strength of a correlation between two variables, derived from statistical techniques such as *Spearman's rank method and the *product moment method. The values of the coefficient run from +1 (perfect positive correlation) through 0 (no correlation) to −1 (perfect negative correlation). Correlation coefficients may also be calculated for multiple *regressions.

corridor A limb of one state's territory that cuts through that of another, usually to allow access. The most famous is the Polish corridor, an extension of Poland created between the First and Second World Wars to give that country access to the Baltic coast ports of Danzig and Gdynia.

corrie See *cirque.

cost curve A graph showing the relationship between the cost of an item and the volume of output. When the volume of production is low, unit costs are high, but they fall as the level of production rises, and *economies of scale come into play which cause unit costs to fall, until they reach a minimum. As production levels rise further, unit costs rise again as *diseconomies of scale come into effect.

cost structure The breakdown of production costs into the expenditure for individual inputs. These include materials, marketing, capital, land, and labour. The cost structure determines the way in which a firm reacts to changes in the industrial environment.

cost surface A three-dimensional 'contour model' representing the variation in costs over an area. There are two 'horizontal' axes: the first from left to right, the second, stretching at 60° from the first to represent the land surface stretching away from the observer. These illustrate distance, while the vertical axis shows the spatial variations in costs, which may be the cost of a single item or of total production. See *revenue surface.

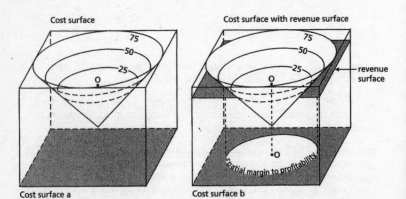

FIGURE 17: *Cost surface*

cost–benefit analysis A technique whereby projected public schemes
are evaluated in terms of social outcomes as well as the usual profit and
loss accounting. The technique begins with an assessment of the costs,
benefits, and drawbacks of the scheme, including *externalities, such as
the generation of noise and other forms of pollution. Financial values are
assigned to these, including qualities like aesthetic appearance, which are
not usually associated with cost. As most major projects are developed over
a long period of time, costs must reflect future conditions. The decision
whether to implement the project is made in the light of the comparison
between the costs of the project and the likely benefits.

This method is far from trouble free. It is difficult to determine which
items should be included, and difficult to put a price on intangibles, such
as aesthetic experience (a new power station might be efficient, but
extremely ugly; and controversy has raged over 'costing' the destruction of
Sites of Special Scientific Interest, such as that at Twyford Down, cut
through by the M3). The discounting of future costs is particularly
problematic, especially as the discount value, which could be based on
current interest rates or on a figure set by the government, is often the key
variable. Even then, moral judgements have to be made—which group in
society should benefit? Should as much weight be given to benefiting the
rich as to the poor? These are political questions which cost–benefit
analysis cannot answer.

cost–space convergence The increasing similarity in accessibility and
costs at any location, and of its products, so that hitherto distant and
expensive products are now in competition with local industries. Thus, it is
not unusual to see French strawberries alongside English ones in British
supermarkets. Of increasing importance in this convergence are changes in
transport: motorways and airways, since the 1950s, and the transfer of
information via the telephone system, computer, and fax.

cottage industry The production of finished goods by a worker,
sometimes together with her/his family, at home. The products may be
sold directly to the public by the worker, or to an entrepreneur who pays
according to the number of goods produced. Cottage industry now exists
in Britain only in the textile trade but is found in some less developed
countries.

coulees *See* *meltwater erosion.

counter-radiation The long-wave radiation emitted from the earth to
the *atmosphere after it has absorbed the shorter-wave radiation of the
sun.

countertrade A trading system under which a country will accept
exports from another country if that country accepts its own goods in
return. Countertrade makes trading easier for those countries lacking
foreign exchange. It allows a nation to export goods for which world
demand is low and is a way of buying in high technology.

counter-urbanization The movement of population and economic activity away from *urban areas. The push factors include: high land values, restricted sites for all types of development, high local taxes, congestion, and pollution. The pull factors offered by small towns are just the reverse: cheap, available land, clean, quiet surroundings and high *amenity. Improvements in transport and communications have also lessened the attractiveness of urban centres, and commuters are often willing to trade off increased travel times for improved amenity. Furthermore, with the ageing of populations in the West, many no longer need to travel to work. Counter-urbanization seems to have diffused from northern to southern Europe, so that by the end of the 1970s only Spain, Portugal, and Greece had failed to show clear signs of the phenomenon.

There is some evidence that counter-urbanization, first identified in the USA in the 1970s, has been a temporary phenomenon, and that cities are now fighting back. While counter-urbanization was still active in northern Italy and Germany in the late 1980s, UK, the Netherlands, and some Nordic states show renewed urban growth; the population of London increased in 1995 for the first time in many years.

country park An area of the country which has facilities for recreation such as picnicking, walking, riding, and fishing; an opportunity for the public to enjoy the countryside at little distance from the city.

country rock A pre-existing rock which has suffered later igneous *intrusion.

county A basic unit of local government in England and Wales based on the medieval feudal earldoms, but much altered in the nineteenth and twentieth centuries.

coupling constraint In *time–space geography a limit to an individual's actions because of the necessity of being in the same space and time as other individuals.

court roll In Britain, the record of the activities of a medieval court.

cover crop A fast-growing crop planted in the rows between the main crop to protect the soil from erosion caused by heavy rainfall.

crag and tail A mass of rock—the crag—lying in the path of a glacier which protects the softer rock in the lee beyond it—the tail. The rock of the tail is a lee-effect depositional landform. A small-scale example of this effect is the formation of morainic ridges in the shelter of individual boulders. The classic example is in Edinburgh, where the castle is situated on the crag, and the Royal Mile extends along the tail.

crater A circular depression around the vent of a volcano. Craters form the summit of most volcanoes. They occur where lava overflows and hardens or where the walls collapse as the *magma sinks down the vent after an eruption. **Funnel-shaped craters** are typical of *stratovolcanoes, such as Mt. Bromo, Java, while **kettle-shaped craters**, like that of Halemaumau, Hawaii, are characteristic of *shield volcanoes.

craton A core of stable continental crust within a continent and composed wholly or largely of Precambrian rocks with complex structures. Two types of craton are recognized: platforms, which are parts of cratons on which largely undeformed sedimentary rocks lie, and *shields.

creep The slow, gradual movement downslope of soil, *scree, or glacier ice. Most creep involves a deformation of the material, i.e. *plastic flow.

Cretaceous The youngest *period of *Mesozoic time stretching approximately from 136 to 65 million years BP.

crevasse A vertical or wedge-shaped crack in a glacier. It can vary greatly in width, from centimetres to tens of metres. The maximum depth of a crevasse is about 40 m because at that depth ice becomes plastic and any cracks merge within the ice. **Transverse crevasses** occur when the ice extends down a steep slope. **Longitudinal crevasses** form parallel with the direction of flow as the ice extends laterally. **Marginal crevasses** occur across the sides of a glacier as friction occurs between the ice and the valley walls. **Radial crevasses** fan out when the ice spreads out into a lobe.

critical group The group most susceptible to damage by a particular form of pollution. If emissions are low enough to avoid damage to the critical group, then the rest of the population will be deemed to be unharmed. Identification of the critical group is not always easy, however.

critical isodapane In A.*Weber's theory of industrial location (1929, trans. 1971) the cost of a good increases with transport costs from the point of production. The critical isodapane is the *isoline where this increase is exactly offset by the savings made by the decreased costs brought about by cheaper labour.

crofting A form of *subsistence farming mostly characteristic of north-west Scotland. Farms and fields are small and usually inherited. One or two cows may be kept and sheep roam over common uplands. Crofting was traditionally supported by fishing and cottage industry—spinning and weaving—and is now linked to the tourist trade.

crop-combination analysis A technique evolved by Weaver to delimit agricultural regions which, he argued, are not regions of simple monoculture as suggested by the names Corn Belt, Cotton Belt, or Spring Wheat Belt, but are areas of combinations of crops.

Theoretical areal values of crop combinations are established so that two-crop combinations take 50% each of the available land, three-crop combinations take 33% of land each, and so on. The real life figures for each crop in the combination are compared with the theoretical figures and the crop combination with the best 'fit' to the theoretical figures is used to classify the area. For example, an American county may produce five crops. Its figures are compared with the theoretical distribution of 3-, 4-, and 5-crop land use. The deviations of the actual figures from the hypothetical are calculated and the area under consideration is assigned to the combination which shows most agreement with the theoretical figures.

crop marks Areas within a field of some plants which are differently coloured, or shorter, and which stand out in aerial photographs. The plants respond to the different soil moisture conditions above ruined buildings, or to the extra soil nutrients in refuse pits. Aerial photographs of crop marks may indicate lines of old buildings, field patterns, original hedgerows, and roads.

crop rotation The practice of planting a succession of crops in a field over a period of years. Rotations can maintain field fertility since different crops use different soil nutrients, so excessive demands are not made of one nutrient. In certain rotations, plants like legumes (peas and beans) are grown to restore fertility.

cross-bedding In a sedimentary rock, the arrangement of beds at an angle to the main bedding plane. The term *current bedding is also used.

cross-section 1. A 'snapshot' of society and its landscape at one moment in time to reconstruct a past geography.
   2. A representation of a vertical section across and through a landscape or landscape feature.

cross-valley profile A section of a valley drawn at right angles to the course of a river at a given point.

cruciform village A village which has developed around an intersection of two routes.

crude rate A *vital rate which is not adjusted for the age and sex structure of a population, for example a **crude death rate**, or **crude birth rate**.

crumb In soil science, a spheroidal cluster of soil particles, i.e. a type of *ped.

crust The outer shell of the earth including the continents and the ocean floor. This is the *lithosphere, formed of *sial and *sima. Sial overlies sima in the continental crust but sima forms most of the ocean floor, i.e. the oceanic crust.

cryofront *See* *cryotic.

cryogenic, cryergic A term used as a synonym for *periglacial. These terms may also refer to the processes carried out by ground ice. The development of cryogenic phenomena is connected with freezing of soil and phase transformation of water. *See* *frost heaving, *frost wedging, *cryoplanation.

cryopediment In *periglacial environments, a low-angle, concave, piedmont footslope developed by slope retreat, brought about through *frost weathering, and the sapping of slopes, and by *nivation surfaces. Cryopediments may coalesce to form a **cryopediplain**.

cryophilous crop A crop which will not fully flower and seed unless it

has experienced low temperatures earlier in its growth. Examples include
some varieties of wheat, peas, potatoes, and apples.

cryoplanation The lowering and smoothing of a landscape by *cryergic
processes. A **cryoplanation terrace** is a *terrace formed under *periglacial
conditions as a plain is dissected by widening *nivation hollows, separated
by rock steps known as frost-riven cliffs, or frost-riven scarps. Such features
are common on Exmoor sandstones. As the hollows merge, a cryogenic
planation surface, or cryoplain may form.

cryosphere The ice at or below the earth's surface, including *glaciers,
*ice caps and *ice sheets, *pack ice and *permafrost.

cryostatic pressure The pressure exerted on rocks and soil when
freezing occurs. As the *freezing front advances, the pressure of the soil
moisture increases since it is trapped. Such pressure can separate
individual grains of soil, forming a mass of fluid mud. This may be driven
near to the surface where it domes up the ground, or where it forms mud
blisters.

cryotic Having temperatures below 0 °C. Non-cryotic areas have
temperatures above freezing point. The boundary of two such areas is the
cryofront or *freezing front.

cryoturbation This term is variously used. It can represent all the
weathering processes that prevail in a *periglacial landscape or be
extended to include the churning up of rocks and soil. Some writers
reserve the term for irregular displacements of soil horizons, while using
the term periglacial *involutions for more regular disturbances.

crystal growth The growth and expansion of crystals of salt or ice along
cracks and fissures in a rock. This expansion causes pressure and splits up
the rock. It is thus a form of weathering.

cuesta A ridge with a *dip slope and a *scarp slope. Cuestas occur in
gently dipping strata which have been subjected to erosion. *See*
*escarpment.

cultural distance A gap between the culture of two different groups,
such as that between the culture of rural societies and that of the cities.

cultural ecology This term, coined by J. Stewart (1955), describes the
study of the relationship between nature and culture in human societies.
One extreme view, that of environmental *determinism, sees nature as the
major control; the other extreme postulates the dominance of culture over
nature, and there are many intermediate views. Cultural ecology is thus
the study of the interactions of societies with one another and with the
natural environment, and as such is a branch of cultural anthropology.
More recent perspectives have stressed that societies are composed of
individual persons acting within given structures, such as societal
constraints.

cultural geography The study of the impact of human culture on the landscape. Themes which have been explored include the effects of plant and animal domestication, fire, hydrological techniques, farming methods, and settlements.

cultural hearth The location in which a particular culture has evolved. C. O. Sauer (1952) controversially chose South Asia and the northern Andes as the cultural hearths for the development of agriculture based on vegetative propagation. Central Mexico and Asia Minor are seen as hearths for the much later seed planting. Urban hearths are relatively easy to date, but the developmental sequence is disputed.

cultural landscape The landscape which results from many generations of human occupancy. Many features of present landscapes were fashioned by past societies who effected more or less permanent changes. The cultural landscape is evolved from the natural landscape by a cultural group.

cultural region A region characterized by a common culture. A distinction can be made between the ethos of London and that of the Western Isles of Scotland, for example.

culture 1. Learned behaviour which is socially transmitted, such as customs, belief, morals, technology, and art; everything in society which is socially, rather than biologically transmitted. The word has many connotations and a geographer might define it differently from, say, an archaeologist, who would distinguish between artefacts—**material culture**—and practices and beliefs—**adaptive culture**. Culture is the primary factor affecting the way in which individuals and societies respond to the environment. The **cultural landscape** is the man-made landscape as an expression of the response of a culture to its natural surroundings.
2. In *Geographic Information Systems, human, as opposed to physical *features.

cumec A measurement of *discharge. One cumec is one *cubic *metre of water per *second.

cumulative causation The unfolding of events connected with a change in the economy. These changes apply to a whole set of variables as a consequence of the *multiplier effect. Thus, the location of a new factory may be the basis of more investment, more jobs both in that factory and in ancillary and service industries in the area, and have a better *infrastructure which would, in turn, attract more industry. The figure illustrates cumulative causation as envisaged by Gunnar Myrdal (1957).

The momentum of change is self-perpetuating, and investment should continue to be attracted to the area. A further part of Myrdal's ideas is that this process of improvement is made at a cost to some other part of the economy; that regions prosper as others feel the loss of investment, and the out-migration of the fittest of their population.

Three stages of regional economies occur: the pre-industrial phase with few regional inequalities; a time when cumulative causation is working, where regional inequalities are greatest because of the *backwash effect;

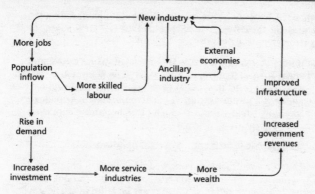

FIGURE 18: *Cumulative causation*

and a third stage where the spread effect stimulates growth in the periphery.

Cumulative causation, like the multiplier, also works 'backwards'—as a major factory closes, the effects are felt throughout the local economy.

cumulative frequency curve Also known as an *ogive, this is a curve drawn by plotting the value of the first class on a graph. The next plot is the sum of the first and second values, the third plot is the sum of the first, second, and third values, and so on. Cumulative frequency graphs are useful in indicating class groupings for a *choropleth.

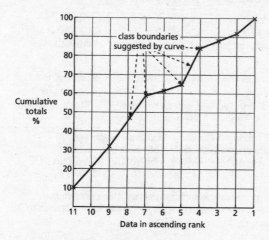

FIGURE 19: *Cumulative frequency curve*

cumuliform In the shape of a *cumulus cloud. Cumuliform convection is vertical convection. Compare with *slope convection.

cumulo-nimbus A low-based, rain bearing *cumulus cloud, dark grey at the base and white at the crown, which spreads into an *anvil shape, as it is levelled by strong upper-air winds.

cumulus An immense, heaped cloud with a rounded, white crown and a low, flat, horizontal base, extending as high as 5000 m. Updraughts within the cloud are strong—up to 10 m s^{-1}, and cloud growth is often rapid, through *entrainment and the saturation of surrounding air. **Cumulus congestus** is a swelling, small cumulus which often becomes *cumulonimbus.

currency In *Geographic Information Systems, up-to-dateness.

current bedding In a sedimentary rock, bedding which is oblique to the 'lie' of the formation as a whole. The structure is original, and not due to tilting or folding. It develops when sandbanks are built up in shallow water, or where sand dunes accumulate from wind-blown sands; here the pattern of bedding reproduces all or part of the outline of the dunes.

currents The rate of flow of a river current varies with depth because friction operates along the bed and sides. The *thalweg is located in the deepest part but all currents change as discharge increases. **Tidal currents** are associated with the rise and fall of the sea. The velocities of ebb and flow vary with the morphology of the coast and any outflow of fresh water. *Rip currents form in the nearshore zone and balance the inflow of seashore currents. They may form a loosely circular pattern of flow as they pass through the surf zone.
   *Ocean currents are driven by the planetary winds.

curvature In meteorology, the wind speed divided by the radius of curvature of the bending air stream. Conventionally, *cyclonic curvatures have positive, and anticyclonic curvatures negative, radii of curvature in the Northern Hemisphere.

cusp A small hollow in a beach, U-shaped in plan with the arms of the U pointing seawards. Beaches tend to have a series of cusps which are formed when outgoing *rip currents and incoming waves combine to set up nearly circular water movements along the beach.

cuspate foreland An accretion of sand and shingle, such as Cape Kennedy, Florida, or Dungeness, Kent, which has been moulded by *longshore drift and constructive waves emanating from two different directions.

customs union A *common market encompassing two or more states within whose boundaries there is free trade; with no tariffs or barriers to the movement of goods.

cut-off low *See* *cold low.

cwm *See* *cirque.

cycle of erosion The notion, first introduced by the American geographer W. M. Davis (*GJ* 1899), that a high-level land surface would be eroded until the whole surface was lowered. During the first stage—*youth*—rivers would be ungraded, waterfalls and rapids common, slopes steep and irregular, and valley cross-profiles V-shaped. The interim stage of *maturity* would show the successive widening and lowering of valleys as rivers and *mass movement shaped the landscape. By the ultimate stage—*old age*—relief would be generally low, approaching a *peneplain, and the landscape covered in *alluvium and *regolith. Davis's theory was based on fluvial landscapes; other cycles have been suggested, for example, in arid lands. The concept of a cycle of erosion is not now generally accepted.

cycle of industry The cycle of industry recognizes times of industrial development and, perhaps, decline. Initially, in its infancy, a region is concerned with *cottage and primary industry. Few industrial towns exist, and urban centres are market towns. The stage of youth sees the emergence of a factory system based on localized resources and/or innovations, accompanied by the development of an *infrastructure. The mature region has experienced large-scale development of manufacturing industry and economic development. The system of industries and services is highly complex, and centres are interconnected by public transport. The mature region may well, however, have derelict buildings and slums. Some regions experience continued maturity while others decline into senility as problem regions where growth is slow, there is overdependence on one or a few industries, the infrastructure is declining, and unemployment is high. Attempts may be made to rejuvenate these areas, usually primarily by government action.

cycle of occupation The growth in numbers and density of a population which then declines, to be followed by a second cycle of population growth.

cycle of poverty A vicious spiral of poverty and deprivation passing from one generation to the next. Poverty leads very often to inadequate schooling and then to poorly paid employment. As a result, the affordable housing is substandard and it may be that crime will increase in these areas of deprivation. Stress is increased and health levels are poor. The children growing up in such areas start off at a disadvantage, and so the cycle continues.

cyclogenesis The formation of *cyclones, especially mid-latitude depressions (also known as frontal wave depressions). Cyclogenesis occurs in specific areas, such as the western North Atlantic, western North Pacific, and the Mediterranean Sea, and is favoured where thermal contrasts between air masses are greatest.
  Cyclogenesis is primarily the result of *convergence of air masses. But cyclones are areas of low pressure. How does this fit with the occurrence of air masses piling up together? Quite simply, because cyclogenesis occurs when *divergence in the upper *troposphere removes air more quickly than it can be replaced by convergence at ground level. The net result is low pressure.

The significance of upper-air movements in cyclogenesis is also indicated by the link with *Rossby waves. Surface depressions develop below the downstream, or eastern, limbs of Rossby waves, where the airflow is divergent. Furthermore, the routes of mid-latitude cyclones, known as *depression tracks closely parallel the movements of the upper-air *jet stream.

cyclomatic number In *network analysis, the number of circuits in the network. It is given by:

$$\mu = e - v + p$$

where e = number of edges, v = number of vertices (nodes), p = number of graphs or subgraphs.

A high value of the cyclomatic number indicates a highly connected network. There exists a significant relationship between the level of economic development of a region and the cyclomatic number of its major transport networks.

cyclone A *synoptic-scale area of low *atmospheric pressure with winds spiralling about a central *low. Cyclonic circulation mimics the rotation of the earth in each hemisphere; thus, it is anticlockwise in the Northern, and clockwise in the Southern, Hemisphere. As air near ground level flows into a cyclone, its absolute *vorticity increases, and it is therefore subject to horizontal *convergence; this infers the ascent of air. This rising causes cooling, which often leads to *condensation, so that *precipitation is associated with cyclones. Since cyclonic circulation and low *atmospheric pressure generally coexist, the term is usually synonymous with *depression. But *see also* *tropical cyclone, for a discussion of hurricane/ typhoon.

cyclone wave The wave-like distortion of flow in the middle and upper *troposphere associated with a mid-latitude *depression.

cyclostrophic Referring to the balance of forces in a horizontal, tightly circular flow of air. **Cyclostrophic flow** is a form of gradient flow parallel to the isobars where the centripetal acceleration exactly offsets the horizontal *pressure gradient.

The **cyclostrophic wind** is the horizontal wind velocity producing such a centripetal acceleration. It equals the real wind only where the *Coriolis force is small, or where wind speed and curvature, and hence centripetal acceleration, are great.

D

daily urban system The area around a city within which daily commuting takes place.

dairying The production of milk, almost entirely from cows, but also from goats and sheep. It is an agricultural system with very high labour and plant costs but using relatively little land. Income is earned throughout the year but there are some difficulties of overproduction. Dairying once circled cities and large towns, but is now possible at any distance from the city by means of specialist, refrigerated transport, or if the milk is subject to UHT processing.

Dalmatian coast Also known as a Pacific, or concordant coast, this is a drowned seashore with the main relief trends running more or less parallel with the coastline, named after the Adriatic coast from Split to Dubrovnik. The old mountain peaks appear as islands above the flooded valleys. Compare with *Atlantic-type coast.

dam *See* *barrage, *hydroelectric power, *irrigation.

Danelaw The areas of northern and north-east England settled by Scandinavians in the ninth and tenth centuries and over which Danish law prevailed.

Darcy's law This states that where the *Reynolds number is very low, the velocity of flow of a fluid through a saturated porous medium is directly proportional to the *hydraulic gradient. For example, the flow of *groundwater from one site to another through a rock is proportional to the difference in water pressure at the two sites:

$$V = \frac{h}{P_l}$$

where h is the height difference between the highest point of the water-table and the point at which flow is being calculated (the **hydraulic head**), V is the velocity of flow, P is the coefficient of permeability for the rock or soil in question, and l is the length of flow. Darcy's law is valid for flow in any direction, but does not hold good for well-jointed limestone which has numerous channels and fissures.

dart leader *See* *lightning.

data capture In *geographic information systems, the encoding of data. This may be done by direct recording, *digitizing, or electronic survey. **Data compression** in *Geographic Information Systems, is the encoding of data in order to reduce its overall volume.

data model An abstraction of the real world; a group of *entity sets together with the relationships between them which represent a human conception of reality, usually tailored towards a given problem or application. The term is common in *Geographic Information Systems, but a map is also a data model.

data quality The fitness for use, precision, accuracy, and completeness of data.

data structure In *Geographic Information Systems, a representation of the *data model as a diagram, list, or array, showing the human implementation orientation of the data.

database A body of information recorded *digitally; an integrated, organized collection of stored data, available for appropriate uses, and reached (accessed) by different logical paths. A **database management system** is a collection of software used for organizing information, and generally has input, verification, storage, retrieval, and combination functions.

dataset In *Geographic Information Systems, a collection of locally related *features, such as woodland or water features, arranged in a prescribed manner.

datum In *Geographic Information Systems, the fixed starting point of a scale or co-ordinate system. *See also* *Ordnance Datum.

dead cliff A sea cliff which is no longer subject to wave attack, either because sea level has fallen, as in the case of much of the Norwegian coast, or because a broad beach has formed which protects the cliff. The dominant process is now *subaerial erosion, which produces a gentler slope. *See* *bevelled cliff.

dead ice Also known as *stagnant ice*, this is static glacier or ice-sheet material. Ice stagnates either after a surge in an Alpine glacier has formed an extension, which is then abandoned further down the valley, or at the end of a glacial period. As stagnant ice melts, the remnants become covered in *ablation moraine. The presence of dead ice in valleys may lead to the formation of *kame terraces.

deadweight tonnage The total weight of all the effects of a ship: cargo, passengers, fuel, etc., when loaded to her safe load line.

death rate The number of deaths in a year per 1000 of the population as measured at mid-year. This crude death rate may be expressed as:

$D/P \times 1000$

where D = the number of deaths, P = the mid-year population.

This is a crude rate because no allowance is made for different distributions of age and sex. For example, Sri Lanka had a crude death rate of 5 per 1000 during the 1970s compared with 9–12 per 1000 in north-west Europe. The low rate for Sri Lanka is a reflection of the youth of its population and does not imply lower mortality in the higher age groups. Within the *EU, 1992 death rates varied from a low of 9.1/1000 in Ireland to a high of 11.9/1000 in Denmark.

A **standardized death rate** compares the rate with a real or assumed population which is held to be standard. Thus the standardized death rate for an age and sex category which is labelled *sa* is

$$\frac{\sigma\left(P_{sa} \times D_{sa}\right) \times 1000}{P}$$

where P = standard population, P_{sa} = number of population of sex and age category *sa*, D_{sa} = specific annual death rate of sex and age category *sa*.

debris Material such as scree, gravel, sand, or clay formed by the breaking-up of minerals and rocks. Through the air, debris is transported by *saltation and *deflation, in water, debris moves in solution, in suspension, by rolling, and by saltation. In ice, debris can be carried on the glacier—supraglacially—within the glacier—englacially—or subglacially. Debris movement downslope is *mass movement.

debris fall *See* *fall.

debris flow The very rapid downslope movement of *debris of a high water content, with a course guided by stream channels. This form of mass movement is less deep seated and rather rarer than a landslide.

decalcification The *leaching out of calcium carbonate from a soil *horizon by the downward movement of soil water. It is an early stage in the formation of a soil, and is generally accompanied by *humus formation and followed by *acidification.

decentralization, deconcentration A process counteracting the growth of *urban areas, and known also as *counter-urbanization. Even while the city is still growing, it has many negative *externalities such as congestion, noise, pollution, crime, and high land values. Such problems are a spur to spontaneous movement away from the cities which has been compounded by the increasing locational freedom of shops, offices, and industries to move to out-of-town shopping centres, office parks, and industrial estates, respectively, together with the increase in numbers of white-collar workers and the consequent rise in incomes, and mass car ownership. Research in the late 1970s indicated that a number of city regions in the UK and north-west Europe were undergoing absolute or relative decline in their cores while growth continued in their *hinterlands, and by the mid-1980s similar trends were observed in Mediterranean cities, especially in Italy.

On a national scale, governments may favour decentralization to restore the fortunes of declining regions which are suffering from out-migration to the extent that services and *infrastructure may be under-used. Governments may attempt to decentralize by discouraging new investment at the centre and encouraging growth in the *depressed areas. Incentives for such relocation include grants, loans, tax concessions, and the provision of industrial premises. *See* *metropoles d'équilibre.

deciduous forest In the cold season of temperate latitudes, a tree's water supply is restricted when the temperature falls below 0 °C. In order to lessen losses of water, deciduous trees shed their leaves until the spring brings more available moisture.

decision-making The choice of one particular strategy to achieve some end. Human geographers have studied decision-making in the context of industrial location, residential choice, migration, response to environmental hazards, and shopping behaviour, and have used two major concepts: *movement minimization and *place utility.

In the earliest models of decision-making, *economic man was seen as the decision-maker, but this concept has generally been replaced by that of the *satisficer who is not blessed with perfect knowledge, who works with *bounded rationality, who seeks only a satisfactory solution, and is therefore *sub-optimal. This may be illustrated for the location of businesses by the use of *spatial margins to profitability; the decision-maker is free to locate within the margins, although not necessarily at the *optimal location. This approach narrows the choice, but does not point to the actual location selected. Decision-making is of great interest to *behavioural geographers, who express the variety of incoming information and the range of the individual's abilities to use that information in a *behavioural matrix. A theoretical basis to decision-making has been attempted with reference to *risk, *uncertainty, and *game theory, but with limited success. More successful generalizations have been based on case-studies. These may show a stimulus such as higher demand for a product. The response of the industrialist may be to expand. From this decision comes the need for an extended site or a new plant. This location decision then demands a host of smaller decisions, and experiencing the results of the decision will then lead to feedback which may affect future decisions. The case-study approach also uncovered the importance of personal factors, which are difficult to build into a theoretical model.

Much of the work on decision-making has been unsatisfactory because it has been based on the crude idea that humans act in mechanistic response to stimuli, and new research is moving to *qualitative investigations based on the meanings people give to different aspects of their lives; for example, the decision to travel to a facility, such as a shopping centre can depend on the significance of shopping to the individual—a chore to be got over with quickly?—a pleasure?—a day out for the family? as well as on an individual's culture and past experience. These may matter as much as, if not more than, movement-minimization.

declining region A region suffering the economic decline associated with closure of factories, outmoded industry, and high unemployment. Services also decline as incomes fall and governments may be concerned with halting such uneven development. UK examples include Tyneside and Merseyside.

declining slope retreat *See* *downwearing.

décollement The movement of underlying strata which is moved during folding as it adheres to the upper layers. This layer, together with the overlying beds, rides easily over an assemblage of older rocks.

decomposers Simple organisms which obtain their nutrients from dead plant or animal material by breaking it down into basic chemical compounds. A decomposer chain can run from a relatively large organism, such as a fungus, to smaller organisms such as bacteria. Decomposers play a major part in the maintenance of *nutrient cycles.

deconcentration *See* *decentralization.

deduction Deducing or inferring the general from the particular; using particular statements as premises from which future developments are shown to proceed logically. Put more simply, deduction begins with a theory from which a hypothesis is derived and then tested. An outstanding example is Burgess's *concentric zone theory, based, as it was, on an observation of the Chicago of the late 1920s and early 1930s. Deduction provides laws from which outcomes may be predicted but such prediction plays little part in human geography because of the extreme complexity of the systems involved. Compare with *induction.

deep weathering The creation of a thick weathered layer (*regolith) through strong and/or sustained chemical weathering. It is particularly associated with gentle slopes in the warm, moist tropics, where it can extend to a depth of more than 50 m, and where transport of the weathered layer is restricted by the lack of gradient and the binding effect of vegetation. Deep weathered layers are also found in desert and temperate regions, but may be *relict features; one school of thought suggests that *tors are an exhumed form of deep weathering. Others suggest that the deep weathering of the tropics does not reflect higher rates of chemical weathering resulting from high temperatures and humidity, and ample humic acid in those regions; rather, that chemical weathering has been active for longer in the tropics.

deepening In meteorology, a decrease in the central *atmospheric pressure of a pressure system, usually a low pressure system.

deer forest A stretch of moorland, usually without trees, managed for deer stalking.

defensible space The environment used by its inhabitants to build their lives in and to feel secure.

deficiency disease A disease brought about because some vital element is missing in the diet. The food lack is generally of protein, vitamins, or minerals. Common conditions are:

LACK OF:	DEFICIENCY DISEASE
vitamin B	kwashiorkor
vitamin C	scurvy
vitamin D	rickets
iron	anaemia

It is claimed that over half the population of the world suffers or has suffered from malnutrition, leaving them debilitated.

deflation The action of the wind in removing material from a surface and lowering that surface. Resulting landforms include small hollows and

*blow outs. The world's largest **deflation hollow** is the Qattâra Depression of the western Egyptian desert, but its formation may also be structural. Deflation is most effective where extensive areas of non-cohesive deposits are exposed, as in *loess or in dry lake beds.

deflocculation *See* *flocculation.

deforestation The complete clearance of forests by cutting and/or burning.

deglaciation The process by which glaciers reduce in thickness, and recede. Deglaciation usually results from climatic change, so that *accumulation decreases and *ablation increases, but it may be brought about by a rise in sea level which increases *calving. The major processes are **backwasting**—the retreat of the glacier—and **downwearing**—the thinning of the ice, and there is much debate over the relative importance of each. Deglaciation is generally accompanied by the formation of recessional *moraines and the release of *meltwater, together with its associated landforms.

deglomeration The movement of activity, usually industry, away from areas of concentration. Deglomeration occurs when the advantages of *agglomeration are outweighed by its disadvantages: high land costs and rents, constricted sites, congestion, and pollution. The 1980s, for example, saw the movement of a number of major firms out of New York city.

deglomeration economies Forces such as congestion and high land values which lead to *decentralization. *See also* *diseconomies of scale.

degradation The lowering and flattening of a surface through erosion, especially the erosion carried out to maintain or restore the graded profile of a river.

degree-day A measure of the difference between daily mean temperature and a given standard—such as 0°. One degree-day represents each degree (Fahrenheit or centigrade) of difference for one day.

degrees of freedom A number which in some way represents the size of the *sample or samples used in a statistical test. In some cases, it is the sample size, in others it is a value which has to be calculated. Each test has its specific calculation, and the correct value for each test must be calculated before the result of the test can be checked for statistical significance.

de-industrialization, deindustrialization The decreasing significance of primary and secondary industry, both in terms of employment and national production, to developed capitalist economies. Only a few countries, the United Kingdom among them, have experienced a decrease in manufacturing capacity as well as in manufacturing employment, but de-industrialization has affected many once strong industrial regions, such as the Belgian coalfield. In other cases, in spite of a decreased labour force, manufacturing still contributes significantly to the national economy in terms of output and exports.

dell In *periglacial environments, a small *dry valley running directly down the direction of slope, with a flat floor and gently inclined sides, several tens of metres wide, and hundreds of metres long. Its development is thought to be due initially to the localized degradation of *permafrost and the thawing of *ice wedges, further modelling being brought about by *solifluction and *sheet wash.

delta A low-lying area found at the mouth of a river and formed of deposits of *alluvium. Deposition occurs as the river's speed, and hence its silt-carrying capacity, is checked when it enters the more tranquil waters of a lake or sea. Furthermore, clay particles *flocculate in salt water, become heavier and sink. Post-glacial rises in sea level have increased the slowing effect of the sea and some deltas grow by as much as 200 m a year.

Other, older deltas may be eroded. The morphology of a delta is the result of the interplay of the following factors: the input of sediment from the river, the density and depth of the sea water, waves, and currents, and any *tectonic activity in the region. **Cuspate deltas** have a pointed seaward end. **Lobate deltas** have a curved seaward end and **digitate** or **bird's foot deltas**, like that of the Mississippi, have long 'fingers' of alluvium extending into the sea. **Inland deltas** form in hot, arid areas of inland drainage where water is gradually lost through evaporation.

demand In economics, the volume of goods which purchasers are able and willing to buy. This depends on their income and preferences, the price of other products, and the price of the product concerned. When preferences alone are isolated, demand can be seen on a **demand curve**, a graph with prices on the y-axis and demand on the x. Typical demand curves slope downwards from left to right, showing demand falling as prices rise.

The specific slope of the demand curve as it relates to any commodity depends on the **elasticity of demand**, that is, the sensitivity of the consumer to changes in price. Together with supply, demand determines prices in competitive markets.

demand cone A depiction of the falling away of demand for a good with distance from the market which is due in principle to the increasing cost of transport. The concept is used in Lösch's model of market area analysis.

demersal Of marine life, living near the sea bed.

demographic regulation The notion that a population will restrict its fertility when a reduction of mortality causes population to grow beyond the ability of the environment to sustain it. This is not a simple response and is accompanied by changes in society in an attempt to maintain an equilibrium.

demographic relaxation theory The view that overpopulation leads to war. A major example is the professed desire for *lebensraum*—space in which to expand—shown by the German government in the late 1930s and which led to the invasion of Czechoslovakia and Poland. The resulting wars destroy populations. In this way, war is seen as relieving overpopulation.

demographic transition An account, but not a complete explanation, of changing rates of fertility, mortality, and *natural increase. These changes are held to occur as a nation progresses from a rural, agrarian, and illiterate state to a predominantly urban, industrial, literate one. Four stages may be recognized:

1. The High Stationary Stage. Birth and death rates are high and the death rate fluctuates from year to year.

2. The Early Expanding Stage. Fertility remains high, but improved conditions mean falling death rates. Population therefore increases.

3. The Late Expanding Stage. Death rates are low and fertility is declining but population is still increasing.

4. The Low Stationary Stage. Birth and death rates are low and the birth rate fluctuates.

Some writers suggest that there is a fifth stage where birth rates fall below death rates so population levels fall. The 1970s did see population fall in Austria, West Germany, East Germany, Sweden, and Luxembourg, which may uphold the theory, but there is some suggestion that, in the 1990s, birth rates are rising in the USA.

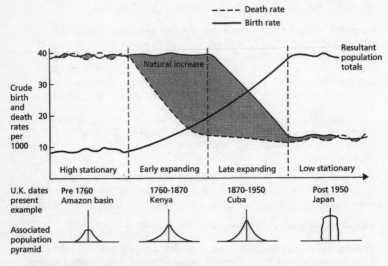

FIGURE 20: *Demographic transition model*

The reasons for falling death rates are improved conditions and health care. The reasons for falling fertility are less clear. Certainly, fertility rates have fallen in countries such as Denmark before reliable contraceptives were developed. Falling fertility has been explained by:

1. The breakdown of the *extended family which means more stress for parents.

2. In a 'modern' industrial society the labour value of children is low whereas in peasant society children contribute to the labour force from an early age.

3. With the provision of pensions, it is no longer necessary to have children as a support in old age.

4. More women are in work.

5. As standards of living rise, more wealth is needed to bring up children.

6. Where infant mortality is low, fewer babies are needed to ensure the survival of the family unit.

There is considerable debate as to whether population growth in the intermediate stage of the transition was a stimulus to the Industrial Revolution or a consequence of it, and whether the demographic transition will follow a similar course in *less developed countries.

While the model seems to hold good with reference to the more developed countries, the first and last stages have almost certainly been over-simplified, and it cannot be applied wholesale to the less developed world; it is by no means certain that the less developed countries will follow the same transition.

demography The observed statistical and mathematical study of human populations, concerned with the size, distribution, and composition of such populations. The main components of this study are *fertility, *mortality, and *migration. *See also* *population studies.

dendritic drainage *See* *drainage patterns.

dendrochronology The technique of dating living wood by counting the annual growth rings. Recently, the study of *isotopes within the rings has yielded information about past temperatures and the width of the rings gives information about times of drought.

density *See* *population density.

density dependent factors The checks to population growth which are the result of overcrowding, such as *competition. **Density independent factors**, like fire and drought, will occur whatever the state of the population.

density gradient The rate at which the intensity of land use or the density of population falls with distance from a central point, a phenomenon illustrated in the models of *von Thünen and Alonso. Population density declines exponentially with increasing distance from the *central business district, and it has been shown that older cities have a steeper density gradient than younger ones.

denudation The laying bare of underlying rocks by the processes of weathering, transport, and erosion. The term may be used more narrowly to describe the removal of weathered rock and the exposure of the material beneath by *mass wasting processes. **Denudation chronology** is the now somewhat outdated study of the long-term formation of specific landscapes.

dependence The condition in which a society is only able to develop, in part or in full, by a reliance on another nation for income, aid, political protection, or control.

A degree of dependence can only come about when one society comes into contact with another. Nearly all today's 'underdeveloped' societies were once viable and could satisfy their own economic needs but their economies were often torn apart after contact with a colonial power. It is therefore argued that Europe did not 'discover' the dependent, underdeveloped countries; on the contrary, it created them. *See* *dependency theory.

Other economists argue that dependence is created by *capitalism, because *accumulation is the key component of the capitalist system. Accumulation depends on the extraction of surplus value, and A. G. Frank (1966) has argued that, in spatial terms, this entails a flow of surplus value from the poor *periphery—the ex-colonies—to the metropolitan *core—the *more developed countries, which dominate the global economy. Others argue that each depends on the other; they claim that without the resources, labour, and markets of the *South, the *North could not survive.

dependency ratio The ratio between the number of people in a population between the ages of 15 and 64 and the **dependent population**: children (0–14) and elderly people (65 and over). It is used as a rough way of quantifying the ratio between the *economically active population and those they must support, but the age limits are somewhat arbitrary as, in the UK for example, 30% of over-16-year-olds go on to higher education, and are therefore still dependent, either on the state or, increasingly, on their parents or bank managers.

dependency theory A group of hypotheses which assert that low levels of development in *less developed countries spring from their dependence on the *advanced economies. It has been argued that the less developed world is doomed to remain economically disadvantaged because the surplus it produces is commandeered by the advanced economies, for example, through *transnational corporations. In this case, the argument continues, the only effective growth strategy for the less developed countries is to cut ties with the more developed countries, and follow self-reliant, socialist, systems. These theories also emphasize the advantages to the West of its economic strength, and the financial and technical power of the West to maintain its advantages.

Dependency theory has been criticized for over-emphasizing economic factors, and has been challenged by the success of *newly industrializing countries such as South Korea, which showed that late industrialization was not impossible.

dependent variable A variable which depends on one or more variables which may control it or relate to it; in other words, that may be seen as a function of another variable. In a study of pedestrian flow and distance from the CBD, pedestrian flow is the dependent variable, and, if graphed, would be plotted on the ordinate, or *y*-axis. *See* *independent variable.

depopulation The decline, in absolute terms, of the total population of an area, more often brought about by out-migration than by a fall in fertility or excessive mortality. In the early 1980s, the areas of Europe experiencing the greatest losses of population were interior Spain and Portugal, southern Italy, northern Scandinavia, Scotland, and western Ireland, with less dramatic losses in the old industrial areas of Lorraine, Limburg, and the French Nord. *See* *rural depopulation.

deposition The dropping of material which has been picked up and transported by wind, water, or ice. *See* *sediment, *sedimentary, *sedimentation.

depressed area An area, usually within a developed nation, where capital is scarce and labour, plant, and *infrastructure are underemployed. Depressed areas develop because their economic activity has been outmoded, often because of competition from cheap labour in less developed countries, and because of world recessions. Competition also comes from newly industrialized areas. Victorian cities decline as the better-off move to rural and suburban areas. Currently, unemployment in the inner cities of the UK runs at four times the national average. *See* *development areas.

depression An area of low pressure (roughly, below 1000 mb; *see* *cyclone). *See* *mid-latitude depression, *cold low, *lee depression, *monsoon depression, polar lows (*see* *mesometeorology) and *thermal low. *See also* *cyclogenesis.

depression tracks The usual paths taken by *mid-latitude depressions. These are influenced by the courses of the *jet streams, energy sources such as warm seas, and mountain barriers.

deprivation Loss; lacking in provision of desired objects or aims. Within the less developed countries deprivation may be acute; the necessities of life such as water, housing, or food may be lacking. Within the developed world basic provisions may be supplied but, in comparison with the better-off, the poor and the old may well feel a sense of deprivation. This introduces the concept of **relative deprivation** which entails comparison, and is usually defined in subjective terms. It has been suggested that high levels of inequality, and thence relative deprivation, lead to differences in life-expectancy.

The idea of a **cycle of deprivation**, refers to the transmission of deprivation from one generation to the next through family behaviours, values, and practices. This idea has been extensively debated and discussed.

deregulation A cutback in the power of the state to control economic activity, usually in order to encourage competition. Probably the most visible piece of deregulation in Britain has been the opening-up of bus transport to increased numbers of operators: almost every British city has been affected, some would say adversely. When deregulation sweeps away wage controls, or health and safety regulations, costs are certainly brought down, and it may be that jobs are created, but possibly with poorer pay and conditions.

derelict land Land which was once used but has now been abandoned; for example, old railway lines, disused factories, old waste tips.

derived-scale map A smaller-scale map made from a basic-scale map.

desalination The conversion of salt water into fresh by the partial or complete extraction of dissolved solids. The methods used include distillation, electro-dialysis, freezing, and reverse osmosis. The processes are relatively simple but costly, and desalination plants tend to be a feature of oil-rich but water-poor states, such as Kuwait.

descriptive meteorology A branch of *meteorology which describes atmospheric events without explanation or theoretical treatment.

descriptive statistics As opposed to *inferential statistics, which predict the state of a population from a sample, descriptive statistics, as the name suggests, draw on complete surveys of the dataset to summarize a state which exists at the present (or existed in the past), using *means, *medians, *modes, *standard deviations, *correlations, and so on.

  Thus, a shopping centre survey which included every user of that centre, rather than a sample, would draw on descriptive statistics.

desert An arid area of sparse vegetation, such that much of the ground surface is exposed. Scanty vegetation can be due to very high, or very low temperatures (as in cold deserts), or to an excess of potential *evapotranspiration over precipitation. *See* *aridity index, *hot desert.

desert pavement A surface comprised of large angular or rounded rock fragments lying over mixed material. These rocky fragments are thought to have been left behind after wind or water has removed the lighter material. Desert pavements can protect surfaces from *deflation.

desert soil Most desert soils show little development of soil *horizons because the climate is too dry for chemical *weathering or the formation of humus. *Leaching occurs only occasionally after sporadic rain, and this downward movement of soil water is soon reversed by evaporation.

desert varnish A film of iron oxide or quartz on desert rocks. The precise cause is disputed; some recognize desert or rock varnish as being *weathering rind, but the polishing of rock by fine wind *abrasion is also significant.

desert wind A dry, and usually dusty, wind blowing off a desert. Examples include brickfielder and *harmattan.

deserted village A village site once inhabited but now abandoned. Most English deserted villages were abandoned in the late Middle Ages and traces may be seen in the landscape in the form of old building lines, lumps of masonry, clumps of stinging nettles, and isolated churches.

desertification The spread of desert-like conditions in semi-arid environments. Desertification means a long-term change in the characteristics of the biome: plant life, vegetation, and soil are changed and impoverished, and so desertification should not be confused with short-term drought; although drought can be a causal factor.

The causes of desertification are by no means clear; overgrazing, overcropping brought about by a switch to export cropping, badly managed irrigation systems, and deforestation for firewood and timber have all been blamed, and yet some writers argue that true desertification is not occurring on a large scale, and counter the claims that between 33 and 37% of the world's land surface is suffering. It may be that we are merely witnessing short-term climatic change. Others suggest that the pastoral *nomadism, once thought to degrade the environment, may represent the best use of desert areas.

It is technically possible to reverse the effects of desertification, and the UN is now backing agro-forestry projects, and the low-technology system of building stone lines.

desilication, desilification The process by which silica is removed from a *soil profile by intense weathering and *leaching. Desilication is common in wet, tropical regions and leads to the formation of a firm but porous soil with a reduced capacity to store water. See *latosol.

desire line A straight line constructed on a map to symbolize a trip between two locations; not necessarily the actual route followed. One line is drawn for each journey. Compare with *flow line.

deskilling The breaking down of jobs into smaller units, each to be tackled separately, so that low levels of skill are required for restricted tasks. It is associated with the development of the assembly line, standardized production techniques, and automation. This technique replaces skilled craft workers with unskilled, cheaper labour, and some theorists have suggested that this will cause wages to fall and insecure employment to increase, but other studies suggest that while certain skilled occupations are disappearing, others, such as computer programming, are expanding, and that many workers are upgrading their skills. *Fordism, neo-Fordism.

destructive margin In *plate tectonics, the zone where two plates meet and where *oceanic crust is being destroyed by *subduction. The crust narrows at destructive margins, which are associated with: intense compression and the formation of fold mountains, as where the Indian plate underthrusts the Eurasian; the formation of *ocean trenches like the Tonga trench, where the Pacific plate dives below the Indian; volcanoes and *earthquakes, for example, where the Nasca plate sinks below the American plate; and island arcs, as in the Aleutian Islands.

destructive wave A plunging wave, with a short wavelength, a high frequency (13–15 per minute) and a high crest, which breaks so that the water crashes downwards from the wave crest and erodes the beach; *backwash thus greatly exceeds *swash. Plunging waves generally occur on steeply sloping beaches. Compare with *constructive wave.

determinism The view that human actions are stimulated and governed by some outside agency like the environment or the economy. Individuals have no choice in regulating their actions, which may be predicted from the external stimuli which triggered them. This view is currently rejected.

detritivore An animal which feeds on fragments of dead and decaying plant and animal material. Detritivores have a vital part to play in *food webs and nutrient recycling.

detritus 1. In geomorphology, fragments of weathered rock which have been transported from the place of origin.
  2. In ecology, dead plants, and the corpses and shed parts of animals.

development The use of resources to improve the standard of living of a nation; the means by which a traditional, low-technology society is changed into a modern, high-technology society, with a corresponding increase in incomes. This may be achieved through mechanization, improvements in infrastructure and financial systems, and the intensification of agriculture. This definition is based on the more obvious distinctions in living standards between developed and less developed countries, but it may be that a change to 'western' conditions is not in the best interests of a Third World nation.
  Narrowly economic definitions of development, based on indicators such as per capita GNP, production, consumption, and investment have been criticized, and the United Nations Human Development Index (HDI) attempts to get away from purely monetary measurements by combining GNP figures with *life expectancy and literacy in a weighted average. Even so other geographers believe that true development includes improvements in social justice; for example, in a more equitable distribution of income, or in an improvement in women's and minority rights.
  Development indicators as used by the World Bank include details of birth and death rates, fertility, life expectancy, health, urbanization, industrialization, production, consumption, investment, capital, income, education, energy consumption, and trade. These indices of development are simply concerned with statistics and do not indicate social structures and patterns of behaviour; there is no definitive definition of what development should be for each society, and no blue-print for how to achieve it.

development area A depressed area in need of investment in industry and *infrastructure. In the UK from 1947 to 1981, investment was deflected from prosperous areas by the failure to grant them development certificates. Other methods of inducing investment in development areas include grants, tax concessions, loans, and the provision of industrial premises.

development control A measure by central government to regulate the location of new industry in the UK. An Industrial Development Certificate (IDC) must be granted by the Department of Trade and Industry who regulate the availability of the IDC according to the size of the plant and the location selected. IDCs are not required in *assisted areas. Local authorities also control industrial development but may be overruled by central government.

development stages growth theory A theory of growth which suggests that all economic development passes through the same sequence. Initially, there is a subsistence economy, but with better transport, regional specialization develops and village industries emerge. Agriculture becomes more intensive and industrialization takes place. Over time, processing gives way to service industries. *See also* *Rostow's model.

device co-ordinate A co-ordinate system for referencing individual points in the area scanned by a graphic device.

devolution The means by which a state allows a degree of independence to a political unit within its boundaries. Regional assemblies for Wales and Scotland were rejected by the British public in the 1970s, but are again being proposed. With the rise of *nationalism, devolution may be seen as one way to avoid ethnic unrest; power has been devolved to the Basques in Spain, for example, partly as a response to local discord.

Devonian A *period of *Paleozoic time stretching approximately from 395 to 345 million years BP.

dew A type of condensation where water droplets form on the ground, or on objects near the ground. Dew forms when strong night-time *terrestrial radiation causes the ground to cool. At the end of a clear night, air in contact with the ground may be chilled to *dew point. If this cooling brings about temperatures below 0 °C, frost rather than dew will form.

dew point The temperature to which a given body must be chilled for it to become *saturated with respect to water, so that condensation may begin. It may also be seen as the temperature of a chilled surface just low enough to attract *dew from the *ambient air. A **dew point meter**, or **dew point hygrometer**, uses this second way of expressing the concept, consisting of a polished metal surface which can be gradually chilled until condensation forms on the metal. The dew point of an *air mass varies with its initial temperature and humidity.

dew pond A pool made to provide water supply in a dry region such as a chalk or limestone upland. The sides and bottom are lined with impermeable clay to retain the water which generally comes from rainfall rather than from dew.

dialect A variant of a language, which depends not only on vocabulary but also on grammar, syntax, and pronunciation. **Dialectology** is the study of social and linguistic variations within a language; **dialect geography** is the study of local differentiations in a speech area; and **dialectometry** is an objective method of determining the degree of difference between geographical variations in dialects, especially lexical differences.

dialectic A theory of the nature of logic. Dialectic is the logic of reasoning. The determination of truth is arrived at by the assertion of the theory (thesis), its denial (antithesis), and then the synthesis of the two to form a new theory. The doctrine of **dialectical materialism**, which was put forward by Friedrich Engels and modified by Karl Marx, was concerned with the conflict between the capitalists (bourgeoisie) who own the *means of production, and the proletarians who have nothing to sell but their labour services.

In human geography, G. Olsson (1980) has written extensively on the dialectics of spatial analysis.

diatreme A passage, generally sloping upwards, which has been forced through sedimentary country rock by volcanic activity.

differential erosion The selective erosion of surfaces, so that softer rocks, such as clays and shales, or lines of weakness, such as joints and faults, are eroded more rapidly than resistant, competent, and unjointed materials. In coastal areas, headlands tend to be made of igneous, metamorphic, or the more resistant sedimentary rocks, such as sandstone or limestone, while weaker strata form bays. Swanage Bay, Dorset provides an excellent example. In granite uplands, areas of close jointing form basins. *See also* *tors.

differential share *See* *shift share analysis.

diffluent col In glaciology, a low pass at a valley side which has been cut by ice spreading out over a col from its own valley into an adjoining one.

diffuse reflection The chromatically uniform *scattering of all available wavelengths of light by clouds, mist, and haze.

diffusion The widespread dispersal of an innovation from a centre or centres. This innovation may be anything from an epidemic disease to a political belief, or even the wearing of reversed baseball caps.

Tortsen Hägerstrand's model of diffusion (1968) implies the existence of a *mean information field which regulates the flows of information around a regional system. These flows are moderated by barriers which can obstruct the evolution of information into innovation, and thus mould the **diffusion wave**; the ripple of innovation which spreads from one location to another.

Various categories of diffusion have been recognized: **expansion diffusion** is the spread of a factor from a centre with the concentration of the things being diffused also remaining, and possibly intensifying, at the point of origin. One example is the spread of girls' public day schools in the nineteenth century, which started in London. As new schools developed further and further away from London, new schools also opened in the capital. **Relocation diffusion** similarly spreads from a centre but the innovation moves outwards, leaving the centre. An example of this is the movement of a bush fire, which has burned out at the origin but continues to spread at the periphery. **Hierarchic diffusion** passes through a regular sequence of orders as when an innovation in a metropolis spreads out to cities, then towns, and finally to villages. A good example is the introduction of video hire shops. The order of the diffusion may also be based on class—the spread of wine-drinking in Britain is a good example. The innovation may spread up a hierarchy, but **cascade diffusion** is a form of hierarchic diffusion which only moves down an order or hierarchy.

The Hägerstrand model has shortcomings: it does not explain why some adopt an innovation and others do not, and it is based on the notion of a uniform cognitive region through which innovation diffuses, whereas *access to information is socially structured.

diffusion barrier Any obstacle which checks the *diffusion of an innovation. Four types of barrier may be encountered. A **super-absorbing barrier** stops the message and destroys the transmitter—the point of origin. An **absorbing barrier** stops the message but does not affect the transmitter. A **reflecting barrier** stops the message but allows the transmitter to transmit a new message in the same time period. A **direct reflecting barrier** deflects the message to the nearest available location to the transmitter.

diffusion curve The graphical representation of the spread of a new development from its inception to its general use. The curve has a typical S-shape. Initially only a few innovators and early adopters are prepared to experiment with new techniques: about 16% of the population. As the idea gains ground, the majority—the next 68%—accept the innovation. Finally, the laggards take notice.

digital elevation model Synonym for *digital terrain model.

digital mapping The storage and display of map data in digital form.

digital terrain model A digital relief map where grid cells contain elevation values.

digitizing Converting *analogue maps and other data into computer-readable form. **Manual digitizing** is achieved by an operator using a cursor and a *digitizing table; **blind digitizing** is a manual process where the operator has no immediate graphical feedback; automated digitizing requires little or no operator intervention. A **digitizing table** also known as a **flatbed digitizer**, is an electronic draughting table which can recognize x, y co-ordinates in computer-readable form, of points on a table.

dike, dyke A vertical or semi-vertical wall-like igneous *intrusion which is discordant, that is, it cuts across the *bedding planes of a rock. Dikes often form in swarms. **Ring dikes** are a concentric series of vertical circular sheets around a central intrusion. They appear to have been formed through repeated subsidence of the cauldron.

dilatancy theory Emphasizes the role played by variations of stress in *till found at the base of moving ice masses. The layer of till is constantly subjected to stress, and deformed during ice movement. A local reduction in stress causes the till to settle and become more compact.

Dilatancy theory has also been used in *earthquake prediction. B. A. Bolt (1988) has shown that, as crustal rocks crack under the tectonic strain which precedes an earthquake, changes in pore pressures within the rock cause a temporary reduction in the velocity of the P waves. Unfortunately, the time between this change and the ensuing earthquake may be months or even years; the greater the magnitude of the earthquake, the greater the lead time.

dilatation *See* *pressure release.

diminishing returns, law of The principle that further inputs into a system produce ever lower increases in outputs. Any extra input will not produce an equal or worthwhile return. Thus, while early applications of fertilizer may increase yields, further applications will not see a corresponding rise in output, and even further applications may actually damage the crop, as excessive fertilizer can burn plant tissue.

dip The angle of inclination of a rock down its steepest slope, that is to say, the direction at right-angles to the *strike. Dip is the angle between the maximum slope and the horizontal. A **dip slope** occurs where the slope of land mirrors the slope of the underlying *strata.

direct cell *See* *atmospheric cell.

directed link In *Geographic Information Systems, a link between two *nodes, with one direction designated.

dirt cone A cone of ice, as much as 2 m in height, covered with a thin layer of debris, found in the *ablation zone of a glacier. It occurs wherever debris forms on a glacier surface, and comes about because the dirt protects the ice below from *ablation. As the cone steepens, the debris will begin to slide off, and the core will decline.

discharge The quantity of water flowing through any cross-section of a stream or river in unit time. Discharge (Q) is usually measured in cubic metres per second (cumecs) and can be calculated as $A \times V$ where A is the cross-sectional area of the channel and V is the mean velocity.

discontinuous media Forms of the transport of energy which can be used flexibly along different routes and to different locations, as with oil tankers and lorries. The journey may have a number of different links by water, rail, and road. *See* *continuous media.

discordant Cutting across the geological grain of an area, as in the case of a stream cutting across an anticline. *See* *superimposed drainage.

discovery–depletion cycle The progression which unwinds as a non-renewable resource is exploited. The sequence begins with the discovery and early development of the resource, followed by rapid expansion leading to peak production. As reserves are depleted, output falls until the resource is exhausted.

discrete choice modelling The use of a group of statistical techniques to model the way in which people choose between distinct alternatives, such as a university course. The basic concept used is that each alternative has a total *utility to the decision-maker, which is the combination of the weighted utilities of all the attributes of the course; for example, the quality of the university teachers, the course content, the entry requirements, distance from home, and local living costs.

It is then possible to calculate the possibility, P, of choosing one out of j alternatives on the basis of the equation:

$$P = \frac{v_i}{\sum_j v_j}$$

where i is the rank, by utility, of the alternative and v_i its utility, but the amount of data and calculation entailed is enormous.

discrete variable A variable which is broken down into separate size categories where no fractions are possible. For example, the number of cars travelling to a town can only be expressed in whole numbers: fractions cannot exist.

diseconomies Financial drawbacks. **Diseconomies of scale** occur when an enterprise becomes too large, where sites become constricted, where the flow of goods is congested, and, perhaps, where the workforce is alienated. *See* *economies of scale.

dishpan experiment If a shallow pan of water is heated at the perimeter, chilled at the centre, and rotated about its centre, the water in the intermediate ring of the dish will describe a wave pattern. This wave-like response to differential heating and rotation seems to echo the configuration of the *Rossby waves.

dispersal In biology and *biogeography, the scattering of seed as a plant species colonizes.

dispersed city A plan of city structure as envisaged by the US architect, Frank Lloyd Wright (1864–1959), with one-family houses surrounded by open space. Shops and factories lie between housing areas and population densities are low enough to give a rural effect. The whole city is to be served by a network of super-highways.

dispersed settlement In comparison with *nucleated settlement, a settlement pattern characterized by scattered, isolated dwellings. Highlands, poor soils, and ubiquitous water supply help to create dispersed settlement as do cultural factors; lowland Wales has dispersed settlement whereas lowland England inclines to nuclear villages.

dispersion diagram A plot of the spread of values in a distribution. A vertical axis is used and each value is shown as a dot.

dissection The cutting down of valleys by river erosion. Thus, a **dissected plateau** is a level surface which has been deeply cut into by rivers, leaving a close network of valleys with hills in between.

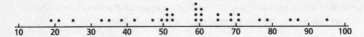

FIGURE 21: *Dispersion diagram*

dissolved load Material carried in solution by a river.

dissolved oxygen Oxygen from the atmosphere and from photosynthesis is dissolved in the upper levels of all bodies of water and is vital for the maintenance of most aquatic life. The amount of oxygen present decreases with depth, rising temperatures, and with the oxidation of organic matter and pollutants. The amount of oxygen used by organisms depends on their *biological oxygen demand. See also *eutrophication.

distance **Absolute distance** is expressed in physical units such as kilometres and is unchangeable. **Relative distance** includes any other kind of distance such as **time-distance**, which is measured in hours and minutes and changes with varying technology. Thus, a location 6 hours away by train is only 90 minutes away by air. **Cost distance** is expressed in terms of currency and varies with the transport mode, the volume and type of traffic and goods, and their destination. **Convenience distance** expresses the ease of travel.

distance decay The lessening in force of a phenomenon or interaction with increasing distance from the location of maximum intensity; the inverse distance effect. Examples include the way in which the intensity of land use declines with distance from the market (*von Thünen), or the number of phone calls made decreases with distance—most people in Britain make more local than long-distance, and more domestic than international calls.

distributary A branch of a river or glacier which flows away from the main stream and does not return to it. *See also* *delta.

distributed data In *Geographic Information Systems, data stored on more than one computer, but accessible to numerous users through communication linkages.

distribution **1.** The physical layout of a feature over an area, such as forest land.
   2. The dispersal of payments to the *factors of production for the output achieved.
   3. The function of *tertiary industry in delivering goods, that is wholesaling, warehousing, and retailing.
   4. The spread of varying observations within a population.

district In behavioural urban geography, a clearly identifiable section of a city, having a distinct *image and geographical extent.

districting algorithm A system, usually using computers, of drawing up electoral boundaries so that the constituencies conform to more or less equal size limits of population. Boundaries may not be manipulated to favour one party.

divergence The spreading out of a vector field; mathematically, *convergence is negative divergence. In meteorology, divergence is the spreading out of an air mass into paths of different directions. In the *atmosphere, **horizontal divergence** predominates, and the word 'horizontal' is understood when this term is used. Divergence is linked with the vertical shrinking of the atmosphere, since, by the conservation of matter, an outflow of air must result.

It is closely related to vertical *vorticity, since it is the principal agency responsible for the vorticity change experienced as a particle flows from a *cyclonic to an *anticyclonic circulation. Divergence in the upper air is associated with depressions at ground level. *See* *jet stream, Rossby waves.

diverging margin *See* *constructive margin.

diversification A measure taken to spread industrial commitment over a large range of activities so that there is no overdependence on one. The term can also refer to the extent to which this takes place. Diversification can take place within a single firm by the taking on of new ventures to spread the risk of any one failing. The development of products which require little adjustment of machinery or skills is **horizontal diversification**. **Concentric diversification** concerns the widening of the use of one product in order to penetrate new markets. **Conglomerate diversification** is the growth of industry into new areas as a result of changes in markets, technology, and products.

Multinationals diversify as they buy up new firms producing different products. In declining areas, there are problems of overdependence of the labour force on one industry, especially in those areas which developed a high degree of specialization in the nineteenth century. Here, **regional diversification of employment** is seen as the answer to overdependence and may also foster economic growth. Governments and local authorities are the usual agents of regional diversification.

diversification curve A diagrammatic technique used to compare *diversification with the minimum and maximum possible, and to compare diversification between regions. For each type of industry, employment is calculated as a percentage of total employment. The percentages are then ranked in ascending order and cumulated (i.e. each percentage is added to the total of percentages before it). The cumulative per cent of the labour force is plotted against the percentage of industrial groups ranked in ascending order. The points are then joined by a line; a 45° line represents the maximum possible diversification.

The **diversification index** is a measure of the difference between a given diversification curve and a perfect diversification line at 45°. The area between the curve and 45° is expressed as a percentage of the total area beneath the 45° line. The higher the index, the more evenly diversified a region is.

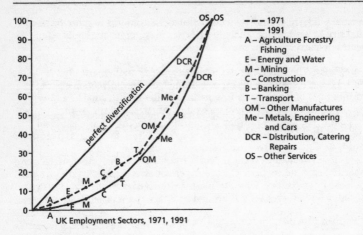

FIGURE 22: *Diversification curve*

diversity The abundance of species within an ecosystem. **Alpha diversity** or between-habitat diversity refers to the range of organisms within a habitat. If an ecosystem is very crowded, species may only survive in the habitats most favourable for them. This is **beta diversity** or within-habitat diversity.

diversivore An organism with a wide range of diet, from plants to other animals. Since the food eaten varies from time to time and place to place, it is difficult to fit a diversivore into the *food web.

divided circle *See* *pie chart.

division of labour The partitioning of a production process into separate elements, with each part assigned to a different worker or set of workers. It is based on the idea that workers can attain a high degree of efficiency if they are restricted to one particular process. Division of labour is one of the hallmarks of the factory system, but can lead to the alienation of the workforce as the workers lose touch with the creative process.

A **social division of labour** divides workers according to their product: steelworkers, miners, and so on. A **sexual division of labour** is a separation of jobs into male and female tasks, and in UK many jobs, such as nursing and bricklaying are highly gendered; in the context of paid work this is known as *occupational segregation*. There are also global patterns; the **international division of labour** describes the pattern of highly paid, more-skilled work in the advanced economies, and low-paid, less-skilled work in the Third World. Increasingly, transnational corporations capitalize on this pattern, and it is not uncommon for a major manufacturer to locate unskilled parts of the production process, such as assembly, in LDCs, while the design and manufacture of the component parts has been carried out in an MDC.

doctor A reviving *tropical wind; the *harmattan brings relief from humidity in West Africa, and the name is also given to the cool *sea breezes of the Caribbean, South Africa, and Australia.

dog days The hottest days of summer, so called because, in classical times, they were thought to be caused by the heliacal rising of Sirius, the dog star.

doldrums Those regions of light, variable winds, low pressure, and high temperature and humidity which occur in tropical and subtropical latitudes.

Doldrums occur over the east Pacific, the east Atlantic and from the Indian Ocean to the west Pacific. They are bounded to the north and south by the *trade winds, and their extent varies greatly with the seasons.

doline, dolina A closed, steep-sided and flat-floored depression in *karst country, such as Dalmatia. The sides are 2–10 m deep and the floor is 10–1000 m in width. **Alluvial streamsink dolines** form when a stream enters the doline and runs down through the rock to form a trough. **Collapsed dolines** form when a cave roof falls in. **Solution dolines** form when solution enlarges a point of weakness in the rock into a hollow. **Subsidence dolines** form when limestone caves develop below insoluble deposits. These superficial deposits may collapse into the cavern below, but the majority of dolines are corrosion forms.

dome An uplifted section of rocks, such as the Harlech Dome of North Wales. The highest part is at the centre, from which the rocks dip in all directions. **Volcanic domes** may be formed from slow-moving, *viscous lava. These domes may be rounded as the result of pressure from lava below. A **plug dome** is a small, irregular dome within a crater. Plug domes may have spiny *extrusions projecting from them.

domestic industry A type of production whereby manufacturers supply materials to workers in their homes, pay them for the finished goods, and sell the products. Remnants of the system still exist, notably in knitwear.

domestication The bringing under human control of a plant or animal.

domino theory The view, which was held by many US administrations and politicians that if one small nation were to 'succumb' to communism, then its neighbours would surely follow. This view has been acted on in a military sense, most tragically in the Vietnam war.

donga A steep-sided gulley resulting from severe *soil erosion. The term is generally used in South Africa.

dormant volcano A volcano which is inactive but not extinct.

dormitory town A settlement made up largely of daily commuters who are employed elsewhere in a larger centre. These commuters have displaced the original residents or live in new housing at the edge of the town or village. Dormitory towns are characterized by a relative paucity of retail outlets since the commuters will use services in the centre of the city or in out-of-town shopping centres.

double water feature Some water *feature, such as a canal whose limits are defined on a published map by two, surveyed lines.

downtown A term used in the USA to denote the heart of the city; the CBD.

downward transition region In Freidmann's *core–periphery model, a region on the periphery characterized by depleted resources, by low agricultural productivity, or by outdated industry.

downwearing, declining slope retreat A model of hillslope retreat where the slope angle decreases over time due to a combination of soil *creep, rain splash, and *sheet wash which causes slope convexities and concavities at the expense of straight hillslope segments. It is now generally accepted that no single model of slope evolution can explain all the features of hillslopes.

drag The braking effect of friction, imparted by a fluid on bodies passing through it.

drainage The naturally occurring channelled flow formed by streams and rivers which removes water from the land surface.

drainage basin The area of land drained by a river and its tributaries. The term is synonymous with river basin.

drainage basin geometry Using A. N. Strahler's 1952 classification of *stream order, (*Bulletin Geol. Soc. Amer.*, 1952) the following relationships hold:
  1. The number of streams of each order falls in an inverse geometrical progression as the stream order rises.
  2. The average slope of a stream falls in an inverse geometrical progression as the stream order rises.
  3. The total length of streams in a drainage basin is linked logarithmically with the *drainage basin order.
  4. The *discharge of a stream is linked logarithmically with the area of the drainage basin feeding that stream.
  5. The average area of a drainage basin rises in direct geometrical progression as the order of the basin rises.
  6. The average length of streams of each different order rises in direct geometrical progression as the order of the drainage basin increases.
  Many of these 'laws' are true of most branching systems and are therefore not notably edifying.

drainage basin order Just as *stream order can be quantified, so can the order of a drainage basin. Thus, a basin serving only a first order stream is a first order drainage basin, and so on.

drainage density The total length of streams per unit area. Any attempt to calculate drainage density is impeded by the difficulty of calculating

total stream length, as the exact point at which a stream starts is
problematical.

drainage network evolution The drainage basin can increase its area
and extend its channels by landslides at the edge of the network, by
*headward erosion, by the extension upslope of underground *pipes, and
by the formation of *rills.

There are two main theories to account for the nature of drainage
network evolution:

1. That the drainage system develops at random, but that random
development will bring about a coherent network.

2. That stream networks develop by the growth of new, first-order
channels at a rate which is proportional to the size of the drainage basin.

drainage patterns The pattern of a drainage network. This pattern is
strongly influenced by geological structure. **Anastomotic drainage** shows a
division of a river into several channels, and develops on nearly horizontal,
coarse sediments. **Annular,** or **radial drainage** shows the major rivers
radiating out from a centre with the tributaries arranged along a series of
nested arcs. Annular drainage develops on domes, particularly where belts
of resistant rock are separated by belts of weaker rock. **Centripetal drainage**
shows a movement into a centre created by a crater or depression.
Dendritic drainage shows a branching network similar to that of a tree,
and is most commonly found on horizontally bedded or crystalline rocks
where the geology is uniform. **Parallel drainage** develops in slopes of
moderate angle. **Rectangular drainage** shows the tributaries running at
right angles to the major river and occurs on rocks which have
intersecting, rectangular joints and faults. **Contorted drainage** is a form of
rectangular drainage on complex metamorphosed rocks. **Trellised drainage**
resembles a trained fruit tree. It usually occurs on dipping or folded
sedimentary or weakly metamorphosed rocks.

drains Channels cut in naturally wet sites constructed to remove excess
water. Porous clay pipes are buried beneath the surface to form **tile drains**
and **mole drains** are unlined channels in the subsoil, formed by pulling a
bullet-shaped plug through the soil.

dreikanter A stone with three clearly cut faces, like a Brazil nut, formed
by sand-blasting in desert environments.

drift 1. A horizontal tunnel driven into the hillside for mining purposes;
an *adit.

2. Any category of *glacial or *fluvio-glacial deposit.

drift-aligned beach A beach which has developed parallel to the line of
longshore drift, normally at an angle of 40–50° to the direction of wave
approach. If the wave direction changes to become more right-angular,
drift slows down, and the beach realigns itself. Compare with *swash-
aligned beach.

drifter A fishing vessel which operates by lowering weighted nets at night and drifting with the winds until morning, when the nets are drawn up again.

dripstone In *karst scenery, underground streams carry calcium carbonate in solution. If the water pressure of the stream falls as it enters a cave, or if the amount of dissolved carbon dioxide decreases, the carbonates will be deposited to form dripstone in hanging *stalactites or in stubbier, cave-floor *stalagmites.

drizzle Rain made of very small droplets, up to 0.2 mm across, and with a fall speed of around 0.8 m s^{-1}. Dense drizzle is not to be confused with light rain, which has at least 3 mm droplets.

drought A long, continuous period of dry weather. Major causes of drought in Britain are the persistence of warm *anticyclones, and the displacement of *mid-latitude depressions by *blocking anticyclones. Droughts in Africa, for example in the *Sahel in the 1970s or in Zimbabwe in the early 1990s, result from the failure of the *inter-tropical convergence zone to move sufficiently far from the equator. *See* *absolute drought.

drove road A broad, unmetalled track used by herders walking their animals, usually cattle or sheep, to market. Drove roads were used for hundreds of years until superseded by the railways.

drumlin A long hummock or hill, egg-shaped in plan and deposited and shaped under an ice sheet or very broad glacier while the ice was still moving. The end facing the ice—the *stoss—is blunt, while the lee is shallow and its point indicates the direction of the ice flow.

Most drumlins result from the reworking of *lodgement *till. It may be that, under high pressure, ice squeezes the till, making it stiffer so that it lodges on the valley floor forming a stoss slope. At points of low pressure, down-glacier, the till is less viscous and may be streamlined, as at the lee of the drumlin. **Drumlin swarms**, or **drumlin fields** are not uncommon, for example in the Eden valley of the English Lake District, or the Eberfinger drumlin field of Bavaria.

Rock drumlins are more commonly known as *roches moutonnées.

dry adiabatic lapse rate, DALR *See* *adiabatic.

dry-bulb thermometer A simple mercury thermometer which is usually housed in a Stevenson screen. *See* *wet-bulb thermometer.

dry farming Farming without irrigation, using techniques which conserve water for the crop. Strategies include mulching, frequent fallowing, working the soil to a fine tilth, and frequent weeding.

dry-point settlement In a marshy, damp, or frequently flooded area, raised sites such as low mounds or gravel terraces attract settlement. The 'ey' portion of names such as Pevensey in Sussex or Hackney in London refers to the dry island upon which these settlements were founded.

dry valley A valley, usually in chalk, or *karst. A dry valley has no permanent watercourse along the valley floor. The theories of dry valley formation are many. Although present-day processes such as *river capture and *superimposed drainage can account for the formation of some dry valleys, most researchers believe that these valleys were cut during *periglacial phases in the quaternary period. Under these conditions, the *permafrost would prevent the river water from soaking through chalk or limestone and would allow *dissection by *meltwater channels.

The periglacial origin of dry valleys seems to be borne out by fieldwork, but there are arguments for falling sea levels and tidal scour as being causal factors. Examples are common on English chalk lands, such as the South Downs, or in the limestone areas of Thuringia and Swabia.

dry weather flow A synonym for *base flow.

dual economy An economy comprising two very different systems, and found in many developing countries where an advanced economy co-exists with a traditional economy and the two have very little contact with each other. In many less developed countries, a *subsistence system operates side by side with, but quite independently of, cash cropping. This is a result of *uneven development. The concept also applies to *more developed economies, where *core regions with large-scale, thriving industries compare with their less developed peripheries: northern Italy and the Mezzogiorno are often quoted as an example.

Certain geographers dislike the concept, arguing that it is too simplistic; why should there be only two elements and are they necessarily completely distinct?

dual labour market The labour market has two sectors: primary and secondary. The primary labour market is typified by high incomes, fringe benefits, job security, and good prospects for upward mobility. This sector may be subdivided into upper primary workers—long-term workers who advance their position as their years of service increase—and lower primary workers who are usually blue collar workers in stable, skilled labour.

The secondary labour market is typified by low incomes, little job security, and little training. There are no rewards apart from wages. Mobility between the primary and secondary markets is very low. These differences are held to derive from the structures of capital. This analysis tends to oversimplify a complex reality; see the end comments for the previous entry.

dumping The off-loading of goods at below cost, usually as exports.

dune See *sand dune or *coastal dune.

duricrust A hard capping found at the surface of the soil in tropical uplands, such as the Djebel Qarah of Saudi Arabia.

Duricrusts are thought by some to be a *plinthite *horizon, originally formed in the B horizon, but now revealed by erosion since plinthite hardens on exposure to the air. The duricrust would therefore be the remains of a fossil soil. An alternative explanation is that duricrusts are formed through 'core weathering' processes, during which rock degradation progresses from the subsurface to the surface of bare rock. Other processes in the formation of duricrust include evaporation of a lake, of *groundwater, or of subsurface and surface waters moving across *alluvial fans and *pediments.

Dust Bowl The name given in 1935 to parts of Colorado, Kansas, New Mexico, Texas, and Oklahoma, which were severely afflicted by *drought and dust storms. It was caused by the loosening of soil, through the removal of natural vegetation and tilling, in combination with a long period of low rainfall. The term has been extended to apply to other regions with similar problems.

dust devil A whirlwind made visible because it is carrying dust.

dust storm Desert dust storms form when the wind picks up small, light particles such as silt. Many tonnes of light soil can be transported in storms some 500 km in diameter. *Accelerated soil erosion caused the dust storms and dust bowl of the Great Plains of America.

duyoda In a *periglacial landscape, a steep-sided, shallow, and often circular depression formed as *baydzharakhs collapse. A duyoda is smaller than, but can develop into, an *alas.

dyke *See* *dike.

dynamic density E. Durkheim's (1893) name for the urban population, who are held to be more responsive than their rural counterparts to new forms of economic and social organization as a result of economic specialization and innovations in transport and communications.

dynamic equilibrium In a landform, a state of balance, in spite of changes taking place within it. Thus, a *spit may appear to be unchanging, although it is fed by deposition from its landward end, and subject to erosion at its seaward end. In the same way, a glacier is fed by *accumulation and depleted by *ablation, and the debris on a slope is produced by *weathering but removed by transport. In all these examples of *open systems, inputs and outputs are in equilibrium.

dynamic meteorology The study of atmospheric motions as expressed by the fundamental hydrodynamic equations, and other systems of equations specific to special situations, such as *turbulence.

E

earth fall *See* *fall.

earth flow A form of *mass movement, where water saturated, weak-slope material flows under the action of gravity at speeds between 10 cm s^{-1} to 10 cm day^{-1}. Earth flows differ from mud flows in that the latter are composed mainly of clay-sized particles. They are common at the foot of landslides, where water content is high.

earth pillar An upstanding, free column of soil that has been sheltered from erosion by a natural cap of stone on the top. They are common where boulder-rich moraines have been subject to gully erosion, as in parts of the southern Tyrol.

earthquake A sudden and violent movement, or fracture, within the earth followed by the series of shocks resulting from this fracture. The point of origin of an earthquake is known as the *focus (but *see* *epicentre). Earthquakes occur in narrow, continuous belts of activity which correspond with the junction of *plates.

The scale of the shock of an earthquake is known as the **magnitude**; the most commonly used scale is the *Richter scale, while the **intensity** of an earthquake is measured by the *Mercalli scale.

Earthquake waves are of three basic types: P, primary, push waves travel from the focus by the displacement of surrounding particles and are transmitted though solids, liquids, and gases. S, secondary or shake waves travel through solids. L, long or surface waves travel on the earth's surface. The monitoring of these waves indicates that the earth's core is molten since S waves do not pass through it. *See* *seismic waves, *seismology.

Fully credible **earthquake predictions** are not yet available; one of the most hopeful avenues entails the application of *dilatancy theory.

easement The granting of a right to lay apparatus across privately held land where there are no statutory rights.

easterlies Winds blowing *from* the east, like the north-east and south-east *trade winds, and the *equatorial and *polar easterlies. Easterlies prevail in low latitudes in both high and low *tropospheres, and in high latitudes in the lower troposphere.

easterly waves In *tropical areas, weak troughs of low pressure, linked with cloud systems.

ecodevelopment Economic development which is sensitive to such environmental issues as the use of scarce, finite resources, pollution, and the destruction of habitats.

ecological balance The achievement of a steady state by an *ecosystem, as in a *climax community. This theoretical concept is rarely accomplished in practice as the controlling forces, such as climate and soil, rarely remain constant, and ecosystems take some time to adjust to change.

ecological climatology, eclimatology A branch of *bioclimatology, which studies the interactions between organisms and their environment. It includes the geographical distribution of plants and animals in relation to climate, and their adaptations to it.

ecological crisis A state of human-induced ecological disorder that could lead to the destruction of ecological conditions on this planet to such an extent that human life, at least, will be seriously impaired for generations, if not destroyed. There is some debate over whether such a crisis has already been reached; evidence to suggest it has included *deforestation, increasing levels of atmospheric carbon dioxide, and current rates of energy use. Others argue, however, that what we call a crisis now is simply a reflection of increased environmental awareness.

ecological dominance The predominance of one or a few species within an *ecosystem such that they have more importance than is possible in a purely random process.

ecological economics A branch of economics which attempts to establish rules for possible sustainable economies on a finite earth with finite resources, so that each economy is integrated into an ecological framework. Compare with *environmental economics.

ecological efficiency The ability of the organisms at one *trophic level to convert to their own use the potential energy supplied by their foodstuff at the trophic level directly beneath them.

ecological energetics The study of the flow of energy from the sun through and up the *trophic levels, expressed in calories. This energy may be fixed in the form of food. With movement up each trophic level, there is a very great loss of the energy available as food. The study of energetics highlights the importance of micro-organisms within ecosystems and stresses the danger of destroying these minute forms of life.

ecological explosion The sudden and dramatic increase in the numbers of a particular species. This may be due to unusually favourable conditions or may happen when an organism is introduced to a new habitat where there is no natural predator, for example, cane toads in Queensland.

ecological fallacy The danger of making an analysis at one level apply at other levels, for example, of inferring individual characteristics from group characteristics. The most famous exposé of this type of reasoning was made by W. S. Robinson (*Am. Soc. Rev.*, 1950), who showed that there was a correlation between the number of black people in American states and literacy levels. He then showed that the assumption leading from this—that black people were inclined to illiteracy—was by no means certain.

ecological imbalance The destabilization and destruction of a fragile environment, often as a result of economic development.

ecological invasion A term used by social geographers to describe the influx of a social group which is better adapted to its new environment than are the existing inhabitants. It is a term used by the exponents of *human ecology, who borrowed extensively from the concepts of mainstream *ecology.

ecological psychology The study of the psychology in a behaviour setting, such as a school, which gives rise to common or regularized forms of behaviour.

ecology This term, coined by Ernst Haeckel in 1866, describes the study of the interrelationships between organisms and their surrounding, outer world; the study of animals and plants in relation to each other and to their habitats. Life forms, including man, are intimately linked with their environment. **Production ecology**, or **community ecology**, is the study of the structure of *communities in terms of the throughput of energy and chemical compounds; key concepts are *primary production, *trophic levels, and *nutrient cycles.

'Ecology', in everyday use, is linked with *environmentalism, or the Green movement, perhaps because most ecological thinking is in favour of *preservation of ecological processes, habitats, and species.

econometrics Economic analysis which applies economic theory to real-life, rather than theoretical, data, with the aid of statistical techniques such as regression analysis. The aim is to make forecasts of, for example, money supply, and to determine economic policy in the light of such forecasts.

economic base theory The view that economic activity in a city can be broken down into two components: activity which meets local, internal demand and activity which meets non-local demand. The former is non-*basic; it serves the city but does not cause it to grow. The latter is basic and city forming because it is the demand from beyond the city which causes the city to grow. This may be expressed in the equation:

$$E = S + B$$

that is, total employment in the city, E, is made up of non-basic S, and basic, B, employment.

The relationship between the local population, P, and employment is:

$$P = \alpha E$$

and between the non-basic sector and population is:

$$S = \beta P.$$

α and β are coefficients which can be obtained by a *regression based on observations from one city or region over time, or a *sample of cities.

The theory is that because growth, or decline, in the city's population and employment is regulated by changes in the basic sector, the impact of a change in that sector on employment, population, and the non-basic sector, respectively, can be predicted from the three economic base equations:

$$E = \left(1 - \alpha\beta\right)^{-1} B$$
$$P = \alpha\left(1 - \alpha\beta\right)^{-1} B$$
$$S = \alpha\beta\left(1 - \alpha\beta\right)^{-1} B$$

A unit increase in B gives rise to $\alpha/(1 - \alpha\beta)$ units of additional population.

Analyses like these have been widely used, in spite of their very simplistic nature in, for example, the *activity allocation and *Lowry models, but there are problems, which arise from difficulties of deciding which activities are basic and which non-basic, and the presumption that the multipliers are correctly established, and will stay the same over time.

economic determinism The thesis, as advanced by Marx and Engels, that economic factors underlie all of society's decisions. Thus, the social relations specific to a particular *mode of production are said to structure social relations between classes and are held to be the base underpinning the legal and political systems. This implies that all political, cultural, and social life can be predicted from the prevailing relations of production. Not surprisingly, such an extreme view has been severely criticized as it denies the existence of free will and individual independence. A more moderate view sees the relations of production acting as a constraint on the possible ways in which individuals and superstructures can develop. See *historical materialism.

economic distance The distance a commodity may travel before transport costs exceed the value of the freight. Because of lower unit transport costs, a small, valuable commodity can be transported profitably further than a bulky commodity of the same value. See *transferability.

economic dualism See *dual economy.

economic efficiency The ratio by value of input to output. The higher the ratio, the greater the efficiency.

economic geography The analysis of the spatial distribution of the transportation and consumption of resources, goods, and services, and their effects on the landscape.

Early economic geography was concerned with describing *what* was produced *where*; a type of *commercial geography, but, in common with other areas of geography, the subject changed with the *quantitative revolution and began to attempt explanations for the location of economic activity; explanations based on the assumptions of *economic man and *optimizing behaviour, which led to models such as Weber's *isodapanes. The influence of *behavioural geography caused a modification of these ideas, with the introduction, for example, of the concepts of the *decision-maker and the *satisficer.

Contemporary economic geography is characterized by a number of themes. A major concern is a re-examination of the nature and causes of *development and *underdevelopment, putting the *mode of production

at the centre, and stressing the interrelationships between the less and
more developed worlds.

Another considers the link between *economic systems and geography,
particularly in interpretations of the spatial impacts of *capitalism and its
role in the development of the world economy. A third challenges the
stifling nature of this predominantly economic view by emphasizing the
dimensions of class, race, and gender in economic systems, stressing, and
sometimes challenging, the way economic systems depend on
discrimination on the basis of these three categories. *See* *feminist
geography.

Other economic geographers study the impact of technological change
and the construction of 'new' industrial spaces, but the hope is that these,
and other, concerns can be brought together under one conceptual
umbrella.

economic growth The growth in wealth of a nation, as measured by an
increase in *gross national product, or in national income. An economy
can grow without benefiting everyone in it; in the 1970s Brasil's economy
grew at rates of around 10%, but at the end of that era, the gap between
rich and poor had widened. For this reason, many geographers are
reluctant to equate economic growth with *development, arguing that
true development should contain an element of social justice. For theories
of economic growth *see* *Rostow's model and *cumulative causation.

economic man A theoretical being who has perfect knowledge of an
economy and has the ability to act in his or her own interests to maximize
profits. The idea of economic man has proved a useful tool in *neoclassical
economics but other writers suggest that the concept of the *satisficer is
more realistic.

economic overhead capital Investment into the *infrastructure which
should encourage new industrial growth. This is often a major part of a
development programme. For example, the Appalachian Development Plan
of the 1960s and 1970s was centred on road construction.

economic rent Economic rent is not synonymous with profit, since built
in to the concept is the notion of opportunity cost; the real cost of
choosing one alternative good or service in terms of the sacrifice of the
next best alternative. A singer may work regularly for £10 000 per annum
or may sing only in starring roles for £20 000. Let her/his total transport
costs for each type of work be £5 000. The net income for starring concerts
is £15 000 but the economic rent is £10 000, i.e. the extra profit over and
above the second choice.

However, the term is often used much more loosely; if opportunity costs
are ignored, economic rent is the difference between revenue and costs.

Economic rent is all about choices, particularly of land use in
agricultural geography, where it is the extra profit derived from operating
one form of land use rather than another. This is evident from the classic
model where different forms of land use successively 'out-bid' the others,
according to their particular **economic rent gradients**. *See* von *Thünen's
model.

economic sector While most geographers and economists recognize the *primary, *secondary, *tertiary, and *quaternary sectors, the division between *public and *private sectors is also significant.

economic system The organization of activity to produce goods and services for given consumers. The central problems are: what to produce, how to produce it, and whom to produce it for; to allocate human resources to supply wants when there is a scarcity of the *factors of production, most of which have alternative uses. Any economic system should be able to determine the needs of society for goods and services, ensure the correct allocation of the factors of production to industry, provide and maintain investment, distribute goods and services by matching supply to demand, and utilize resources efficiently. *See also* *mode of production.

economic transformation The change in the structure of an economy over time from a *subsistence economy, through *industrialization, to an industrial, or even post-industrial society.

economically active population The total population between the ages of 15 and 65 in any country, used in the calculation of *dependent population.

economics The study of the relation of available scarce means to supply for a proposed end; economists assume that people have wants and needs, and then study how societies are organized to supply them, trying to establish whether one method is better than another. **Micro-economics** explains how demand and supply affect prices, wages, rentals, and interest rates. **Macro-economics** focuses on the aggregate (large scale) demand for goods and services, and especially on the relationship between unemployment and the economy. **Marxist economics** sees the economy as a reflection of the history and sociology of a society. In particular, it focuses on the historical evolution of, and the conflict between, classes.

economies of scale The benefits of producing on a large scale. As the volume of production increases, the cost per unit article decreases. It is suggested that, after a certain volume of production, this fall in cost will be halted as diseconomies arise, but this will happen only at very high levels of production, if at all. **Internal economies of scale** arise from within a plant and include *indivisibility, specialization, and *division of labour. Furthermore, overheads like Research and Development cost less per unit article when production levels are high. Increases in plant size can be important; doubling the capacity of a machine does not necessarily double its cost. In a larger production system, specialist machines can be used to advantage. Buying in bulk reduces costs. These different factors operate with differing force in different industries. *See also* *external economies.

ecosystem A community of plants and animals within a particular physical environment which is linked by a flow of materials through the non-living (abiotic) as well as the living (biotic) sections of the system. Thus, ecosystems can range in size from the whole earth to a drop of water, although in practice, the term ecosystem is generally used for units below the size of *biomes, such as sand dunes, or an oak woodland.

ecotone A region of rapidly changing species between two *ecosystems, for example between coniferous forest and tundra. Within the ecotone, local factors, such as soil and groundwater conditions determine species. An ecotone usually marks a change in soil, in water supply, or in exposure to the elements.

ecotope A defined *niche or niche space within a habitat.

ecumene (oecumene) The inhabited areas of the world, as opposed to the non-ecumene which is sparsely or not at all inhabited. The ecumene of a nation is its more densely inhabited core. These very simplified classifications pose difficulties of delimitation.

edaphic Of the soil; produced or influenced by the soil.

eddy A roughly circular movement within a current of air or water. Eddies may have the circular and intermittent motion of a *vortex, the continuous corkscrew motion termed *helicoidal flow, the cylindric motion of rollers, or surge phenomena which are short-lived outbreaks of greater velocity in any flow.

edge 1. In *network analysis, another term for the *link between two *nodes.
   2. In behavioural urban geography, the edge is an informal boundary imagined by an individual or group separating one clearly identifiable *district from another. The definition of such boundaries is frequently problematic.

edge map In *geographic information systems, to match data, in terms of attribute and position, along edges of adjacent maps.

EEC, EC The European Economic Community; the former name for the *European Union.

efficiency The ratio of the work done to the effort used; usually the relation of output to input. Although output and input are usually measured in financial terms, analyses of agricultural efficiency can be made using input and output in terms of energy, and this analysis makes commercial agriculture look far less efficient than in conventional, financial analyses. The concept of **spatial efficiency** is central to *locational theory, and relates to the organization of space in order to minimize costs but ignores considerations of *social justice.

effluent The flow into a river, stream, or lake of sewage, fertilizers in solution, or liquid industrial waste.

effluent stream A small *distributary. This expression has no connection with the term 'effluent' as in pollution.

EFTA The European Free Trade Association, established in 1960, an economic union of Sweden, Austria, Portugal, Finland (as an associate) Iceland, Norway and Switzerland, and, later, Liechtenstein. Subsequently, all but the last three have transferred their allegiance to the *EU. In 1991 EFTA was linked with the EU to form the European Economic Area, a single, free-trade grouping.

Ekman layer In the *atmosphere, the transition stratum between the surface *boundary layer, where the *shearing stress is constant, and the free atmosphere. In *meteorology, the **Ekman spiral** is a mathematical model of the wind distribution with height in the *planetary boundary layer.

el Niño, el Niño southern oscillation, ENSO A southerly, warm *ocean current off the coast of Ecuador, so called because it generally develops after Christmas. The el Niño effect is an extension of this current to the coasts of Chile and Peru, replacing the usual nutrient-rich cold currents. Deprived of sustenance, many organisms die, and the result is a red tide. This effect, which occurs about fourteen times per century, also brings rain to the dry Peruvian coast. el Niño is also associated with major variations in tropical climates, such as the 1982-3 droughts in Africa, Australia, north-east Brasil, and the USA.

elasticity In economics, a measure of the responsiveness of supply and demand to changes in price. Elasticity is calculated as the change of supply or demand, related to a 1% difference in price. Where the percentage change in demand is greater than the percentage change in price, the demand is said to be **elastic**. Where the opposite applies, demand is said to be **inelastic**. The same terminology is used for supply. **Elasticity of substitution** is an indication of how easily one of the *factors of production can be substituted by one of the others; for example, labour by capital (machinery).

electoral geography The geographical analysis of elections. The study includes the drawing of constituency boundaries (*see* *districting algorithm and *gerrymandering), the spatial patterns of voting and power (*see* *pork-barrel), the influence of sociological and local factors on voting behaviour (see *contextual effect), and the influence of voting decisions upon the environment.

electromagnetic radiation Waves of energy propagated through space at the speed of light. *Solar radiation is of this type as is the energy measured by *remote sensors.

electronic mapping The production of maps and related cartographic products, and the presentation of these in different formats, for electronic media and geographic information systems.

ellipsoid A figure shaped like a sphere, not perfectly spherical but with an oval form.

eluviation The lateral or downward movement in suspension of clay or other fine particles. The clay-depleted horizon thus formed is an **eluvial**

horizon and is found either below or in place of the A *horizon. The eluviated particles may accumulate in the *B horizon to form a clay pan.

emergent coastline Johnson (1919) attempted a classification of coasts based on rising (submergent coast) or falling (emergent coast) sea levels. The Atlantic seaboard of the USA was cited as an example of the latter, and a succession was suggested: low coast, coastal bar, lagoons, which silt up, and the movement onshore of an offshore bar. The concept has been challenged, as nearly all coasts, including the Atlantic seaboard, have experienced a net inundation in the last 10 000 years.

emigration The movement of people from one place to another, usually from one country to another. While the years immediately following World War II witnessed a massive redistribution of population within Europe—6 million Germans emigrated from Poland, Czechoslovakia, and Hungary in 1945–9, for example, current emigration flows are from the less developed world. *See* *migration, *chain migration, *Ravenstein.

emission standard The level of pollution which is allowed, by law, into the environment.

empirical Based on, or acting on, observation or experiment, not on theory. An empirical view regards sense-data as solid information and strives for objectively verifiable measurements so that knowledge can be derived from experience alone. **Empiricism** is the theory that all concepts emanate from experience and that all statements claiming to express knowledge must be based on experience rather than on theory. Valid statements must be based on what can be proved to exist, not on what appears to exist. This is known as ontological privilege since *ontology relates to the being or essence of things. Such statements must be able to be declared true or false without reference to theoretical statements. This is epistemological privilege since *epistemology is the study of knowledge. Knowledge is held to be substantiated by justification derived from observed facts.

Empiricism is the basis of scientific knowledge, in geography as in many other disciplines, and many human geographers search for general principles and laws in the light of the data which they have accumulated.

enclave 1. A small area within one country administered by another country. West Berlin was an enclave within Eastern Germany between 1945 and 1990.
2. A part of a *less developed economy which is regulated by foreign capital and has few *linkages with the national economy. *Free trade zones may be considered as **economic enclaves**.

enclosure The fencing of once common land to bring it into private ownership, most significantly in England in the sixteenth and eighteenth centuries.

end moraine *See* *moraine.

endemic Occurring within a specified locality; not introduced.

endogenetic, endogenic Meaning 'from within'.
1. In geomorphology, this refers to those forces operating below the crust which are involved in the formation of surface features.
2. In human geography, it is those forces acting from within, for example, a society. Compare with *exogenetic.

endogenous variable In economics, a variable explained within the theory studied. Compare with *exogenous variable.

energy The physical capacity for doing work. Nearly all our energy derives from the sun, and technical progress has reflected more and more sophisticated uses of energy, from wind and water, through *fossil fuels, to *nuclear power. In the early stages of industrialization, the consumption of energy is closely related to levels of economic development, and hence per capita *GNP, although mature economies tend to be more energy-efficient, perhaps because technology improves and the emphasis shifts to service industries. Nevertheless, the *advanced economies still account for most of the world's energy consumption. The breakdown of Western Europe's energy consumption in 1991 was: solid fuels 18%, liquid fuels 46%, gases 21% and primary electricity 15%.
World demand for energy has increased so much that an **energy crisis**—a potential shortage of energy—has now been identified and this, together with the adverse environmental effects associated with the burning of fossil fuels (*greenhouse effect, *acid rain) has lead to increased emphasis on energy conservation.
Energy resources are commonly divided into non-renewable (fossil fuels) and renewable (wind, water, and *solar energy). *See also* *geothermal heat.

energy crisis *See* *oil crisis.

enforcement notice A notice requiring the perpetrator of unpermitted development to comply with planning regulations.

englacial Within a glacier.

ENSO *el Niño southern oscillation.

enterprise zone An area of declining or derelict land, usually within an *inner city, which is chosen for rejuvenation. Private enterprise is attracted by such inducements as concessions on land development tax and local council tax, 100% capital allowances, a general speeding up of procedures, and a relaxation of planning regulations. Enterprise Zones were introduced in the 1980s in tandem with the British government's emphasis on small businesses, and these highly localized initiatives were designed to replace in part the much larger Development Area which had been part of British regional policy since the 1930s. *See also* *urban development corporation.

entisol In *US soil classification, young soils, high in mineral content and without developed soil *horizons. *See* *rankers.

entity In *geographic information systems, a thing that exists such as a

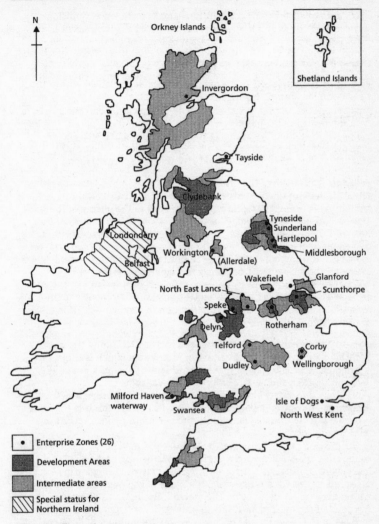

N

Orkney Islands

Shetland Islands

Invergordon

Tayside

Clydebank

Londonderry

Belfast

Workington

(Allerdale)

Tyneside
Sunderland
Hartlepool

Middlesborough

Wakefield

Glanford

North East Lancs

Scunthorpe

Speke

Delyn

Rotherham

Telford

Corby

Dudley

Wellingborough

Milford Haven
waterway

Swansea

Isle of Dogs

North West Kent

• Enterprise Zones (26)

Development Areas

Intermediate areas

Special status for
Northern Ireland

FIGURE 23: *Enterprise zone*

building or a lake, which is distinguishable from another entity, cannot be divided into two or more similar entities, and about which information can be stored, possibly in terms of attributes, position, shape, and relationships. An **entity class** is a specified group of entities.

entrainment 1. In *geomorphology, the picking up and setting into motion of particles, either by wind, water, or *ice. The main entrainment

forces are provided by impact, *lift force, and *turbulence. Collision between particles is an important process where the lifting agent is air.
   2. In *meteorology, the incorporation of buoyant air into a cloud.

entrepôt Also known as a **free port**, this is a point of transhipment between nations where goods are held without incurring customs duties. Examples include Singapore and Rotterdam.

entrepreneur An organizer, singly or in partnership, who takes risks in creating, investing in, and developing a *firm from its inception through to hoped-for profitability as goods and services are marketed. The enterprise of the entrepreneur can be seen as a fourth *factor of production, but other writers would classify it as a form of labour.

entropy A measure of the disorder within a system. Any state of order is actually a state of unequal distribution, and is virtually certain to randomize as time passes. (Entropy can be seen as the enemy of the houseproud; the 'ordered state' of tidiness is *sure* to break down). As an isolated system tends towards equilibrium, entropy increases; thus, it is the tendency of a system to move from a less probable (ordered) to a more probable (less ordered) state. As the amount of entropy in a system increases, the amount of free energy in that system decreases.
   Geographers study entropy levels in different population distributions and settlement patterns and use *entropy-maximizing models to find the most probable pattern of spatial distribution in a system which is subject to restrictions.

entropy-maximization procedure A method of finding the most probable pattern of spatial distribution in a system which is subject to restrictions, that is to say, of making the best estimate of a probability distribution from the limited information available.
   *Entropy may be seen as a measure of a system's disorder, and maximum entropy is maximum disorder within a system; it is the most probable state within a system subject to constraints, since everything tends to disorder.
   The method can be illustrated by looking at a method of calculating commuting flows within a city without investigating each individual's movements. Consider a matrix showing individuals taking a trip along a variety of routes. On a micro-scale, the name of each individual is recorded within each cell of the matrix. The macro-scale shows only the column and row total of the matrix. Most macro-states will correspond to a large number of micro-states, and entropy assesses the number of different micro-states which can correspond to a particular macro-state. Maximum entropy shows the greatest correspondence between macro- and micro-states. The macro-state with the maximum entropy value is the most likely pattern to occur.
   The mathematics of all this is complicated, to say the least, but the entropy-maximization procedure is superior to many models—such as the *gravity model—because it is not *deterministic and can be applied to complex situations. The procedure has not been without its detractors who

note that, as in the gravity model, no attention is paid to individual evaluation.

enumeration district A unit of census survey; in the UK, about 150 households.

environment The surroundings. The **natural environment** includes the nature of the living space (sea or land, soil or water), the chemical constituents and physical properties of the living space, the climate, and the assortment of other organisms present. The **phenomenal environment** includes changes and modifications of the natural environment made by man. The effect of the environment on man is modified, in part, by the way the environment is perceived and human geographers distinguish this—the **subjective environment**—from the **objective environment**—the real world as it is. The objective environment is of less importance to the individual than his or her perceived image of it. A division may also be made between the **built environment** and the **social environment** which is made up of the various fields of economic, social, and political interactions.

environmental determinism The view that human activities are governed by the environment, primarily the physical environment. According to this view, individuals build up knowledge by encountering the world through their senses, and are unable to transcend their responses to the environment; they are at the mercy of environmental stimuli. This rather crude view of human behaviour has come under fierce criticism and has been, in part, displaced by *environmentalism, *possibilism, and *probabilism.

environmental economics An economic viewpoint which holds that, if environmental goods, such as habitats, are monetized—expressed and valued in money terms—then they will be subject to market forces, and the *ecological crisis will be solved. This view is derided by some ecologists.

environmental hazard Sources of danger to humans and to their built environments which arise in the environment. Most of these are natural hazards, such as hurricanes and other high winds, lightning, floods, droughts, earthquakes, and volcanoes. It has been noted that human beings consistently underrate natural hazards; the growth of San Francisco did not halt after the earthquake of 1906, nor, for that matter, after the shocks of 1989. Some geographers see this response—or lack of response—as an example of *cognitive dissonance. Other hazards are man-made: pollution, oil spills, and pesticides, for example.

environmental impact A change in the make-up, working, or appearance of the environment. These changes may be planned, like aforestation, or accidental, like the introduction of Dutch elm disease into Britain. Most accidental impacts bring about undesirable change, and deliberate actions may have an unexpected impact, as when the construction of a sea wall leads to the destruction of a beach. Sometimes the damage is irreversible, like the introduction of DDT into food webs.

Environmental Impact Assessment, EIA seeks to consider the probable consequences of human intervention in the environment so as to restrict environmental damage. The US National Environmental Policy Act of 1969 required such an assessment to be drawn up for all major federal developments, giving information about the technology and location of the development, an appraisal of the likely environmental effects, both positive and negative, an outline of possible alternatives, and an estimate of any irreversible commitment of scarce resources.

environmental lapse rate The fall in temperature of stationary air with height, averaging 6 °C per 1000 m. *See* *lapse rate, *adiabatic change. This fall in temperature varies greatly with the time of day, and with the nature of the *air mass concerned and is, for example, much more rapid in a polar maritime air mass than in a mass of tropical maritime air. This fall of temperature with height in stable air is due to a fall in the density of the air.

environmental perception The way in which an individual perceives the environment; the process of evaluating and storing information received about the environment. It is the perception of the environment which most concerns human geographers because decision-makers base their judgements on the environment as they perceive it, not as it is (*see* *mental maps). The nature of such perception includes warm feelings for an environment, an ordering of information, and an understanding, however subjective, of the environment.

The concept of the 'perceived environment' has been used to challenge the concept of *economic man, which lies at the heart of *neoclassical economics (*see* *bounded rationality, *satisficer) and to explain supposedly irrational behaviour, such as moving to a flood- or earthquake-prone location (*see* *cognitive dissonance).

It is suggested that environmental perception can be seen as a five-stage model:

1. An emotional response.
2. An orientative response with the construction of *mental maps.
3. A classifying response as the individual sorts out the incoming information.
4. An organizing response as the individual sees causes and effects in the information.
5. A manipulative response as the individual seeks to change the environment.

environmental psychology A study concerned with the ways in which man perceives his environment. Man can interact with the environment by interpretation, evaluation, operation, and response. Much of *behavioural geography is concerned with the first two processes as in the description of *images, *milieus, and *mental maps.

environmentalism A concern for the environment, and especially with the bond between man and the environment, not solely in terms of technology but also in ethical terms: we are reminded of the necessity for sharing and *conservation. Man is seen as having a responsibility for his environment.

The term may also be used as a synonym for *environmental determinism, but stressing the influence of the environment rather than control by the environment.

Environmentally Sensitive Area, ESA A fragile *ecosystem area where the conservation or preservation of the natural environment is sustained by state controls and/or grants.

eolian *See* *aeolian.

epeirogeny Broad, and generally large-scale, vertical movements of the earth's crust which do not involve much alteration in the structure of the rock, hence **epeirogenic**; caused by the relatively gentle raising or lowering of the earth's crust. Compare with *orogeny.

ephemeral Short-lived. *r-strategist plants are ephemeral in that they grow and reproduce rapidly when conditions are favourable, dying within a short space of time. **Ephemeral streams** flow only during and after intense rain. Such streams are typical of arid and semi-arid areas.

epicentre The point of the earth's surface which is directly above the *focus of an *earthquake, usually 0 to 50 km below it. Most epicentres occur near *plate margins.

epilimnion The upper layer of a body of water. Light penetrates the epilimnion so that photosynthesis can occur. This zone is warmer, and contains more oxygen, than the layers below. The lower limit of the epilimnion is the thermocline.

epiphyte A plant growing on another plant but using it only for support and not for food. Epiphytes are most common in areas of *tropical rain forest.

epistemology The philosophical theory of knowledge which considers how we know what we know, and establishes just what ought to be defined as knowledge. One view is that justification distinguishes genuine knowledge, and there are two main types of justification: rationalism, which uses formal logic and mathematics to construct human knowledge by 'pure' reasoning; and *empiricism, which takes the impressions of sense-data as the foundation of all knowledge. Different types of knowledge may be recognized: knowledge-how, knowledge-of, and knowledge-that.

In *geography, the term is used to indicate the examination of geographical knowledge—how it is gained, sent, changed, and absorbed, in other words, how do we know what we know about geography, what do we choose to call geography, and how have the ways people think about geography changed?

epochs The subdivisions of the units of geological time known as *periods. Thus, the Pleistocene epoch is part of the Quaternary period.

equal area map A map so drawn that a square kilometre in one portion of the map is equal in size to a square kilometre in any other portion. Equal area maps of the whole globe tend to be elliptical in shape, and severely distort the shapes of regions far from the equator.

equation of motion In *meteorology, a form of Newton's second law of motion for a body with constant mass:

$$a = \frac{F}{M}$$

where F is the force acting on a body, M is the mass of the body, and a is its acceleration.

equator The imaginary *great circle around the world at latitude 0°. The equator is equidistant between the North and South Poles. It has a length of 40 076 km: about 25 000 miles.

equatorial rain forest See *tropical rain forest.

equatorial trough A narrow zone of low pressure, between the two belts of *trade winds, arising from high *insolation, especially in the centre of continents in summer. Also known as the *Inter-Tropical Convergence Zone, the equatorial trough is not constant in position, breadth, or intensity; from time to time it disappears completely. This zone also includes the *doldrums.

equifinality In the study of systems, the recognition that different initial states can lead to similar end states. This is an important understanding, not least for geographers, because it works against a tendency to monocausal (single explanation) thinking, especially in geomorphology, where very similar landforms, such as *tors, may be formed by different processes; such as frost weathering in a cold climate and chemical weathering in a warm, humid environment.

equilibrium A state of balance. In **dynamic equilibrium**, inputs are balanced by outputs so that the *status quo* remains. A good example is a beach, where longshore drift is responsible both for inputs and outputs of sand. If these are in balance, the extent of the beach will be unaltered, yet the beach is in a constant state of movement.

Neoclassical economists see the *market mechanism as being a force which will re-establish an equilibrium, and define an **equilibrium price** as the price established by the interaction of supply and demand. Similarly, some economists and politicians have believed that market mechanisms can bring about **spatial equilibrium**—equality between regions, usually expressed in terms of wages. This balance, they believe, will be brought about by the interaction of labour moving to high wage areas and capital moving to regions where wage rates are low. In practice, this mobility is severely restricted: workers may not be able to sell their houses, and capitalists may not get planning permission to move into cheaper areas, for example.

equilibrium line In glaciology, the point at which expansion of the glacier by accumulation is outstripped by losses of ice through *ablation.

Snow does not remain below the equilibrium line throughout the warmer season.

equilibrium species These species show characteristics which are consonant with a stable *niche. Persistence of individuals enables the species to survive. Dispersal is less important and perseverance is more significant than recovery from adverse conditions. The survival of the young is more important than high fecundity. Large desert plants exhibit these properties. They grow rapidly in the rains and, unlike *opportunist species, put most of their energy into growth and conservation of resources. Reproduction is a rare event; some species may set seed only once in several years as with cacti. *See* *r-selection.

equinox Equinoxes, those times when day and night are of equal length, occur twice a year. The **spring**, or **vernal equinox**, is on 21 March and the **autumn equinox** is on 22 September. On these dates, the sun is directly overhead at the equator. The changes in day length result from the changes in the tilt of the earth with respect to the sun, or to what is known as the *apparent movement of the sun*, although it is, of course, the earth which moves.

era The largest unit of geological time. The approximate datings of the eras are:

ERA	DURATION IN MILLIONS OF YEARS BEFORE PRESENT
Precambrian	4600–570
Paleozoic	570–225
Mesozoic	225–65
Caenozoic	65–0

erg Arid, sandy desert, particularly within the Sahara, as in the Grand Erg. The Saharan ergs occupy basins, and the sand in them has been transported, probably by running water, from surrounding, and now sand-free areas of *hammada, and *reg.

erosion The removal of part of the land surface by wind, water, gravity, or ice. These agents can only transport matter if the material has first been broken up by weathering. Some writers use a very narrow interpretation of the word, claiming that erosion refers only to the transport of debris and that *denudation includes the weathering as well as the transport of rocks.

erosion surface A relatively level surface produced by erosion. Much of Africa is composed of extensive plains, often cutting *discordantly across varied geological structures, and these have therefore been identified as erosion surfaces. In south-eastern Britain the presence of a number of chalk summits between 200 and 300 m is regarded as evidence of a Tertiary erosion surface, but some geomorphologists question the assumption that *accordant summits are remnants of an erosion surface. In Africa, remnants of a Gondwana surface have been identified at high altitudes at such places as the Nyika Plateau of Malawi. Not all erosion surfaces are flat (but *see* *planation surface).

The whole concept is bound up with the theory of the *cycle of erosion in which the erosion surface would represent the end-point. Erosion surfaces are not easy to date because they rarely exhibit deposits which were associated with their formation.

erosivity The ability of a soil to be eroded by a given geomorphological force.

erratic A large boulder of rock which has been transported by a glacier so that it has come to rest on country rock of different *lithology; all erratic blocks in northern Germany, for example, originated in Scandinavia. L. Agassiz used the presence of erratics, such as rocks originating in Scotland but now found in Anglesey, as evidence of glaciation in the British Isles.

error Geographical surveys, like all surveys, are prone to error. **Sampling error** is due to faults in the process by which data are selected. **Interviewing error** occurs when the interviewer introduces bias—perhaps by asking 'leading' questions. Any writing based on a geographical project should contain an assessment of likely errors and a suggestion of how, with infinite resources, errors could have been avoided.

ESA *See* environmentally sensitive area.

escarpment A more or less continuous line of steep slopes, facing in the same direction and caused by the erosion of folded rock. Some writers use the term as a synonym for *cuesta.

esker A long ridge of material deposited from *meltwater streams running subglacially, roughly parallel to the direction of ice flow. Eskers range in size from tens of metres, as in north-east Scotland, to several hundred kilometres, as in Finland, and wind up and down hill across the landscape because *subglacial streams are under great hydrostatic pressure, and can flow uphill. Since eskers were formerly river beds, they have typical stream channel *bedforms, with ripples and dunes. Nearly all eskers have bedding which is slumped at the sides, indicating that the stream was contained within ice walls which then melted. Some have an *anastomosing pattern, while others are single features. In many cases, mounds occur along the length of the feature, perhaps where a temporary delta formed. Such a feature is a **beaded esker**.

estancia A large farming estate in Spanish-speaking Latin America, particularly in Argentina.

estuary That area of a river mouth which is affected by sea tides. An estuary differs from a *delta in that the former debouches into the sea whereas the latter *progrades seaward.

eta index, η An expression of the relationship between a network as a whole and its *edges:

$$\eta = \frac{C}{e}$$

where $C =$ total length of all the edges in the network, and $e =$ number of edges in the network.

etchplain A tropical *planation surface where deep weathering has etched into the bedrock. The removal by streams of this weathered rock may lay the etchplain bare.

ethnic group, ethnicity A group within a larger society which considers itself to be different or is considered by the majority group to be different because of its distinctive ancestry, culture, and customs. Ethnicity in a group generally becomes pronounced as a result of migration (forced or voluntary), and a group may only achieve the status of an ethnic association as a result of migration; for example, the group labelled as 'Pakistanis' by white Britons may not only contain very diverse individuals who would have little in common in Pakistan in terms of class or even language, but is often widened to include Nepalis or Indians. Ethnicity is often the basis for social discrimination, and **ethnic unity** tends to increase as a result of such discrimination.

Human geographers have been greatly concerned with the development of **ethnic segregation** in the city, and have identified causes both external and internal to the ethnic minority. The external causes—imposed by the majority *charter group—include discrimination, low incomes which direct them towards inner-city locations, and the need for minorities to locate near the CBD since much of their employment is located there. Internal causes—springing from the ethnic group itself—include a desire to locate near facilities serving the group, such as specialized shops and places of worship, desire for proximity to kin, and protection against attack.

*Indices of segregation have been developed in order to measure the extent of segregation, as well as policies to further integration.

ethnocentricity 1. Giving priority to one's own ethnic group.
2. Making assumptions about other societies which are based on the norms of one's own society. This may result in the development of global models or theories which are based, for example, on Western conditions.

ethnography, ethology In some classifications, these are synonyms for anthropology. In English either term denotes descriptive anthropology.

euphotic zone The upper layer of a body of water receiving light and thus where *photosynthesis is possible. Unlike the *epilimnion, the euphotic zone is defined solely by light input, and not by temperature.

European Coal and Steel Community, ECSC An organization set up in August 1952 with the aim of jointly managing the coal, iron and steel production of Belgium, France, West Germany, Italy, Luxembourg and the Netherlands. The ECSC was the precursor of the *EEC, subsequently the *EU.

European Economic Area A single, free-trade grouping, comprising the member states of the *European Union and *EFTA.

European Monetary System In the late 1980s, the EU governments attempted to stabilize the currencies of member states by fixing their values (in terms of exchange rates) within fairly narrow limits, as a prelude to the establishment of a single currency. For some member states, including Britain, the experience was disastrous, perhaps because the exchange rate had been wrongly set, and led to doubts about the wisdom of the entire enterprise. None the less, by the mid-1990s, most member states seemed still to feel that the advantages of a single currency, such as ease of payment and lack of uncertainty over exchange rates, outweighed the disadvantages. Others feel that the establishment of a pan-EU currency will lead to a loss of national self-determination. In fact, the *ecu/euro*—the proposed European currency unit—already exists, but its value within each nation varies according to fluctuations in exchange rates.

European Union, EU A *free trade area comprising Belgium, the Netherlands, Luxembourg, France, Germany, Italy, UK, Denmark, Eire, Finland, Spain, Portugal, Greece, Sweden, and Austria. Designed initially as an economic unit, the European Union is now attempting uniformity in social as well as economic policies. *See* *Common Agricultural Policy, *European Monetary System.

eustasy A world-wide change of sea level, which may be caused by the growth and decay of ice sheets (**glacio-eustasy**), by the deposition of sediment, or by a change in the volume of the oceanic basins.

eutrophic Fertile, productive; usually of lakes. **Eutrophication** is the process by which *ecosystems, usually lakes, become more fertile environments as detergents, sewage, and agricultural fertilizers flow in. The response to this enhanced fertility in a lake is *algal bloom, which inhibits the penetration of light into the water, thus restricting photosynthesis. The consequent loss of an oxygen input into the water causes widespread death of all species unable to survive in an anaerobic environment. In an attempt to reverse the effects of eutrophication in Cockshoot Broad, the Norfolk Broads Authority has dredged the broad to remove phosphates in the mud. Some increase in oxygen content of the water and species diversity has been observed.

evaporation The changing of a liquid into a vapour, or gas, at a temperature below the boiling point of that liquid.
Evaporation occurs at the surface of a liquid, and energy is required to release the molecules from the liquid into the gas. The use of this energy, known as latent heat, causes the temperature of the liquid to fall.

evaporite A deposit formed when mineral-rich water evaporates. The most common evaporites are gypsum (hydrated calcium sulphate) and halite (sodium chloride).

evapotranspiration The release of water vapour from the earth's surface by evaporation and transpiration. Transpiration is the biological process whereby plants lose water vapour, mainly through pores in their leaves. This water is usually replaced by a continuous flow of water moving upwards from the roots. Rates of evapotranspiration vary with factors such as the temperature and humidity of the air, wind speed, plant type and the nature of the land surface. Since evapotranspiration is so variable, physical geographers prefer to use the concept of **potential evapotranspiration (PE)**. This is the greatest amount of water vapour which could be diffused into the atmosphere given unlimited supplies of water.

*Lysimeters may be used to measure PE, and various formulae have been devised to predict it.

evergreen A plant which keeps its leaves throughout the year instead of losing them seasonally. Most evergreens have some defence against water loss in the winter in the form of needle-like or waxy leaves.

evodeviant Describing life conditions which are different from those which prevailed in the natural life habitat of a species. These may cause disturbances in physiology or behaviour; examples include 'diseases of civilization' such as typhoid and coronary disease.

evolution The change in attributes of a species over a long period of time such that a different species emerges. *See* *natural selection.

evorsion The erosion of rock or sediments in a river or stream bed. *Hydraulic action and *fluid stressing are the predominant processes in this form of erosion.

exceptionalism The study of the unique and particular; the *idiographic. In the nineteenth century it was argued that geography was exceptionalist, since it was concerned with the differences between regions, and in the 1960s regional geography was attacked for its perceived idiographic nature. Model building became the *paradigm. Recent thinking has combined the search for 'laws' about spatial patterns with an understanding of the importance of the regional context. *See* *contextual theory.

exclave A portion of a nation which lies beyond national boundaries, as with West Berlin between 1945 and 1990. This type of territory is also an *enclave in terms of the host country.

exfoliation Also known as *onion weathering, this is the *sheeting of rocks and their disintegration, thought to be due to *thermal expansion, at least on small structures. An **exfoliation dome**, such as the Sugar Loaf (Pão de Açúcar), Rio de Janeiro, is a large, single, dome-shaped body of massive rock revealed through sheeting, probably as a result of *pressure release.

exhumation The removal of young deposits to reveal the underlying structure of older rocks.

existentialism A doctrine which emphasizes the difference between human existence and that of inanimate objects. Later supporters of this philosophy saw human beings as self-created; they are not initially endowed with characteristics but choose their own characteristics by 'leaps'. Thus a person may be said to believe in God because he or she has chosen to do so. Other existentialists see that the only certainty for each one of us is death, and that the individual must live in the knowledge of that certainty.

In *geography, existentialism sees individuals as striving to build up a self which is not given, either by nature or by a culture. Human beings are thus not rational decision-makers but the subjects of their experiences. Landscapes are seen through the eyes of the beholder. Such a view runs counter to the 'mechanistic' views of, say, environmental determinism or economic determinism, which would seem to deny human beings any freedom of action.

exogenetic 1. Applying to processes which occur at or near the earth's surface.
2. In human geography, as a result of outside, environmental influences. Compare with *endogenetic, endogenic.

exogenous variable In economics, a variable explained outside the theory studied. Compare with *endogenous variable.

expanded-foot glacier A small *piedmont glacier, formed when a valley glacier spreads out from the mouth of the valley making a broad mass of ice stretching out on to the lowland.

expanded town A town which has agreed to house *overspill population from large cities in order to relieve pressure on housing there. Expanded towns were introduced in the Town Development Act of 1952 so that local authorities with housing problems could co-operate with the receiving towns in rehousing some of their people. The rationale was that the receiving town would also benefit from the operation in that the new supplies of labour would stimulate its economy. Certain government grants were available to help with the building programme.

exponential growth Geometric growth, where each increment is twice the size of the last, expressed as $Y = 2^x$. *Malthus postulated that, unchecked, population would grow exponentially.

exponential growth model This presupposes that population will grow exponentially through time until it reaches a ceiling beyond which population exceeds resources. At this point population growth becomes a problem, as it is highly unlikely that *zero population growth will occur. More likely are the checks proposed by *Malthus including famine, disease, and war. It is possible that society, becoming aware of the approaching crisis, will make a progressive adjustment and population growth will slacken.

extended family A family unit which consists of relatives by blood and by marriage as well as two parents and their children. Compare with *nuclear family.

extending flow The extension and thinning of a glacier, often marked by an *ice fall. Extending flow occurs near the *equilibrium line, and where the velocity of the glacier increases, for example, down a rock *step. Extending flow can transmit material from the surface of a glacier to its base, thus increasing its powers of *abrasion. It is also responsible for *crevasses and is typical of the zone of *accumulation of ice. Compare with *compressional flow.

extensive agriculture Farming with low inputs of capital and labour, and generally with low yields per acre. It is associated with regions of cheap available land where high revenues are unimportant. Certain forms of commercial agriculture may be extensive; such as cattle farming in the Australian outback.

external economies The cost-saving benefits of locating near factors which are external to a firm, such as locally available skilled labour, training, and research and development facilities.

externality A side-effect on others following from the actions of an individual or group. This effect is not bought by those affected and may be unwished for. Thus, while the acquisition of a car may benefit one household by improving mobility, it generates pollution and creates congestion for others.

  Externalities may be positive—the building of a hall of residence may bring new business to a local shop—or negative—a new road scheme may create *planning blight for home owners. Two types of externality are recognized: **public behaviour externalities** covering property, maintenance, crime, and public behaviour, and **status externalities** resulting from the social and ethnic standing of the household.

extrapolation The prediction of a value made by projecting the trend shown by a set of data into the future. Extrapolations may vary; in the early 1990s, the UN made three extrapolations of world population by the year 2000.

extra-tropical cyclone A cyclonic disturbance outside the tropics; for example, a *mid-latitude depression.

extruding, extrusion flow *Glacial flow which is slow in the brittle, upper layers, but rapid 30 m below the surface. No adequate explanation, or, indeed, evidence, for extruding flow has been advanced.

extrusion A formation of rock made of *magma which has erupted onto the earth's surface as lava and has then solidified. The crystals in extrusive rocks are small, since the lava solidifies rapidly, giving little time for crystal growth. Extrusions emerge from *fissure eruptions and *volcanoes.

exurb American for *dormitory settlement.

ex-works pricing *See* *f.o.b. pricing.

eye The calm area at the centre of a *hurricane (tropical cyclone).

fabric In geomorphology, the physical make-up of a rock or sediment. **Fabric analysis** may be carried out to determine the *dip and orientation of particles in the sediment and thus to uncover information about its origin.

facies The characteristics of a rock, such as fossil content, or chemical composition, which distinguish it from other formations and give some indication of the way it was formed.

factor analysis A *multivariate statistical technique, usually used to make analysis more simple. Sets of original, unique data are replaced by a smaller number of sets. This method can be illustrated by considering a number of characteristics which often go together. For example, if we think of a typical left-wing woman we would expect her to be in favour of abortion, and gender and racial equality, and against capital punishment and stringent immigration policy. Where these views do coincide, they are said to be a **factor**. Factor analysis can be used to see how closely these aspects are related to the individual, or it can be used to see how far all these variables can be reduced to a smaller set; if we can be sure that all pro-abortionists are anti-capital punishment then we can save a great deal of time in an analysis.
    Factor analysis attempts to determine a possible underlying pattern of relationships so that the data may be reordered and reduced to a smaller set of factors. From this it is, in theory, possible to select the *dependent variables. It is a technique used to change a set of original variables into a number of basic dimensions, explained in non-technical language by P. R. Gould (*Trans. Inst. Br. Geogrs.*, 1967). The most commonly used method of factor analysis is *principal component factor analysis.

factorial ecology A term used to describe those investigations of urban spatial structure which use techniques of *factor analysis. *Factors relating to housing and socio-economic characteristics are worked out and used to divide the city into a number of distinctive, smaller areas. *See* *social-area analysis.

factors of production The requirements for production, usually represented as capital, labour, and land. Capital covers all man-made aids to future production; fixed capital stays put, and includes the physical plant, buildings, tools and machinery, while circulating capital includes raw materials and components.
    Labour includes all human resources. It may be unskilled, semi-skilled, or skilled, and local labour markets vary in the size and nature of the pool of labour. Cheap, unskilled and semi-skilled labour may be an important locational factor for *multinational corporations while skilled labour is significant in high technology industries. Industries may be capital- or labour-intensive. Management skills can be a vital factor of labour or can

be seen as a separate factor of production under the heading of
*entrepreneurship.

Land includes natural resources, as in mining, and is an increasingly
important factor as modern factories extend on one level and require space
for storage and parking.

factory farming A system of livestock farming in which animals are
kept indoors throughout the greater part of their lives in conditions of
very restricted mobility. Pigs, laying hens, broiler chickens, and veal calves
are the animals most often kept under these conditions Factory farming
leads to a standardized product raised *en masse*. *See* *agribusiness.

factory system A concentration of the processes of manufacturing—fixed
capital, raw material, and labour—under one roof, in order to provide the
mass production of a standardized product or products. The factory system
superseded *cottage industry because it made possible *economies of scale,
a standardized product, and the *deskilling of labour, which increased
efficiency and increased the power of the employer. Mass production also
meant that specialized machines could be used to their maximum effect.
See *indivisibility.

fall A form of *mass movement in which fractured rock and soil separates
into blocks and falls away from the parent slope. **Debris falls** and **earth
falls** occur on cliffs as joints weaken or as the slope is undercut. **Rock falls**
occur on high and steep rock slopes and are of major importance in rock
slope erosion.

fallow Agricultural land which is not used for crops but is left unused in
order to restore its natural fertility. **Summer fallow** is the practice of
leaving the ground uncultivated during a long, dry spell. **Three-year fallow**
is part of the *three-field system. *See also* *bush fallowing.

false bedding A synonym for *current bedding.

family In geolinguistics, a group of clearly related languages; a genetic
group of languages with numerous cognates and regular correspondences,
for example, the Indo-European family. These features can be accounted for
as the language has diffused and diverged over a long period of time.

famine A relatively sudden flare-up of mass death by starvation, usually
relatively localized, and usually associated with a sharp rise in food prices,
the sale of household goods, begging, the consumption of wild foods, and
out-migration. It has been estimated that more than 5 million people died
in the famine of 1920–1 in the USSR, and as recently as 1959–61, a possible
30 million peasants died from famine in northern China. Currently,
perhaps the most stubborn problems of famine occur in the *Sahel.

The actual causes of the food shortages include drought and other
natural disasters, but many would argue that the prime causes of famine
are social. Civil war is clearly a cause, but there seems to be a class basis to
the effects of famine, and the main sufferers are those who lose access to
food. This may be simply illustrated by reference to the Irish famine of the

1840s, for while perhaps three-quarters of a million people died of hunger, food was still being exported from Ireland.

farm fragmentation The division of a farmer's land into a collection of scattered lots. Fragmentation is usually the result of inheritance but may also reflect present processes like *bush fallowing and past processes like the *three-field system. Farms may be composed of up to twenty different plots, restricting mechanization and decreasing *efficiency. Consolidation programmes have been initiated by many central governments. *See* *remembrement.

farm rent Regular payments from a tenant farmer to a landlord. Rent varies with the quality of the land, the size and location of the holding, and the length of the tenancy.

fast reactor A nuclear reactor fuelled by plutonium which has been produced as a by-product from a *nuclear power station. The reactor is cooled by liquid sodium which uses the heat from the reaction to boil water. The resulting steam drives a turbine to create electricity.

fault A fractured surface in the earth's crust along which rocks have travelled relative to each other; the fault bounding the Teton Range, Wyoming, for example, is nearly 50 km long, with an estimated vertical *throw of about 6000 m. Usually, faults occur together in large numbers, parallel to each other or crossing each other at different angles; these are then described as a **fault system**. The slope of the fault is known as the *dip. Where rocks have moved down the dip there is a **normal fault**; where rocks have moved up the dip, there is a **reverse fault**. A **thrust fault** is a reverse fault where the angle of dip is very shallow and an **overthrust fault**

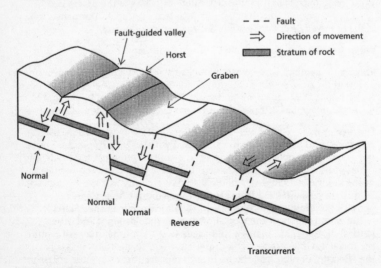

FIGURE 24: *Fault*

has a nearly horizontal dip. A **fault plane** is the surface against which the movement takes place. A **tear fault** is where movement along the fault plane is lateral. This latter type of fault may be termed a **strike-slip fault**.

Regions, such as the Harz of Germany, that are split by faults into upland *horsts or depressed *rift valleys are said to be **block faulted**.

fault block A section of country rock demarcated by faults and which has usually been affected by *tectonic movement. Single-block landscapes are rare; the best example is part of a lava and tuff plateau in South Oregon.

fault breccia A zone of angular rock fragments, located along a fault-line, and formed by the grinding action associated with movement either side of the fault. Fault breccias are relatively easily eroded.

fault scarp A steep slope resulting from the movement of rock strata down the *dip of a normal fault. The Teton Range is one of North America's major fault scarps, part of a nearly 50 km long fault, with an estimated vertical throw of 6000 m.

A **fault-line scarp** is coincident with a fault, yet not actually formed by faulting, but through erosion of weaker rocks which have been brought alongside stronger ones through movements along a fault.

faunal realms The simplest groupings of the animals of the world. The holarctic realm covers two subdivisions: the nearctic (almost all of North America, with Greenland) and the palearctic (Asia north of the tropics, Europe, and Africa north of the Sahara). The neotropical realm covers Central and South America, and the ethiopian embraces Africa south of the Sahara and Arabia. The oriental realm is Asia south of the tropics with an ill-defined boundary between it and the Australian realm which includes New Zealand, Australia, Oceania, and some of South-East Asia.

feature In *Geographic Information Systems, a geographical *entity. A **feature serial number** is a unique code designating a unique feature; a **feature code** is an alphanumeric classifying or describing a feature.

fecundity The potential of a woman or of women in a society to bear live children. This contrasts with *fertility, which is a measurement of actual childbearing. Fecundity in a population is, of course, closely linked to the proportion of women of childbearing age, but explanations of the difference between the numbers of children that *could* be born and the numbers that *are* born continue to fascinate *demographers.

federalism A two-tier system of government. The higher, central government is usually concerned with matters which affect the whole nation such as defence and foreign policy, while a lower, regional authority generally takes responsibility for local concerns such as education, housing, and planning although, of course, the division of responsibilities varies from case to case.

Federations were designed to preserve regional characteristics within a united nation, or to contain *nationalist elements which were agitating for independence. Thus, the Nigerian federation was substantially amended after the Biafran war of 1967; the number of states was increased from twelve to nineteen, and each state was given more power.

feedback The response within a *system to an action or process. **Negative feedback** causes the situation to revert to the original. One theory illustrating negative feedback (*Malthusianism) suggests that as population expands, its food supply per individual is diminished; the result is that the level of the population begins to fall. **Positive feedback** causes a change; one example is the growth of motor traffic. Initially, few individuals had the money to buy cars, but later the growth in private car ownership made possible a settlement structure marked by increasing distances between home and work. This structure, in turn, motivated more people to own and drive a car.

felsenmeer A surface of broken rock fragments found in *periglacial environments. The fragments are the result of the *frost shattering of exposed bedrock.

feminist geography A geography which emphasizes the oppression of women and the gender inequality between men and women. An important theme was making women visible—writing a geography of women—which would emphasize the way in which women's perceptions and experiences differ from those of men. The way that cities are designed overwhelmingly reflects the male viewpoint, as an Oxford example will illustrate. Within the city, a cycle track has been hidden from the main road by a bank with a hedge on the top. From a male point of view this makes the track quieter, and less polluted. The track is desperately unsafe for women, and a number have been attacked along it.

Some feminist geographers argue that cities should be restructured, in order to reduce gender inequalities, since it is argued that women's access to a range of goods and services is more restricted than men's.

Socialist feminist geographers are concerned with the way in which the structuring of space perpetuates traditional gender roles and relationships, and note the way in which spatial variations in gender relationships can affect industrial location; the availability of cheap female labour is a major attraction to employers, and the quantity of this type of labour varies regionally, nationally, and globally. There are those who draw analogies between women and colonized people, but others suggest that the commonalities between women and Third World people are far outweighed by the differences between them.

Other topics of interest include studies of the ways in which *environmental perception and the representation of space vary with gender, and claims that the very language of geography is gendered, and possibly sexist. It is also argued that in social research the gender of the researcher may influence the result.

Further themes concern the geography of women's issues like abortion laws, women as wage-earners, and women's access to education, income, health care, and day care for children and aged dependants.

fence-line In *Geographic Information Systems, a physical, surveyable boundary to a parcel of land.

ferrallitization The combined effect on a soil of the strong *leaching and intense weathering found in the tropics. Ferrallitic soils are a characteristic of the tropical rain forest. Destruction of organic residues is very rapid, so that there is little *humus. Most of the bases and some of the silica is leached away. The B2 *horizon is massively impregnated with *sesquioxides which form a hardpan rich in clays. This pan is often exposed by erosion of the upper horizons. In *US soil classification, ferrallitic soils are oxisols.

Ferrel cell An *atmospheric cell lying between the two thermally direct cells: the Polar cell and the Hadley cell. It transfers warm air to high latitudes and shifts cold air back to the subtropics, where it is warmed.

Ferrel's law This states that a body moving over the earth will be deflected to its right in the Northern Hemisphere and to its left in the Southern Hemisphere. This occurs as a result of the earth's rotation and applies particularly to movements of the atmosphere.

ferricrete A soil *horizon made up of the *cementation of iron oxides at or near the land surface. Ferricrete forms a very hard layer which has been used as building bricks.

ferruginous Of or containing iron or iron rust. A **ferruginous soil** is a *zonal soil developed in warm temperate climates without a dry season, or in the tropical *savanna or bushlands. These soils develop to a great depth because of the intensity of tropical weathering. The A *horizon is a dark red-brown with a weak crumbling structure. The B horizon is stained red by the ferruginous gravel present. In *US soil classification, a ferruginous soil is an ultisol.

fertility The level of childbearing; in an individual, but more often in a society or nation. Crude *birth rate is the simplest and commonest measure of fertility, but it does not relate the number of births to the number of women of childbearing age. For this reason, other measurements are used, such as the **general fertility rate,** or **fertility ratio,** which shows the number of births in a year per 1000 women of reproductive age, generally given as being between 15 and 45 (sometimes this is given as 15 to 49). **Cohort fertility rates** survey the number of births to women grouped either according to their year of birth or their year of marriage.

Fertility rates vary widely globally; the rate for the UK, 1990, was 1.8, as compared with 7.0 for Malawi. Reasons for variations in fertility include availability of land in rural societies, death rate, public health programmes, access to birth control, income, female employment opportunities, industrialization, *modernization, and customs of age at marriage, celibacy, and inheritance. Fertility has been declining in industrial societies since the late nineteenth century; a decline which preceded easily available artificial contraception. There is a very strong global correlation between fertility rates and per capita *GNP.

While Western Europe witnessed a fractional increase in fertility in the late 1970s and early 1980s, fertility rates have declined consistently since then.

fetch The distance that a sea wave has travelled from its initiation to the coast where it breaks. While the extent of fetch substantially controls the energy and height of a wave, such that in the North Sea the greatest fetches, and the biggest waves, come to the British Isles from the north and north-east, wind speeds are of greater importance; within the tropics, even when there is a fetch of 1500 km, wave heights rarely exceed 3.5 m (unless there is a *tropical cyclone). The longest fetch, and hence the dominant wave direction, will affect the direction of *longshore drift and the orientation of beaches.

feudalism A system, common in Europe in the Middle Ages, where access to farm land was gained by service to the owner: the **feudal lord**. Initially, no money was involved in transactions between the serf and the lord, although the payment of cash in lieu of service became common in the later Middle Ages.

fiard, fjärd An inlet of the sea with low banks on either side. These are not to be confused with fiords, since they lack the characteristic steep walls. They are common along the Gulf of Finland, and were formed by the post-glacial drowning of the low-lying, glaciated, rocky terrain of the Fenno-Scandian shield.

field In *Geographic Information Systems, a group of one or more characters incorporating map information.

field capacity The volume of water which is the maximum that a soil can hold in its *pores after excess water has been drained away; the state of a soil in this condition, when the only water that remains is water retained by the soil particles through surface tension.

field drainage *See* *drains.

field system The layout and use of fields. Different communities have given rise to different systems. The extent and use of fields varies with the natural environment, the nature of the crops and livestock produced, and aspects of the culture of the farming community such as inheritance rights and available technology. A major element of field system study has been the pattern and evolution of medieval field systems with the distinction between communally organized common fields and open fields which were not available to the community as a whole.

fill-in migration As an individual moves from a small to a large city, a vacuum is created in the smaller centre which is filled by someone moving out of an adjacent rural area. *See* *stepwise migration.

filter-down process In the early stages of an industrial development specialist skills are usually required, and consequently activity is concentrated in places of industrial sophistication. As the production process is rationalized and made into a routine, less skill is needed. Wage rates are now too high for the low levels of skill needed and industries therefore seek out industrial backwaters where the cheaper labour can handle the lesser demands of the simplified process. Thus, innovation filters down from the more advanced to the less sophisticated regions. Some writers believe that the evidence for a filter-down process does not appear to be very strong.

filtering down The movement of progressively less affluent individuals into housing stock. It is suggested that the rich move away from the city to newly built houses because their old houses are out of date, difficult to maintain, or surrounded by types of land use which are not appealing. The next social and occupational class moves into the houses vacated by the rich. Homes are subdivided and passed on to successively poorer groups. This would be a major cause of *inner city decline.

However, it is by no means true that only the rich move into new homes, as public housing schemes have attested. Many higher status housing areas have managed to withstand infiltration by poorer social groups. It is also the case that the well-off invade run down areas, in the reverse process of *gentrification.

fines The name given to smaller stone particles, whose poor thermal conductivity may aid the development of *patterned ground in *periglacial environments.

finger lake A long, narrow lake occupying an over-deepened basin in a glacial trough. British examples include Derwent Water and Ullswater.

finger plan The development of new towns or suburbs along routes, road or rail, radiating from the city centre. Planners see commuting as a fact of life which should be made as efficient as possible.

fiord, fjord A long, narrow arm of the sea which is the result of the 'drowning' of a glaciated valley. Most fiords, including the fiord coastline of Norway, are located on the west coast of continental masses, and it is thought that this is connected with the westerly winds which prevail in these locations. Fiords are distinctive both because of their great depth, and because of the over-deepening of their middle sections which are deeper than the water at the mouth. Søgnefjord, for example, is 1200 m deep, but its mouth is only 150 m below sea level. The shallow bar at the seaward end of the fiord is thought to represent the spreading and thinning of ice as it was released from its narrow valley and spread out over the lowland. Glaciation may not be the only process in the formation of fiords; the configuration of the Norwegian fiords may be *tectonically controlled.

fire ecology The study of the effects of fire on *ecosystems. These are not always deleterious; it is suggested that some ecosystems, like those of the *coniferous forests, depend on fire to evolve fully. Many ecosystems like *savanna or *garrigue have resulted from fire, usually man-made, although fires do occur naturally from lightning strikes.

firm An independent unit which utilizes the *factors of production to produce goods and services. Revenue is kept high enough to cover costs and to generate profit. It should be noted that whereas the creation of profit is the key objective of a firm, businessmen have other motives in addition to profit, such as increased managerial satisfaction. It should also be noted that in a purely capitalist society it is numbers of individual firms who make decisions rather than an industry as a whole.

firn Ice formed when falls of snow fail to melt from one season to another. As further snow accumulates, its weight presses on earlier snow, compacting and melting it to a mass of globular particles of ice with interconnecting air spaces. Further snow fall, and further compaction, drives out the air spaces and turns the firn to pure ice. Where temperatures are around 0 °C, snow can turn to firn within five years. The process takes much longer in very cold conditions. An alternative term for firn·is névé. The **firn line** is the line at which firn forms, and is close to the equilibrium line. This line varies with *aspect; it is notable that former glaciers descended lower on the eastern than on the western slopes of Mount Ruwenzori, for example.

First World Western Europe, Japan, Australia and New Zealand, and North America. These were the first areas to *industrialize. Synonyms (which are much more often used) include 'the developed world', 'the North', 'the more developed countries' (MDCs) and 'the advanced economies'. *See also* Third World.

fiscal migration Migration for financial, usually tax reasons, common in pop stars. *See* *betterment migration.

fish farming The rearing of fish in man-made pools or tanks. Fish farming has been practised for thousands of years using ponds and fenced-off enclosures of rivers to rear fish. Manuring has traditionally taken place by raising ducks and allowing their droppings to fall into the water. Careful control of the fish is required to prevent losses from disease which would spread rapidly in the confined conditions of the tanks. Recently this has been achieved by routine dosing of the water with antibiotics and other chemicals.

fission track dating Minerals and glasses of volcanic origin contain traces of a radioactive *isotope of uranium. This isotope decays by spontaneous fission and the resulting fragments tear into the surrounding material, leaving tracks of about 10 μm in length. The number of such tracks indicates the age of the volcanic matter.

fissure A long, narrow opening made by cracking or splitting.

fissure eruption A volcanic eruption where *lava wells up through fissures in the earth's crust and spreads over a large area. Fissure eruptions are usually of very fluid *basic lava; several were responsible for the formation of the Antrim plateau of Northern Ireland. *See* *plinian eruption.

fixation line Certain controls to the growth of a town which structure the plan of the town. Old town walls act as very influential fixation lines; in the case of Barcelona, for example, expansion beyond the city walls was forbidden until the 1880s. Growth of the town may then spread out in a radial pattern based on the fixation lines, as in Amsterdam, where the fixation lines were the canals, or, as again in the case of Barcelona, a radically new town plan may be drawn up.

Flandrian The time succeeding the most recent glacial stage. During this time, there has been a global rise in sea level known as the **Flandrian transgression** due to the melting of *ice sheets and *glaciers.

flashy In hydrology, applied to a natural watercourse which responds rapidly to a storm event, hence **flash flood**; a very sudden, brief, and dramatic flood event. Rising and falling limbs are steep, and the period of peak flow is short. Flash floods are, surprisingly, major hazards in deserts, for, although rainfall is rare, the hard-baked ground may be impermeable. Furthermore, there is little or no vegetation to intercept the rainfall or slow the flow of floodwater. A typical flash flood occurred in Big Thompson Canyon, Colorado, 1976, when 139 people drowned after 300 mm of rain fell in less than six hours.

Flash flooding is the major process in the formation of *wadis—which, incidentally, should therefore be avoided as camp sites. Flash flooding is by no means confined to semi-deserts; recent estimates suggest that in tropical countries 90% of deaths due to drowning are the result of intense rainfall on small, steep catchments upstream of poorly drained urban areas.

flint A form of silica, found in bands in the Upper Chalk, and formed when water rich in silica from marine organisms percolated through the chalk. When hammered, flint splits into the sharp-edged flakes so useful to *Palaeolithic and *Neolithic peoples.

flocculation In soil science, the process whereby very small particles aggregate to form *crumbs. The term is usually applied to clays. In certain *subsoils of arid areas, downward *translocation of soluble salts leads to the breakdown of these crumbs in the process of **deflocculation**.

flood Floods occur when peak discharge exceeds channel capacity; and this may be brought about naturally by intense *precipitation, snow- and ice-melt, *storm surges in coastal regions, and the rifting of barriers, such as ice dams; or by the failure of man-made structures, by deforestation, urbanization, which reduce *infiltration and *interception, and by engineering works, such as land drainage and the straightening and embankment of rivers. **Flood prevention** and **flood control measures** include aforestation, the construction of relief channels and reservoirs, the provision of water meadow areas in which to divert flood water, and a ban on building in flood-prone environments, such as *flood plains. While flood insurance is compulsory in many areas of the United States, geomorphologists have argued that this is a flawed strategy, since it is costly, and leads to an increase in building in unsuitable environments.

flood frequency analysis The calculation of the statistical probability that a flood of a certain magnitude for a given river will occur in a certain period of time. Each flood of the river is recorded and ranked in order of magnitude with the highest rank being assigned to the largest flood. The *return period here is the likely time interval between floods of a given magnitude and can be calculated as:

$$\frac{\text{number of years of river records} + 1}{\text{rank of a given flood}}$$

flood plain The relatively flat land stretching from either side of a river to the bottom of the valley walls. Flood plains are periodically inundated by the river water; hence the name. The flood plain may be thought of as an area of *alluvium which is introduced to the valley, stored, subjected to weathering, and then transported downstream. Flood plains are often ill-drained and marshy, and characteristic *fluvial features include meanders, levées, and ox-bow lakes.

floral, floristic realms These may be recognized and mapped as areas characterized by the indigenous plant species and not by the *biome. Thus, the Brazilian rain forest is part of the Latin American floral realm.

flow 1. The movement of goods, people, services, and information along a *network.
   2. The *mass movement of material held in suspension by water. Flows are classified by the size of the particles: debris flow refers to coarse material; earth flow to soil; and mud flow to clay. Flows may be the result of very high water pressure in the *debris and can occur in clay if the particles have absorbed a great deal of water before they are *entrained.

flow line graph, flow chart A map which depicts the movement of people or *commodities by a line whose width is proportional to the volume of flow, and drawn in the general direction of movement. Compare with *desire line.

flowage The movement of solids such as ice or rock without fracturing. In such cases, the term *plastic flow is equally applicable.

fluid stressing The erosion of weak, cohesive rocks, such as muds, by the force of water in a river. The effect of this force depends, among other factors, on the strength of the bed, the percentage of clay in the bed, the velocity of the water, and the *turbulence of the water.

flume A man-made channel which conveys water for some specific purpose. In practical hydrology, a flume is an apparatus placed across a watercourse in order to measure its *discharge.

fluvial, fluviatile Of, or connected with, rivers.

fluvial erosion Erosion by streams or rivers. This involves the destruction of bedrock on the sides and bottom of the river, the erosion of channel banks, and the breaking down of rock fragments into smaller fragments.

fluvio-glacial Of, or concerned with, streams and rivers formed from melting glaciers.

flysch A series of alternating sediments: clays, shales, and sandstones.

f.o.b. (free-on-board) pricing Pricing a commodity to include the cost of loading onto freight vehicles at the point of sale but excluding the cost of transporting the goods from the point of sale to the buyer. *Ex-works pricing* is a synonym.

focal area In geolinguistics, an area of relative uniformity, as indicated by sets of shared linguistic features which acts as the central area of a *dialect: the central area of industrial Scotland has been identified as one such area.

focus The point of origin of an earthquake. *See* *epicentre.

foehn *See* föhn.

fog A cloud of water droplets suspended in the air, limiting visibility to less than 1000 m. Fog forms when a layer of air close to a surface becomes slightly supersaturated and produces a layer of cloud, that is, when vapour-laden air is cooled below *dew point. In **advection fog**, this cooling is brought about as warm, moist air passes over cold sea currents, such as the Labrador current. **Radiation fog** forms during cloudless autumn nights when strong *terrestrial radiation causes ground temperatures to fall. Moist air is chilled by contact with the ground surface. The fog lingers until it is dispersed by warm sunlight.
   Where cold air streams cross warm waters, **steam fog** forms. This is common when relatively warm surface air over lakes in *frost hollows convects into the cold *katabatic airflow above it, and is also the mechanism behind *Arctic sea smoke. Frontal fog forms when fine rain falling at a warm front is chilled to *dew point as it falls through cold air at ground level. *Hill fog is not true fog.

föhn When moist air rises over a mountain barrier, it experiences *adiabatic temperature changes, and cools at the slow saturated adiabatic lapse rate. Precipitation is common. Once past the mountains, the air, now much drier, descends. It warms at the dry adiabatic lapse rate, higher than the saturated rate by some 3 °C/1000 m. A dry, warm, gusty wind, which can reach gale force, results. Effects in summer can be so desiccating that bush fires are a serious risk; in winter, snow melt can be rapid. *See also* *chinook.

fold A buckled, bent, or contorted rock. Folds result from complex processes including fracture, sliding, *shearing, and *flowage.
   An arch-like upfold is an anticline which may be symmetrical or

asymmetrical. This is also true of downfolds or synclines. A complex anticline is an anticlinorium; a complex syncline is a synclinorium. In an overturned fold the upper limb of the syncline and the lower limb of the anticline dip in the same direction. In recumbent folds the beds in the lower limb of the anticline and the upper limb of the syncline are upside down. *See also* *nappe.

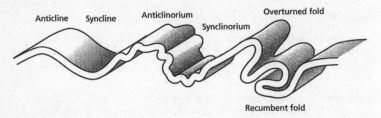

Anticline　Syncline　Anticlinorium　Overturned fold

Synclinorium

Recumbent fold

FIGURE 25: *Fold*

fold mountain An upland area, such as the Alps or Andes formed by the buckling of the earth's crust. Many fold mountains are associated with destructive or collision margins of *plates. **Young fold mountains**, such as the Caucasus and Alps, were formed by the Alpine orogeny of 65 million years BP, and reach elevations of 10000 m. Structurally, they are characterized by *nappes and overfolds, as are **old fold mountains**, such as the Grampian mountains of Scotland, which were created by earth movements pre-dating the Alpine orogeny but have been extensively eroded. Some old fold mountains have been uplifted and re-eroded.

food chain A linear sequence representing the nutrition of various species from the simplest plant through to top *carnivores, as in: rose → greenfly → ladybird → sparrow → sparrowhawk. This direct pathway is too simplified. Plants and animals are usually linked together in a *food web. Plants (primary *producers) and *consumers at various *trophic levels are interconnected in their diet and in their role as sources of food.

food conversion ratio The ratio of the number of calories of a prey required to produce one calorie for a predator.

food web A series of interconnected and overlapping *food chains in an *ecosystem.

footloose industry An industry whose location is not influenced strongly by access either to materials or markets, and which can therefore operate within a very wide range of locations. Any form of 'direct line' business, operated almost entirely through telephone and fax lines, would be an example.

foraminifera Usually marine micro-organisms of plankton and *benthic animals with calcite skeletons, found over much of earth's ocean beds. Foraminifera are very sensitive to temperature, and their presence in the fossil record may be used to reconstruct past environments.

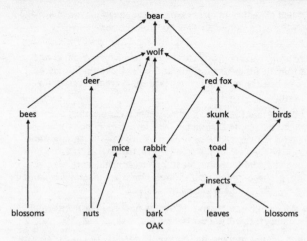

FIGURE 26: *Food web*

förde An elongated bay in an area of glacial deposition, formed when glacial tongue basins are drowned by rising water levels. The term comes from the type locations in northern Germany and Denmark.

Fordism A term coined by A. Gramsci (trans. 1971) to describe a form of production characterized by an assembly line (conveyor belt factory system) and standardized outputs linked with the stimulation of demand brought about by low prices, advertising, and credit. Fordism, exemplified by the mass-production systems based on the principles of *Taylorism used by the car maker, Henry Ford (1863–1947), gave workers high wages in return for intensive work. Many commentators believe that Fordism was characteristic of Western industry from about 1945 to some time in the 1970s, and that it was linked with the rise of major car manufacturing regions in the Western world, such as the West Midlands of Britain. Fordism is associated by geographers with a distinctive spatial pattern of economic activity, or spatial *division of labour; that is, with the spatial separation of the development of the product, at the centre of research and development, and the actual sites of the production of a standardized product. *See* *post-Fordism.

foredune A ridge of irregular sand dunes, typically found adjacent to beaches on low-lying coasts, and partially covered with vegetation.

foreign aid *See* *aid.

foreign exchange The foreign capital earned by a country's exports. Since the currency of many less developed countries is not accepted by international markets, it often becomes necessary to earn foreign exchange in order to buy imports.

foreset bed

foreset bed A bed of sediments laid down at the seaward, inclined edge of a delta or an advancing sand dune. *See* *bottomset beds.

forest park An area of forestry which, as well as supplying timber, may be used for recreation. Forests provide excellent cover for car parks and picnic sites and may be used for camping and pony trekking. It may be that some areas of forestry are more important for recreation than for their economic value.

forestry The management of woodland to provide timber for sale. New areas are ploughed and planted while cut-over forests may be replanted. The trees are given fertilizer and arxe protected from pests, diseases, and major fires. They are felled when the trees are mature, when there is overcrowding, or when the trees die. Most forests in Britain are planted with fast-growing, softwood conifers although *hardwoods may be planted at the periphery to soften the environmental impact.

formal game An exercise in *game theory, as opposed to an informal game which simulates real-world events. *See* *game theory.

formal region A region marked by relative uniformity of characteristics, such as the Scottish Highlands. The variations within the region are less than variations between the region and other areas. *See* *functional region.

formal sector All those types of employment which offer regular wages and hours, which carry with them employment rights, and on which income tax is paid.

formation, great plant formation A major vegetation system, determined by climate, e.g. tropical grassland. Thus, a formation is the plant community of a *biome.

fossil fuel Any fuel which is found underground, buried within sedimentary rock: coal, oil, and natural gas. Reserves of fossil fuels are dwindling and some writers suggest that all fossil fuels will be used up by AD 2800.

fossil water *See* *connate water.

fractus *See* *cloud classification.

fragmentation *See* *farm fragmentation.

free atmosphere That part of the atmosphere, above about 500 m, which is generally free of the influence of the earth's surface.

free face, fall face An outcrop of rock which is too steep for the accumulation of soil and rock debris.

free port A port, such as Hong Kong, or part of a port, where customs duties are not payable. Costs are, therefore, lower, as is insurance and administration.

free trade Trade between countries which takes place completely free of restrictions. Such trade allows specialization in member states of free trade areas, and lowers costs because, together with competition, the markets

are increased. Within a free trade area there are no barriers, such as
*tariffs and *quotas. However, there is not necessarily a common policy on
trade with countries outside the free trade area. *See* EFTA.

free trade zone A designated area, often within an *LDC, where normal
tariffs and quotas do not apply. In addition, it is common for the
conditions of employment to be more repressive; such zones are often
mandatorily union-free, and working conditions can be harsh.

freehold The complete and unrestricted title to land.

freeway In the USA, a multi-lane inter-state motorway.

freeze–thaw The weathering of rock which occurs when the water,
which has penetrated the joints and cracks, freezes. This process is,
therefore, probably effective only in well-jointed rocks. Water expands by
9% when it freezes, and it has been suggested that this expansion causes
the rock to shatter. Freeze–thaw is most active where there is a maximum
number of temperature oscillations around 0 °C; it is therefore more
frequent in *periglacial, rather than *polar environments.

   In *glacial geomorphology, some writers distinguish between the freeze–
thaw active within a glacier (see *bergschrund), and a similar process
acting above the glacier, which may be termed **frost shattering**. This effect
has shaped the Cuillin Hills of Skye, for example, above the snowline.
Some geomorphologists believe that this force is insufficient to break up
any but the softest rocks, and that what has been called frost-shattering is
really *hydro-fracturing. Alternative terms are: congelifraction,
frost-shattering, gelifraction, and gelivation.

freezing front The edge of frozen or partially frozen ground. In areas of
seasonal freezing with no *permafrost, the freezing front moves downward
through the earth. In areas of permafrost, the front can also move upwards
from the frozen layer below.

freezing nucleus A *nucleus on which a water droplet will freeze to
form an ice crystal. Ice nuclei thus formed are much less common than
*condensation nuclei, but their effectiveness rises as the temperature falls
below 0 °C. They grow by *sublimation if the *ambient air is saturated
with respect to water.

freight rate This cost of transporting goods reflects a number of factors
besides basic transport costs, such as the nature of the commodity.
Non-breakable, non-perishable items, like coal, are carried most cheaply as
they can be carried in bulk on open wagons. The more careful the
handling required, the more expensive is the freight rate. Sophisticated
manufactured goods can bear high freight rates because of their greater
value. Raw materials are carried for less so that they can be moved over
greater distances.

   Distance is an important factor. Many freight rates are *tapered; that is,
the rate per tonne-mile or tonne-kilometre drops as the distance increases,
but this change in rates is expressed in a series of distance 'bands' so that,
on a graph, the relationship between cost over unit distance and distance

would appear as a series of downward steps rather than a smooth diagonal line.

Competition between alternative modes of transport can also cut freight rates. Thus, because of competition with the New York State Barge Canal, rail freight rates from Chicago to New York are less than from Chicago to Philadelphia, even though the latter journey is slightly shorter.

frequency distribution The range of values of any data set shown by the number of occurrences in a series of classes. The distribution may be shown as a *histogram or the mid-points of each class may be joined up with a line to make a frequency curve or frequency polygon.

friable Easily crumbled; usually referring to soils.

friction The force which resists the movement of one surface over another.

Friction between the surfaces of two mineral grains is related to the hardness of the mineral, the roughness of the surface, and the number and area of the points of contact between the grains. Friction is of major significance in any study of the movement of sediment since the forces moving the sediment must be greater than the resistance provided by friction.

friction of distance As the distance from a point increases, the interactions with that point decrease, usually because the time and costs involved increase with distance. In this context, however, distance need not be reckoned solely in spatial terms; the frictional effect of distance 'on the ground' is far less in a lowland area with good communications than in an upland area of difficult terrain, and has slackened with improvements in transport and communications. *See* *distance decay, *space-time convergence.

frictional force In meteorology, the roughness and irregularity of the earth's surface which reduces wind speeds. The **friction layer**, where this effect is strongest, roughly comprises the lowest 100 m of the atmosphere.

Friedmann's core–periphery model *See* *core–periphery model.

fringe belt At the edge of a town or built-up area, a zone of varied land use: Victorian hospitals and cemeteries, located beyond the city for reasons of public health; recreation facilities, such as playing fields, riding stables, and golf courses; and utilities, such as water- and sewage-works. Many of the functions of the fringe belt have been squeezed out from the town centre due to congestion, high land prices, the need for a special site, or disturbances in the central area. Sometimes further urban expansion leap-frogs the fringe belt.

front The border zone between two *air masses which contrast, usually in temperature. A **warm front** marks the leading edge of a sector of warm air; a **cold front** denotes the influx of cold air. Fronts are intensely *baroclinic zones, about 2000 km long and 2000 km wide, moving at around 14 km per day. In *mid-latitude depressions, fronts develop as part

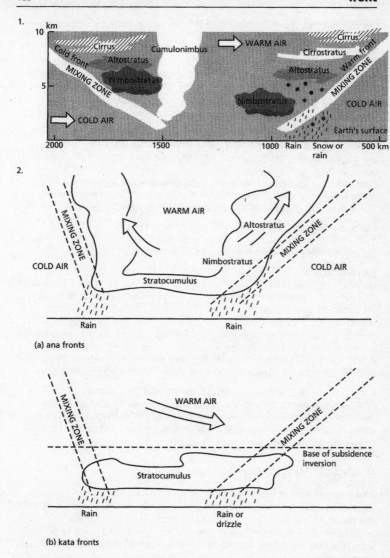

FIGURE 27: *Fronts*

of a horizontal wave of warm air enclosed on two sides by cold air. These
frontal wave forms move from west to east in groups known as **frontal
wave families**.

The basic classification into warm fronts, with a slope of 1 in 100, and steeper cold fronts is further divided by the type of air movement at the front. In **ana-fronts**, the warm sector air is rising, and a succession of cloud types and precipitation results. In **kata-fronts**, the warm sector air is descending, clouds are few and precipitation is reduced to a drizzle.

Traditionally, much weather forecasting has been based on the correct interpretation of fronts, but recent research would indicate that *divergence in the upper air is probably more important than the convergence of air masses in the lower atmosphere.

frontier That part of a country which lies on the limit of the settled area. It differs from a boundary because the term frontier indicates outward expansion into an area previously unsettled by a particular state. Some frontiers have occurred where two nations advance from different directions, leading to boundary disputes. A **settlement frontier** marks the furthest advance of settlement within a state while the **political frontier** is where the limit of the state coincides with the limit of settlement.

frontier region A sparsely populated, little developed, *peripheral region, often with a hostile physical environment, such as Arctic Siberia. These are the **resource frontier regions** of *Friedmann's model.

frontier thesis A theory put forward by F. Turner who held that the westward expansion of the USA was due to the existence of 'free' land in that direction. Expansion was seen as a series of waves moving to the west: the Native American, the fur trader, the hunter, the rancher, and the arable farmer. Each wave moved further away from Europe and contributed to a new society. It is claimed that the process of expansion led to the evolution of an individualism which promoted democracy. This theory has been severely criticized, not least because it is utterly silent on the constrictions on female freedom imposed by the American frontier.

frontogenesis The development of *fronts and frontal wave forms. Frontogenesis occurs in well-defined areas; for example, Atlantic Polar fronts form off the east coast of North America, while the Arctic front occurs across Canada at around 50° N. Fronts are less common in the tropics where contrasts between *air masses are less marked. *See* *frontolysis.

frontolysis The change and decay of *fronts.

frost Frozen dew or fog forming at, or near ground level. **Black frost**, as the name suggests, is a thin sheet of frost without the white colour usually associated with frost.

Air below 0 °C is **air frost. Hoar frost**, or rime, is a thick coating of white ice crystals on vegetation and other surfaces. **Ground frost** occurs when the air at ground level is chilled below freezing point.

Frost hazard on roads is most common in maritime climates, such as in Scandinavia and the British Isles, where rainfall is regular and winter temperatures oscillate unpredictably around 0 °C, and while salt is an effective de-icing agent it damages both road surfaces and vehicle bodies. Frost is also an agricultural hazard; in 1971 the annual frost loss in the USA was estimated at $1.1 billion. Measures such as direct heating of orchards with oil burners and the use of wind machines may prevent frost damage. The most effective response, however, is the choice of a frost-free site. *See* *frost pocket.

frost cracking This can occur within seasonally thawed ground, but is more a feature of permanently frozen areas. When the frozen ground reaches very low temperatures, it contracts, splitting up to form a pattern of polygonal cracks. The edges around the polygon are lifted when the ground expands, perhaps as the result of slightly warmer temperatures. These polygonal areas are known as sand-wedge or ice-*polygons according to the nature of the material within the crack.

frost creep The downslope movement of debris, firstly through the growth of needle-like ice which lifts a thin surface crust at right angles to the ground, followed by thawing which washes the loosened debris downslope. Opinions vary about the efficacy of this process.

frost heaving The upward dislocation of soil and rocks by the freezing and expansion of soil water. **Frost push** occurs when cold penetrates into the ground. Large stones become chilled more rapidly than the soil. Water below such stones freezes and expands, pushing up the stones. **Frost pull** can alter the orientation of a large stone causing it to stand upright. This occurs when ice creeps downwards from the surface. The growth of ice crystals on the upper part and the drying of the soil around the lower part cause the stone to be pulled into a more vertical inclination.

frost hollow *See* *frost pocket.

frost pocket A concentration of cold air in a hollow or valley floor. This occurs when night-time *terrestrial radiation is greatest on valley slopes. Air above these slopes becomes colder and hence denser, therefore it flows downslope. Minimum temperatures in the pocket may be tens of degrees below the surroundings. For this reason, fruit growers try to avoid frost pockets.

frost-riven cliffs, frost-riven scarps *See* *cryoplanation terraces.

frost shattering, frost spalling *See* *freeze–thaw, but note that some writers distinguish between the freeze–thaw active within a glacier and the similar process acting above the glacier, which may be distinguished as frost shattering. Others apply the term frost shattering, or **frost weathering** to ice formation in fractures on the surface of a rock.

frost thrusting The lateral dislocation of soil and rock by the freezing and expansion of water.

frost wedging *See* *freeze–thaw.

Froude number The ratio of the velocity (v) of a river to its *celerity where celerity is the product of the acceleration due to gravity (g) and the mean depth of flow (d). The Froude number (Fe) is calculated from the equation:

$$F_e = \frac{v}{\sqrt{gd}}$$

where F_e is less than 1, deeper flow is tranquil. Where F_e exceeds 1, the flow is turbulent.

fugitive species *See* *opportunist species.

fumarole A vent in a volcano through which steam and volcanic gases are emitted.

fumigation In meteorology, polluted air may be trapped beneath an *inversion. Plumes of air may then rise because of *convection, causing downdraughts which return the polluted air to ground level.

functional classification of cities A categorization of cities according to the functions they discharge: administration, defence, culture, provision of goods and services, communications, and recreation. In reality, any one city may fulfil a number of functions.

functional linkage The link between industries, like headlight manufacturers and car makers, or between an industry and the public. A firm's linkages are classified by their direction of movement: **backward** or **input linkages** are received by the firm; **forward linkages** are supplied by it to another undertaking. Links include information, components, raw materials, finished goods, and transport links. They create the chains which bind firms together so that the difficulties faced by one firm may have severe repercussions on others. Functional linkages are at the heart of *agglomeration economies.

functional region A type of region characterized by its function such as a city-region or a drainage basin.

functionalism Basically, an anthropologist's view of society as an expression of its biological and social needs. The way in which things function must be studied in order to understand their effects. Schools of thought strongly grounded in functionalism include anthropogeography and certain aspects of *cultural geography.

fungi Important in soil science, fungi are a group of simple parasitic plants. Fungi are lacking in chlorophyll and therefore cannot photosynthesize. They attack a wide range of organic residues, such as the woody tissue of plants, and are a major element of the soil-forming processes.

G

gabbro A coarse-grained, basic, igneous rock composed chiefly of calcium-rich plagioclase and pyroxene. Gabbro is formed through the crystallization of basaltic magma, usually as a large igneous *intrusion deep within the earth's crust. The Black Cuillins of Skye furnish a British example.

Gaia hypothesis The concept, formulated by J. Lovelock (1988), that the entire planet earth—the *atmosphere, ecosphere, geosphere, and *hydrosphere—is a single ecosystem, or indeed, organism, which regulates itself by feedback mechanisms between the *abiotic and *biotic components of the system. To some extent, therefore, the ecosystem can moderate the effects of any changes made to it; it is, in some degree, self-regulatory, and tending to equilibrium.

It is argued that human agency is overriding this regulatory mechanism, tipping the *biosphere out of equilibrium. This hypothesis stresses the overriding importance of the entire planetary ecosystem, rather than the health of any individual species.

game Not to be confused with formal *game theory, this is the technique of mimicking real-life processes in a game in order to teach the participants an understanding of a particular aspect of reality.

game theory This deals with the question of making rational decisions in the face of uncertain conditions by choosing certain strategies in order to outwit an opponent in a formal game. In geography, this strategy is often used to overcome or outwit the environment, and when the environment is unpredictable man has only highly probabilistic notions based on past experience to work on.

Consider a very simplified case. Farmers in the Middle Belt of Ghana can choose either hill rice or maize as a staple crop. The climate may be wet or dry. A pay-off matrix can be set out, showing the likely yield of the crops:

	WET	DRY
maize	61	49
hill rice	30	71

The difference between each pair is calculated and, regardless of sign, is assigned to the alternate strategy:

	WET	DRY	
maize	61	49 = 12;	$\dfrac{41}{12-41} = 77.4\%$
hill rice	30	71 = 41;	$\dfrac{12}{12-41} = 22.6\%$

The pay-off matrix shows that more maize than hill rice should be grown but does not say whether the percentages indicate areas of both crops each year or whether maize should be planted in preference to hill rice in seven years out of ten. A further source of difficulty for all pay-off matrices lies in the assigning of values for each cell of the matrix.

gamma index, γ In *network analysis, a measure of the *connectivity in a network. It is a measure of the ratio of the number of *edges in a network to the maximum number possible: $\frac{1}{2}n(n-1)$.

$$\gamma = \frac{e}{\frac{1}{2}n(n-1)}$$

where e = number of edges, and n = number of nodes (vertices).

The index ranges from 0 (no connections between nodes) to 1.0 (the maximum number of connections, with direct links between all the nodes).

gap town A town located in a pass to an upland area which benefits from being a focus of routes. Corfe Castle, in Dorset, is a minor, British example; Lincoln a more important one.

garden city A planned settlement, as conceived by Ebenezer Howard (1850–1928), offering the benefits of urban living without the crowding and squalor of the Victorian city. Housing densities were to be low; parks, open spaces, and allotments were to be plentiful. The maximum city size was to be about 30 000. In 1903 work began on the building of Letchworth, in England, the first garden city. Using Howard's plans, roads, parks, and factory sites were laid out and private developers were invited to build carefully regulated houses on prepared sites. In 1919 Welwyn Garden City was founded.

Howard also founded the Garden City Association which, in 1918, became the Town and Country Planning Association, which is still an important pressure group. His ideas were also echoed in the construction of *new towns in the UK.

garden suburb A planned suburban development with open spaces and low-density housing inspired by the ideas of Ebenezer Howard. Garden suburbs were built in the late nineteenth and early twentieth centuries as in Bedford Park in 1875, and Hampstead Garden Suburb in 1907.

garrigue, garigue *Xerophytic and evergreen vegetation, for example, rosemary and thyme, found as the result of grazing, browsing, and burning in areas of Mediterranean climate.

Gästarbeiter Originally a German expression, literally meaning guest worker, this describes foreign migrant workers, usually engaged in manual labour, working in European cities.

gatekeeper An individual—or possibly a group—who is able to control access to goods and/or services. For example, urban planning departments have tremendous power in deciding who has permission to develop land or property, and can be seen as gatekeepers who control the evolution of the city. Other gatekeepers include bank managers, building society officials and estate agents. *See*, particularly *redlining.

GATT, The General Agreement on Tariffs and Trade, set up in 1947, but subject to almost continuous subsequent negotiation, is an arrangement between the states of the free world to encourage the gradual abolition of trade barriers. The number of signatories has increased from 23 to almost 100, but the agreement reserves the right of each state to protectionism if it seems to be necessary, and this clause has proved to be a fundamental weakness.

gavelkind The equal distribution of inherited land amongst the male heirs. Land may be left equally to daughters if there are no male heirs.

GDP, gross domestic product The total value of the production of goods and services in a nation measured over a year. (This includes production by non-nationals; compare with *GNP.) This is an unduplicated measurement, that is to say, if vinyl, for example, is used to press a record, the value of that vinyl is not registered in addition to the value of the record itself. In other words, components for a finished product are not taken into account; only the finished articles are recorded. The decision as to what constitutes a finished product varies from one country to another. GDP is an imperfect measurement of a nation's economy because certain forms of production, especially *subsistence production, are not recorded.

geest A heathland area of glacial sands and gravels; especially in north-central Europe.

gelifluction The downslope flow of soil in association with ground ice. This occurs in *periglacial environments, where water cannot percolate downwards because of the *permafrost. Spring melts of ice and *ice lenses provide enough lubricant to cause downslope flow. Gelifluction only occurs in areas of permafrost, in contrast to *solifluction, and may give rise to **gelifluction flats** and **gelifluction terraces**.

gelifraction Synonymous with *freeze–thaw.

gemeinschaft society A community bound together in a tightly knit pattern which is socially homogeneous and based on a clear-cut piece of territory. Gemeinschaft is a world dominated by face-to-face contacts, where each person is aware of his or her status, each is attached to a particular place, and the community is well regulated. It is said to be a feature of villages and small towns, and contrasts with *gesellschaft society, said to be a feature of an urban, industrialized populace. Both concepts are idealized; in real life there are ingredients of both, in varying proportions, in all societies. For example, within many cities there are neighbourhoods which operate as urban villages.

gendarme On an *arête, an abrupt rock pinnacle which has resisted *frost shattering.

gender While it is generally accepted that sex is biologically determined, gender is the role fabricated for us by society, which constructs appropriate behaviour for each sex. Not only is the ordering of space strongly gendered, it may also reinforce gender stereotypes; when space is constructed so as to make women feel unsafe, (secluded woodlands, dark alleyways, ill-lit multi-storey car parks) they are much more aware of their vulnerability and lack of physical strength, and this will further constrain their movements, so fulfilling the stereotype that women are less adventurous than men.

general circulation of the atmosphere The world-scale systems of pressure and winds which persist throughout the year or recur seasonally. Such winds transport heat from tropical to polar latitudes, thus maintaining the present patterns of world temperatures.

This global circulation is driven by intense differences in *insolation between the tropical and polar regions, and is strongly influenced by the *Coriolis force. Air moves vertically along the *meridians and horizontally with the wind systems, both at ground level and in the upper atmosphere.

generator cell A localized cyclonic development, responsible for *meso-scale rainfall events, and caused by fluctuations in the polar front *jet stream.

gentrification The rebuilding, renewing, and *rehabilitation of depressed areas of the inner city as more affluent families seek to live near to the city centre, trading off space and quiet for access to the goods and services of the city centre. The process has been facilitated by those local authorities which have provided home improvement grants as part of an *urban renewal programme. They are repaid by an increased rate, or council tax income. The original inhabitants move out as leases fall in, houses are sold, or landlords harass their tenants into moving. There is often a change of tenure from renting to home ownership.

In London, Islington is a classic example of a gentrified area; in Paris, gentrification has extended from the Marais eastwards. See *Alonso model.

geocode A code, for example a co-ordinate pair or postcode, expressing the spatial character of an *entity.

geodesy The science which deals with the shape and size of the earth.

geographic information Information which can be related to a *location on earth, particularly to a culture, human resource, or natural phenomenon.

Geographic Information Systems, GIS are integrated, spatial, data-handling programmes which will collect, store, and retrieve, *spatial data from the real world. They are powerful tools in decision-making, as they can incorporate co-ordinated data. It should be noted, however, that GIS only contain selected data; solely the properties which investigators have considered relevant, so that many variables will not be fed into the systems.

geography The discipline has many interpretations, which might best be understood if they are taken chronologically. Whereas the Greeks were concerned with what we now call *chorography, or *areal differentiation, throughout most of human history, geography was concerned with exploration, in order to increase a knowledge of the earth. Indeed, the Royal Geographical Society still sponsors 'voyages of discovery'. Linked with this reading of the subject was the drive to record phenomena in map form; at first, the mapping of relief features and settlements, but later the thematic mapping of anything from rainfall totals to the distribution of malaria. This tradition is still strong within the discipline.

In the nineteenth century, the doctrine of *environmental determinism took hold; the belief that human actions were moulded by physical conditions. The major proponents of this doctrine were Mackinder, with the concept of the *heartland, and Ratzel, who believed that the lot of a *Volk* was inextricably linked to the territory, or *Raum*, it occupied. (*See* *lebensraum*.) In the United States, Semple used this idea to explain the development of American history, and Huntington saw climate as the mainspring of civilization.

At the same time, there were those who emphasized the ability of humans to transform their environment. Herbertson and Fleure, for example, stressed the interplay of human and natural forces in the development of the *natural region.

By the turn of the century, the French school of geography, linked most strongly with de la Blache, saw the environment as a limiting factor rather than as a deterministic force. This school stressed the concept of *possibilism, and may be linked with the emphasis given by the Berkeley school, and its chief proponent, Sauer, on the importance of culture in the making of the landscape. Both schools saw geography as a study of *regions, and immense effort has been given to the delineation of *natural and human regions. Throughout, there has been an emphasis on the interaction between the human and the physical, and geography has long been promoted as the bridge between humanity and nature; a tradition which flourishes to this day.

In the 1950s a new practice developed. F. Schaefer (*AAAG*, 1953) initiated a move to seek 'laws' which would explain geographical phenomena, particularly within the field of human geography, initiating the shift to *logical positivism and the concept of geography as a science of *spatial distribution. This movement to locational analysis was seen as a revolution in geography, and was accompanied by extensive *quantification. In time, *positivism came under fire, for it was felt that this approach reified human beings. In consequence, there has been an exploration of the role of *perception, at the heart of *cultural geography, and of the deep structures, most notably capitalism, which underlie human actions. This is the field of 'structuration', proposed by Giddens, but modified by geographers, who stress that the results of the reactions between social structure and human agency depend also upon location.

It would take considerable temerity to find a unifying definition throughout the twists and turns that the discipline has taken. One possible approach would be to adopt Linton's proposal that geography is 'the study of the landscape'; a view that goes far in uniting the many concerns of the geographer.

geological column Also known as the **stratigraphical column**, this is the separation of geological time into *eras and periods, as below:

ERA	PERIOD	EPOCH	END DATE, MILLION YEARS
Quaternary		Holocene (recent)	
		Pleistocene (glacial)	
Tertiary		Pliocene	2
(Cainozoic)		Miocene	
		Oligocene	
		Eocene	
Secondary	Cretaceous		65
(Mesozoic)	Jurassic		135
	Triassic		190
Primary	Permian		225
(Palaeozoic)	Carboniferous		280
	Devonian		345
	Silurian		395
	Ordovician		440
	Cambrian		500
	Precambrian		570

geomagnetism The magnetic field of the earth, also known as terrestrial magnetism. The axis of this field emerges from the earth's surface at the *magnetic poles. The position of these poles varies over time, and sometimes the positions of the north and south magnetic poles switch places. The pattern of the magnetic field at any one time will be preserved in any contemporary *extrusions of volcanic rock. The study of past magnetic fields, **palaeomagnetism**, can yield information about the creation of new material at the oceanic ridges, about continental drift, and about the dating of certain deposits.

geomorphology The study of the nature and history of landforms and the processes which create them. Initially, the subject was committed to unravelling the history of landform development, but to this evolutionary approach has been added a drive to understand the way in which geomorphological processes operate. In many cases, geomorphologists have tried to model geomorphological processes, and, more recently, some have been concerned with the effect of human agency on such processes.

geopolitics The view that location and the physical environment are important factors in the global power structure; the state may be seen as a realm in space. Early proponents of the study were Halford Mackinder (*see* *heartland) and Friedrich Ratzel, and the 1970 domino theory may be seen as a branch of geopolitics.

Other aspects of geopolitics include studies of relationships between states, especially with reference to the growth and decline of great powers, and the importance of location in the ability of states to compete in the world economy; ideas of core and periphery are important here. Geopolitics must not be confused with *geopolitik.

geopolitik A view of *geopolitics developed in Germany in the 1920s. Individuals are subordinate to the state which must expand with population growth, claiming more territory—*lebensraum—to fulfil its destiny. These ideas are not synonymous with national socialism, but were used as a quasi-science by the Nazis to justify their territorial demands.

geosophy The study of geographical knowledge. One aspect may be the development of geography as a branch of knowledge; the other relates to human beings as a whole. Thus, all human beings have some geographical knowledge, although the standpoint of this knowledge may be different from that of the professional geographer, and, even though the knowledge may be false, it may be acted on in any case. Alternatively, the mind may erect barriers to certain places, even though there are now no unexplored parts of the earth. *See* *mental maps.

geostationary satellite A satellite with remote sensors, in an orbit 5.6 times earth's radius, so that it 'hangs' over the same spot as the earth turns; capable of viewing nearly a full hemisphere, although with much lateral distortion at the edges.

geostrophic wind A theoretical wind, occurring when the force exerted on the air by the *pressure gradient is equal to the opposing *Coriolis force (assuming straight or nearly straight *isobars; when the isobars are strongly curved, the effect of centrifugal force should be added in). The net result is a wind blowing parallel to the isobars, with speeds proportional to the *pressure gradient. Except in low latitudes, where the Coriolis force is minimal, the actual wind direction is the same as that of the geostrophic wind.
 Supergeostrophic flow describes wind speeds greater than the expected geostrophic wind. It occurs at a *jet entry, where winds are experiencing linear acceleration. **Subgeostrophic flow** describes wind speeds less than the expected geostrophic wind.

geosyncline A thick, rapidly accumulating body of sediment formed within a long, narrow, subsiding belt of the sea which is usually parallel to a plate margin. The sediments may form gently tilted strata of uniform *dip in which case it is a **geocline**.

geothermal flux, geothermal heat Heat from the earth's interior generated by early gravitational collapse, and later radioactive decay. **Geothermal energy** may be extracted by pumping water down through an injection well and forcing it through joints in the hot rocks. When the water returns to the surface it may be converted to steam, for use in a generator, or run through a heat exchanger. This is potentially an important energy source in any volcanically active area, and is used in Iceland and New Zealand.

gerrymandering Redrawing constituency boundaries in order to gain a political advantage. One method of doing this is to concentrate most of the opposition's vote into a few electoral districts so that, although they have major support the number of successful candidates they gain is small. Alternatively, opposition votes may be spread over a large area where they will have little electoral impact. The expression comes from the boundary drawing of Governor Gerry of Massachusetts, who redrew the boundaries of an electoral district (so that it looked like a salamander) in order to help his party in the elections of 1812.

gesellschaft society A society characterized by formal and aloof relationships. In such a society, people merely reside in their neighbourhoods and are free from social bonds and ties. Gesellschaft society is said to be a feature of the city, although this is strongly debated. *See* *gemeinschaft society.

gestalt theory In human geography, a theory which suggests that between stimulus and response lies perception, which intervenes between the two. Thus, observed objects are organized into patterns, and behaviour is based on the perceived, rather than the actual, environment. The same environment may have very different meanings to individuals coming from culturally different backgrounds.

geyser A jet of hot water and steam, issuing from beneath the earth, usually as a result of the *geothermal heating of a store of underground water which is connected to the surface by a narrow outlet pipe. When the water is heated above boiling point, some of it becomes superheated, and this fraction expands rapidly, forcing its way up the pipe, and emerging as a fountain of steam and hot water. Geysers often erupt at regular intervals, as pressure builds up and is discharged, and the most famous is 'Old Faithful', in Yellowstone National Park, USA, which erupts, on average, every 65 minutes, to a height of 45 m.

ghetto A part of a city, not necessarily a slum area, occupied by a minority group. The term was first used for the enforced concentration of Jews into specific residential areas in European cities from the Middle Ages, but has now spread to include other *ethnic groups in unofficial ghettos, especially black minorities in the USA. Lifestyles within the ghetto differ distinctly from those of the 'host' population and the prejudices of the host confine the sub-group to particular locations. *See* *redlining. Although ghettos are characterized by social disadvantage, most ghettos display a spread of socio-economic groups and the better-off may move to the affluence of the 'gilded ghetto'. *See also* *segregation.

giant nuclei *See* *nucleus.

Gini coefficient In a *Lorenz curve, a measure of the difference between a given distribution of some variable, like population or income, and a perfectly even distribution. More simply, it tells us how evenly the variable is spread; this might be a measure of how wealth is distributed over the regions of a country, or over the classes in society. A diagonal line shows an even distribution, and the calculation of the Gini coefficient uses the 'gap' between the diagonal and the actual curve. The coefficient, also known as **Gini's concentration ratio**, may be calculated as the ratio of area between the diagonal and the Lorenz curve to the total area beneath the diagonal. The lower the Gini coefficient, the more evenly spread the variable.

glacial 1. Of or relating to a glacier.
   2. An extended length of time during which earth's glaciers expanded widely.

glacial breaching The erosion of a breach between two adjacent valleys when a transfluent glacier flows across the watershed between them.

glacial deposition The laying down of *sediments which have been removed and transported by a *glacier. The sediments—known as *till, or drift—are deposited when the ice melts; that is, when *ablation is dominant. Glacial deposition is predominant in marginal areas of present and past ice sheets, such as the Eden Valley, and much of the English Midlands, but also occurs in uplands, especially in the form of *moraines. *See also* *drumlins, erratics,*kettle holes.

glacial erosion Glacially eroded landscapes are moulded by abrasion, the incorporation of debris, the 'conveyor belt' transport of *moraine on top of a glacier, rock fracturing, *plucking, plastic moulding, *pressure release and the action of *meltwater. The highlands from which ice dispersed, such as the Highlands of Scotland, are primarily areas of glacial erosion.
   Vertical erosion is due to glacial scour; lateral erosion is thought to be due to freeze–thaw and pressure release.
   The features of glacially eroded uplands are striking, and best developed in resistant rocks, such as the Borrowdale volcanics of the English Lake District. In this context, it is noticeable that features of glacial erosion are much better developed in the Scottish Highlands than in the Southern Uplands of Scotland. It might be that the hardness of resistant rocks preserves features of glacial erosion, or that more resistant rocks are better jointed. Furthermore, it should be noted that glacial erosion is very selective, being at its most effective when localized, high-velocity *ice streams flow in part of an otherwise stationary valley glacier.
   In glacially eroded lowlands, relief is confused, as in the Laurentian *Shield. Weathered rock is stripped away, and *abrasion and *quarrying of the bedrock are active. Drainage is rambling as earlier patterns are disrupted by erosion and the deposition of *moraines. Lakes and ponds abound, and perched blocks and erratics may be common. This type of landscape is well developed in the Lewissian gneiss areas of north-west Scotland, and is known as knock and lochan topography.

glacial margin channel A stream running between the side of the glacier and the valley sides. The stream may undercut the side of the valley.

glacial movement Glacier ice will move if the temperature at the base is above the *pressure melting point. The temperature at the base of a glacier is a function of the thickness of the ice, friction, the input of *firn, and the altitude. If these combine to give a *warm glacier, and there is a gradient, the ice will move downslope. *Cold glaciers move much more slowly than warm glaciers.

Ice moves by *compressive and *extending flow; these two forms may alternate down the valley profile. It also moves when the ice crystals change into a series of flat platelets, i.e. it becomes *plastic and then creeps downslope. Further movement takes place by *basal slipping.

glacial surge The swift and dramatic movement of a glacier. Glaciers may surge at up to 100 m per day. Surging occurs for a short period of time only and is associated with the growth of ice up-glacier to unstable proportions and with severe *crevassing. See *kinematic wave.

glacial trough Once termed a U-shaped valley, this is a wide valley floor with steep sides formed by *glacial erosion. Glacial troughs tend to have a straighter course than river valleys. The harder the rock which the glacier has cut through, the steeper the valley walls. The shape of a glacial trough more resembles a parabola than the letter U.

The long profile of a glacial trough is frequently irregular and marked by basins and steps. Explanations for this over-deepening and reverse flow vary: that deeper sections are formed where two glaciers meet, that *plucking is more effective in the weaker or closely jointed rocks, that flow alternates between *compressive and *extensive, or that glaciers may pass through naturally occurring narrows. In this last case, the power of the glacier would be intensified as it pushes up against the valley walls.

glacier A mass of ice which may be moving, or has moved, overland: when enough ice has accumulated, a glacier will start to move forwards. A glacier may be seen to be the result of a balance between *accumulation and *ablation.

Glaciers are classified by their location (*cirque glacier, *expanded-foot glacier, valley glacier, *niche glacier, *piedmont glacier), by their function (*diffluent glacier, *outlet glacier), or by their basal temperature (*cold glacier, *warm glacier).

glacier budget The balance in a glacier between the input of snow and *firn, that is, *accumulation, and the loss of ice due to melting, evaporation, *sublimation, and *calving, that is, *ablation. A glacier grows where the budget is positive and retreats when it is negative.

glacio-eustasy See *eustasy.

glacio-isostasy See *isostasy.

glaze See *hail.

gley soils Soils with mottled grey and yellow patches. These are caused by intermittent waterlogging. **Gleying** occurs where soils are waterlogged because there the air is excluded and the supply of oxygen is reduced. Under these conditions, *anaerobic micro-organisms flourish by extracting oxygen from chemical compounds. This is most conspicuous when the *sesquioxide of iron, ferric oxide is reduced to ferrous oxide by the removal of oxygen. This process gives a greenish-blue-grey colour to the soil. Gley soils are sticky and hard to work.

global circulation *See* *general circulation of the atmosphere.

global energy balance The difference between the total influx of *solar radiation to the earth's surface and the loss of this energy via *terrestrial radiation, evaporation, and the dissipation of sensible heat into the ground.

global warming The increase in global temperatures brought about by the increased emission of *greenhouse gases into the atmosphere. There is no doubt that concentrations of, for example, atmospheric carbon dioxide have risen since the 1950s; what is less certain is the extent to which this has altered the earth's climates, or the extent to which climates will change in the future. *See* *greenhouse effect.

globalization The concept of the interactions of natural and human phenomena on a global scale. *Global warming, for example, is a world-wide phenomenon where human agency may have major repercussions on the *atmosphere, geosphere, and *hydrosphere. *See also* *air–sea interactions, *Gaia hypothesis.

gneiss A highly metamorphosed rock of a granular texture and with a banded appearance. Gneisses may have been metamorphosed from *schists or quartzites, themselves metamorphic rocks. Some gneisses may be produced by the interaction of igneous magma with metamorphic rocks. Gneisses are resistant rocks, but inclined to *exfoliation along the sheet joints.

GNP, gross national product The *GDP of a nation together with any money earned from investment abroad, less the income earned within the nation by non-nationals. Compare with *GDP. **GNP per capita** is calculated as GNP/population and is usually expressed in US dollars. It may be used as an indicator of *development. In the mid-1990s, a per capita GNP figure of $10 000 would indicate a more developed country, while for a *least developed nation, the figure would be around $600. GNP is an imperfect measurement of a nation's economy because certain forms of production, especially *subsistence production, are not recorded.

Gondwanaland A 'supercontinent' occurring as a continuous region of land formed of the now separate units of Africa, Madagascar, Antarctica, Australia, and India. *See also* *continental drift.

gorge A deep and narrow opening between upland areas, usually containing a river. Gorges, such as Cheddar Gorge in the UK, occur in *karst scenery partly as a result of the collapse of caves. A further cause, as in the Elbe Gorge in Saxony, is when the downcutting power of the river is greater than the processes of valley-wall erosion. The latter may be less effective because the water permeates the side walls rather than flowing over and eroding them.

government incentives Measures taken by a government to attract the development of industry in specified areas. These include grants for building, works, plant, and machinery, assistance in encouraging sound industrial projects, removal grants to new locations, free rent of a government-owned factory for up to five years, taxation allowances against investments, loans, and contract preference schemes. These last give preferential treatment to companies in *assisted areas in the UK when tendering for government contracts.

graben *See* *rift valley.

grade In geomorphology, a state of *equilibrium in a system such as a hillslope or river.

graded river A stable stream where slope and channel characteristics are such that the *discharge is enough to provide transport for the *load, but where there is no energy for erosion. A graded river is in a state of *equilibrium. If any of the factors controlling discharge change, the river would respond in such a way as to re-establish *grade. A river may establish grade over one section of its course; this is a **graded reach**.

graded slope A slope of such inclination and character that output, throughput, and input remain in equilibrium. No change will be detected unless the balance of forces alters.

gradient level The lower limit of the free *atmosphere; also the lower level at which the quasi-geostrophic balance between the Coriolis and pressure-gradient forces exists.

granite A coarse-grained igneous rock that consists largely of quartz, alkali feldspar, and plagioclase feldspar. Granite is formed by the slow crystallization of deep igneous intrusions but may also be formed by *metasomatism. Major granite intrusions in the British Isles are found on the Isle of Aran, Dartmoor, and Bodmin Moor. *See* *batholith, *tor.

granitization *See* *metasomatism.

granular disintegration, granular disaggregation A form of weathering where the grains of a rock become loosened. Grains fall out to leave a pitted, uneven surface. Granular disintegration may be the result of *freeze–thaw, *hydro-fracturing, *thermal expansion, or *salt weathering.

graph theory The mathematical study of *networks and topological maps. In this context, a **graph** consists of a set of points, also known as *nodes or *vertices, and the links, also known as arcs or lines, connecting them. Graph theory studies the nature of the links between points, and the location of the points themselves.

graphic In picture form.

graticule A regular grid, used for referencing points on a map.

graupel Spongy *hail.

gravel A loose deposit of rock fragments rounded by river erosion. The lower size limit of gravel is 2 mm but the upper limit is either 10 mm, 20 mm, 50 mm, or 60 mm according to different authorities. While **river gravels** are well rounded as a result of *attrition, **solifluction gravels**, produced by freeze–thaw, are more angular. The **plateau gravels** of southern England are extensive deposits between 0 and 1250 m above sea level, made of flint and chert fragments in a mould of sand and clay. They were deposited by solifluction and meltwater during the *Pleistocene.

gravity anomaly The difference between the actual gravitational force and the calculated force. When a plumb line is set up near a mountain range, it is attracted from the vertical towards the mountains but by far less than would be expected from calculations. This is a negative gravity anomaly and can be explained by the hypothesis that the density of the mountains is less than estimated. If the plumb bob is attracted more than expected from calculation, a positive gravity anomaly is said to exist.

gravity model A model of the interaction between two population centres based on Newton's Law of Universal Gravitation: two bodies in the universe attract each other in proportion to the product of their masses and inversely as the square distance between them. Thus, expected interaction between city i and city j is shown as:

$$k \times \frac{P_i \times P_j}{d_{ij^2}}$$

where P_i = the population of town i, P_j = the population of town j, d_{ij} = the distance between them, and k = a constant.

This original equation has been changed to accommodate features like wages, employment opportunities, and so on, and has been widely criticized, but is still used to predict future interactions. The gravity model may be applied to fields of influence of settlements, trade, traffic flows, telephone calls, and migration. Perhaps the most severe criticism of the model is that it has no theoretical basis, but is based on observation only. Furthermore, planning on the basis of the model will only reinforce differences between places; people will interact more with larger towns if planners are geared to that assumption and plan for it accordingly. *See* *Reilly's law.

gravity slope A slope which has formed at the *angle of repose of the unattached material resting on it.

great circle An imaginary line on the earth's surface which, if projected underground, would pass through the centre of the earth. Any great circle route between two points will represent the shortest line between the two.

great soil groups The primary classification of global soils into groups; a classification similar to the *formations of vegetation. *See* *soil classification.

green belt An area of undeveloped land encircling a town. An early green belt, about 15 km wide, was set up in the 1950s around Greater London in order to limit the spread of suburbs.

Other cities have followed this example by restricting development in the semi-rural areas beyond the built-up zone. The amount of new building is restricted although by no means completely banned. In some cases, development has been switched to areas beyond the green belt which is then sandwiched. Later motives for creating green belts have been the provision of open areas for recreation and the preservation of agricultural land. Some planners advocate the establishment of green 'wedges' which project into the city rather than a green belt.

green manure A leguminous crop not harvested but ploughed into the fields after it matures. Leguminous crops fix nitrogen from the air and thus improve the fertility of the soil.

green revolution The development and use of high-yielding crops (HYVs) in conjunction with improved agricultural technology. New breeds of crops have been developed to increase yields two to four times, to shorten the time required for growth such that more than one crop a year can be produced, and to produce a plant which can withstand extremes of climate or disease. The use of Mexican wheat has doubled yields in the Punjab, and HYV rice has been used to such effect in the Philippines that imports are no longer necessary. The Green Revolution has had most impact in South and East Asia, and in South America, but has not been taken up to the same extent in sub-Saharan Africa.

There have been drawbacks, however. The grain may not be as palatable or as attractive in appearance as the grain it replaces, and it may use up more energy to process. Seeds have to be bought, as the hybrids are not self-fertile, and some varieties are less resistant to drought and disease. Heavy applications of expensive fertilizers and insecticides are required and these are often made from *non-renewable resources.

Herbicides are required because the fertilizer stimulates weed growth as well as crop growth. The high yields and reliance on artificial fertilizers can lead to impoverished soils. Traditional rice exporters, like Burma, have seen the collapse of their markets. Increased yields mean that landowners can use their holdings more profitably and this often means that tenants are dispossessed. Copious, but strictly regulated, irrigation is required.

The green revolution has benefited the most prosperous farmers in the most prosperous areas but its price is too high for many of the peasants who need its help. To that extent, it has only been a partial success.

green village A settlement with houses and a church gathered around a common or village green.

greenfield site Areas beyond the city where development can take place

unfettered by earlier building and where low density, high amenity
buildings can be constructed.

greenhouse effect The warming of the atmosphere as some of its gases
absorb the heat given out by the earth. Short-wave radiation from the sun
warms the earth during daylight hours, but this heat is balanced by
outgoing long-wave radiation over the entire 24-hour period. Much of this
radiation is absorbed by atmospheric gases, most notably water vapour,
carbon dioxide, and *ozone, but also by methane and
*chloro-fluorocarbons. All of these may be called greenhouse gases.
Without this absorption, which is also known as *counter-radiation, the
temperature of the atmosphere would fall by 30–40 °C.

Through human agency, such as the clearance of rain forest, or the
increased rearing of livestock, the concentration of greenhouse gases in
the atmosphere is increasing; measurements taken at Mauna Loa,
Hawaii, show that the concentration of atmospheric CO_2, for example,
increased by 8% between 1959 and 1983, mostly because of the increased
use of fossil fuels. It would follow, therefore, that increased
concentrations of such greenhouse gases would lead to a rise in global
temperatures, and, indeed, global mean temperatures have increased by
0.3 to 0.7 °K over the last century, but the cause of this temperature rise
has not been unequivocally put down to the increase in greenhouse
gases. It may be that the uptake of CO_2 by the oceans actually increases
with higher temperatures. Others argue that increased concentrations of
CO_2 foster improved rates of photosynthesis in plants, so that
faster-growing trees, for example, might partially offset increased
concentrations of carbon dioxide. Thus, general models of the effect of
growing greenhouse gas levels do not give unequivocal predictions of
future trends in climates.

The analogy with a greenhouse is not perfect, since a greenhouse retains
heat through lack of movement in the air as well as by absorbing *counter-
radiation.

greenhouse gas Gases which absorb outgoing *terrestrial radiation,
such as water vapour, methane, CFCs and carbon dioxide. *See* *greenhouse
effect.

greywacke A sedimentary rock consisting of angular fragments of
quartz, feldspar, and other minerals set in a muddy base. Its origin is
problematic, since, according to the normal laws of sedimentation, sand
and mud should not be laid down together, and some geologists attribute
its formation to submarine avalanches or turbidity currents. Deposits of
greywacke are found on the edges of the *continental shelves.

grèzes litées Deposits down a hillslope of *imbricated rock fragments
bedded parallel to the slope. It is suggested that the debris is shattered by
*freeze–thaw and that larger fragments roll downwards under the
influence of gravity. With thawing, the finer debris washes downslope,
forming a fairly smooth layer of sediment on top of the coarser material.

grid A planimetric frame of reference.

grid cell In *Geographic Information Systems, a two-dimensional object representing one element within a regular *tessellation of a surface.

grid plan An urban area in which the basic street pattern is planned as a grid with regular spacing between blocks. Many American cities were built to this pattern.

grike The joints on an exposure of limestone which have been widened through solution by *carbonation. *See* *clint, *limestone pavement.

grit An extremely coarse *sandstone.

gross domestic product *See* *GDP.

gross national product *See* *GNP.

gross reproduction rate The number of female babies born per thousand women of reproductive age. The **net reproduction rate** also takes into account the number of women who cannot or do not wish to have children.

ground failure *See* *avalanche, *debris flow, *landslide, *rock fall.

ground frost *See* *frost.

ground moraine *See* *moraine.

groundwater All water found under the surface of the ground which is not chemically combined with any minerals present, but not including underground streams.

Group of Seven Canada, France, Germany, Italy, Japan, the UK and the USA, as the world's leading economies, formed this group to advise on the running of the global economy, to lend to the IMF, and to further international economic co-operation. The value of the group has been questioned.

grouped data Data which have been arranged in groups or classes rather than showing all the original figures, for example, the data in a *population pyramid.

growth pole A point of economic growth. Poles are usually urban locations, benefiting from *agglomeration economies, and should interact with surrounding areas spreading prosperity from the core to the periphery. Observation of naturally occurring growth poles has inclined planners to create new growth poles; the best-known attempt at creating growth poles took place in the Mezzogiorno (south) of Italy, with industrial complexes planned at Taranto and Bari. Such artificially created growth poles, as in France, have not stimulated regional development as much as was hoped. *See also* *metropoles d'équilibre.

groyne A breakwater running seawards from the land, constructed to stop the flow of beach material moved by *longshore drift.

GST, general systems theory A general science of organization and wholeness. Bertalanffy, the founder of the science, dated its inception from 1940, but it did not influence geographical thinking until the late 1960s. It introduced the application of the *system to geography and claimed that any phenomenon cannot properly be understood until it is seen as a system of many associated parts.

guest workers Temporary immigrants who do not plan to, or are not allowed to, settle permanently in their work place. Within Western Europe, the earliest flow of guest workers was from Italy to West Germany, France, Belgium, and Switzerland; later flows came from Spain, Portugal, Greece, Algeria, Morocco, Yugoslavia, and Turkey. The German *Gästarbeiter*, of whom the greatest proportion are Turkish, are the classic example: they are predominantly male; work in low-skilled, poorly-paid, repetitive work; they have no job security; and most live in very poor housing. If the economy declines, the 'guests' are less welcome.

Guinea current A warm *ocean current off the coast of West Africa.

gully A water-made cutting, usually steep-sided with a flattened floor. Gullying usually occurs in unconsolidated rock and rarely cuts through bedrock. Gullies usually form quickly as a result of destruction of the plant cover. **Gully erosion** is the removal of topsoil and the creation of many steep-sided cuttings in a hillside. It can be stopped by restoring a vegetation cover, by *contour ploughing, and by making terraces and small dams across the hillside.

gust A temporary increase in wind speed, lasting for a few seconds. A typical ratio of gust speed to wind speed in rural areas is $1.6:1$, increasing to $2:1$ in urban areas due to the effects of high buildings and narrow streets.

Gusts may also be associated with heavy rain, as downdraughts sweep eastwards from *cumulo-nimbus clouds. The repeated hammer blows of gusting winds do more damage than does the persistent pressure of steady wind.

guyot A truncated sea-floor volcano occurring as a flat-topped mountain which does not reach the sea surface. Guyots are thought to be associated with *hot spots.

gypcrete A *duricrust composed of hydrated calcium sulphate.

H

habitat In *ecology, the area in which an organism can live and which affords it relatively favourable conditions for existence.

haboob *See* *local winds.

hacienda In Spanish-speaking countries, a large farm, usually a ranch.

Hadley cell A simple, vertical, thermally direct, *atmospheric cell, first suggested by G. Hadley in 1735. Hadley supposed that cells, extending from equator to pole in each hemisphere, were made up of warm air rising from equatorial regions and moving polewards, transporting heat energy. In reality, this simple pattern is complicated by the *Coriolis force, the shape of the earth, relief barriers, ocean currents, and the distribution of land and sea.

Hadley-type cells do exist, however, in the tropics. Air rises at the *inter-tropical convergence zone, (ITCZ), where the trade winds meet, and drifts polewards, cooling a little through a net loss of radiation, being thus responsible for the transfer of heat. As the latitudinal expanse of the earth diminishes with distance from the equator, the moving air has to converge. This combination of cooling and convergence, together with deflection by the *Coriolis force, causes the air to sink around latitude 32°. The air then returns to the equator as surface winds. Three basic components may be identified: weak upper-air easterlies above the ITCZ, subtropical anticyclones associated with subsidence at the descending limb, and easterly *trade winds associated with the return of air to the equator, together with *synoptic scale disturbances, such as easterly waves or *tropical cyclones. *See* *atmospheric heat engine, *general circulation of the atmosphere.

The westerly subtropical *jet is located at the poleward limit of the Hadley cell.

hail A form of snow, consisting of roughly spherical lumps of ice, 5 mm or more in diameter. Hailstones often show a roughly concentric pattern of alternating clear ice (glaze) and opaque ice (rime). They form when a frozen raindrop is caught in the violent updraughts found in warm, wet *cumulo-nimbus clouds. As they rise, they attract ice, and as they fall, the outer layer melts, but refreezes when the droplet is again lifted by updraughts. The onion-like structure of hailstones shows that they must have passed up and down several times. A hailstone will descend when its fall-speed is enough to overcome the updraughts in the cloud. Soft hail is white and of a low density because it contains air. *See also* *cloud seeding for limiting hail formation.

half-life The rate of decay of a radioactive *isotope. The half-life is the time taken for half the original, parent isotopes to decay. At the end of the first half-life, half of the parent isotope is left; at the end of the, equally long, second half-life, one quarter of the parent isotope remains and so on. Hence the amount of residual radiation in a rock can be used to determine the age of the rock.

halo In meteorology, a ring of light around the sun or, more rarely, the moon. It is caused by the refraction of light by ice crystals, thus, in popular lore, a halo round the moon foreshadows snow. A coloured halo is a corona.

halophyte A plant which can grow in saline conditions, which include salt marshes, estuarine environments, and the lower parts of sea cliffs. Halophytes on salt marshes tend to trap sediment at high tide, and this gradually increases the height of the marsh. Common British halophytes include sea aster and marsh samphire.

ham 1. In Anglo-Saxon place names, a home, as in Birmingham.
  2. A water-meadow of rich pasture.

hamlet A small settlement without services or shops and usually without a church.

hammada, hamada In hot deserts, such as the Syrian–Iraqi desert, a pavement of angular or rounded stone fragments. The processes of formation include the removal of finer particles through *deflation, and through water action during infrequent storm events, or the upward movement of stones as a consequence of wetting and drying the surface.

hanging valley A high-level tributary valley from which the ground falls sharply to the level of the lower, main valley. The depth of the lower valley may be attributed to more severe glaciation, because it contained more ice. Some writers suggest that these features are caused by two phases of glaciation separated by a period of *fluvial erosion, or that the erosive power of the tributary stream has been less than the erosive power of the larger stream; glaciation may not be the only process involved.

har Advective, sea *fog.

hard copy A copy of data in tangible form.

hard water Water containing dissolved carbonates of calcium and magnesium which inhibit the formation of a lather with soap or detergent. When hard water boils, the carbonates are deposited out as 'lime scale'.

hardness The hardness of a rock may be indicated by comparing it to the rocks on the **Mohs scale**. On this scale hardness is indicated by the ability of the specimen to scratch the rocks of the scale. A rock which could scratch quartz (7 on the Mohs scale) but is scratched by topaz (8 on the Mohs scale) would have a hardness of 7–8. The complete scale is:

    1 = talc   6 = orthoclase feldspar
    2 = gypsum   7 = quartz
    3 = calcite   8 = topaz
    4 = fluorite   9 = corundum
    5 = apatite   10 = diamond.

Mohs's qualitative scale is now being replaced by more quantitative tests.

hardpan A cemented layer in the B *horizon of a soil, formed by the *illuviation and precipitation of material such as clay (forming a **clay pan**), humus (forming a **moor pan**) or iron (forming an **iron pan**), leached from the A horizon. Hardpans hamper drainage and may make cultivation difficult.

hardware The physical equipment in a computer system.

hardwood Wood obtained from temperate deciduous trees such as oak, or from tropical evergreens such as teak, mahogany, and ebony.

Harmattan *See* *local winds.

Hawaiian eruption A volcanic *fissure eruption where large quantities of *basic lava spill out with very little explosive activity. *See* *shield volcano.

hazard *See* *natural hazard.

hazard perception The view which an individual has of a natural or man-made hazard. A person may have a high perception of a hazard which occurs often, but still may suppress knowledge of such occurrences because of a desire to remain in a particular location. It is not the hazard as such which influences behaviour but the assessment of its likelihood and extent. San Francisco was rebuilt after 1906 partly because buildings were then constructed which could to some extent survive earthquakes, and partly through a desire not to think of the risks. Later events have shown that the risks are very real. *See* *cognitive dissonance.

haze A suspension of particles in the air, slightly obscuring visibility. These particles may be naturally occurring—sea salt or desert dust—or may be man-made, like the smoke formed from the burning of fossil fuels.

head An alternative name for a *solifluction gravel.

headland An area of high land jutting out into the sea.

headward erosion The lengthening of a river's course by erosion backwards from its source. *Sapping is an important process in headward erosion.

heartland A term suggested by Halford Mackinder (*GJ*, 1904) to indicate the wealthy interior of Eurasia. Mackinder maintained that whoever controlled the heartland would eventually control the world as political units became larger and larger. He did not live to see the rise of *nationalism in the late twentieth century.

heat island In a city, air temperatures are often as much as 3–4 °C higher than over open country. These higher temperatures are generated by the combustion of fuels in factory, heating, and transport systems, and, more importantly, the release at night of heat which has accumulated during the day in the fabric of the city, for the bricks and concrete of the buildings act as enormous storage heaters. This effect is compounded by air pollution, which reduces night-time *terrestrial radiation, and by the low humidity which results from the lack of vegetation. A heat island is developed during calm conditions; winds disperse heat.

heathland An uncultivated, open area of land with a natural vegetation of low shrubs such as the family *Ericaceae* (heathers). Heathland tends to develop on poor soils such as *outwash sands and gravels.

heavy industry Manufacturing industry which needs large quantities of often bulky raw materials. These are usually transported by water or rail as in iron smelting or shipbuilding. These industries have a high *material index. Productivity per worker is generally low, and heavy industries are often dirty and noisy. Compare with *light industry.

hegemony Originally, leadership, especially by one state of a federation, in terms of power and politics. More recently, within *Marxist geography, the term has been applied to the ruling class. In this context, it refers to the way in which a ruling class will represent its interests as being everyone's interests. Marx believed that, historically, each ruling class did actually represent universal interests rather better than the one before. The ruling class may keep its grip on society either by **social hegemony**, that is, the use of force to maintain order in society, or, much more ubiquitously, by **cultural hegemony**; by producing ways of thinking and seeing, and especially by subtly eliminating alternative views to reinforce the *status quo*.

helicoidal, helical flow A continuous corkscrew motion of water as it flows along a river channel. It has been explained by the tendency of the *thalweg to flow in a straight line, so that it 'hits' the outer banks of a *meander and causes a 'head' of water to build up at that point. To compensate for this, a return flow develops across the channel. Note, however, that other geomorphologists have suggested that helicoidal flow actually causes the development of meanders. Helicoidal flow is particularly associated with sinuous and meandering river channels.

henge A British circular earthwork dating from the late Neolithic having an encircling bank with a ditch inside.

herbivore Any animal which eats only plant material. *See* *food chain, *trophic level.

heritage coast Stretches of unaltered coastline which are outstandingly attractive and are protected from development. For illustration, *see* *Area of Outstanding Natural Beauty.

hermeneutics The art, skill, or theory of understanding and classifying meaning. It is often applied to the interpretation of human actions, utterances, products, and institutions. A **hermeneutic interpretation** requires the individual to understand and sympathize with another's point of view.

In *geography, the hermeneutic approach was used to challenge *empiricism and *positivism, and to develop the field of *humanistic geography, which stressed human meaning and intentionality. Although this approach is now less fashionable, the importance of open-mindedness and varieties of interpretation has not diminished.

heterogeneous nucleation The freezing of water droplets around a *freezing nucleus. *See also* *homogeneous nucleation.

heterotrophe An organism which has to acquire its energy by digesting food which has been manufactured by other organisms. Thus all organisms are heterotrophes except the primary *producers which can manufacture their own food usually by *photosynthesis.

heuristic In computers and computerized problem-solving exercises, based on trial and error; in education, describing the technique of learning by the 'discovery method'.

hidden lines The lines in a three-dimensional drawing which are obscured by features in the foreground.

hide A medieval unit of land of varying size; usually 40 to 50 hectares, but often much smaller.

hierarchical diffusion *See* *diffusion.

hierarchy Any ordering of phenomena with grades or classes ranked in sequence. *Central place theory posits a hierarchy of settlements from regional capitals to hamlets. It is suggested that the same grades of settlement in the hierarchy are spaced evenly: villages are closer together than towns, which, in turn, are closer together than cities. Research suggests, however, that settlements occur in a continuum rather than a hierarchy. It seems unlikely that the presence of a hierarchy can be established, partly through the difficulty of *ranking towns.

high A region of high atmospheric pressure. In Britain, the term is generally applied to pressures of over 1000 mb. *See* *anticyclone, *col.

high-energy society A society heavily dependent on machines powered by the burning of *fossil fuels. The ecological impact of such societies has been estimated to be 10 000 times greater than the impact of preceding societies, and the signs of this disturbance include the *greenhouse effect, the hole in the *ozone layer, and the effects of *acid rain.

high farming A time in eighteenth- and nineteenth-century Britain associated with buoyant farm prices which stimulated new techniques, such as enclosures and scientific breeding. It also encouraged the owners of estates to engage in cultivation themselves rather than to let farms for a fixed return from their tenants.

high-order goods and services Goods and services with a high *threshold population and a large *range. Examples include furniture, electrical goods, and financial expertise. These goods are usually *shopping goods. Compare with *low-order goods and services.

high technology, high-tech Any industry concerned with advanced technology, such as biotechnology, computers and microprocessors, and fibre optics. While in theory such industries are *footloose, since the products have a high value to weight ratio, favoured locations tend to be near motorways, airports, and *research and development facilities, such as science parks. 'Silicon Fen' near Cambridge is a case in point.

highland clearances The eviction of inhabitants of the Scottish highlands from their land, accompanied by the destruction of their dwellings for the creation of *deer forests. The clearances were intense from 1790 and reached a peak in 1800.

hill farming The extensive farming of an upland area, usually rearing sheep, although some cattle may be kept more intensively. Numbers of cattle are restricted by a lack of winter fodder, and the sheep, grazing at about two hectares per sheep, must be brought to the lowlands for fattening. British hill farming has been supported by government subsidies since the 1940s, but now receives subsidies from the EU. Traditional hill farming has given way in places to improved, sown pasture and reclaimed moorland so that sheep can be stocked at one per 0.25 hectares.

hill fog Not strictly *fog, but low cloud.

hill fort A fortified site on a hilltop, usually with a ditch and ramparts. The earliest date from the Iron Age but some British examples were created as late as the Dark Ages. Favoured sites for such structures in southern England seem to be chalk uplands.

hill wave See *lee wave.

hillslopes *Escarpments and valley sides. Slope studies are generally concerned with hillslopes and are not concerned with *flood plains, *river terraces, or submarine slopes.

hinge line A line either side of which *isostatic readjustment proceeds unevenly.

hinterland The hinterland is the area serving and being served by a settlement. The term was originally applied to ports, and one port may share part of its hinterland with another, but has now been extended to refer to the *sphere of influence of a settlement. Christaller's *central place theory was based on nested, hexagonal hinterlands.

histogram A graph which uses bars (rectangles) to show the frequency of certain classes of values within a dataset. Classes can be descriptive, as in a histogram showing numbers of voters for different parties, or numerical, so that the numbers, or percentages of a population in different age groups (0–4, 5–9, 10–14, and so on) are illustrated by a rectangle (bar). The widths of the rectangles should be proportional to the class intervals just as the heights are proportional to the frequencies of occurrence (numbers, or percentages) within each class.

historical geography The study of past human geographies; of past landscapes. This is usually achieved by teasing out *cross-sections or by making a series of successive sections through time. Historical geographers emphasize the historical perspective in geography, and, increasingly, stress the geographical perspective in history, and both are involved in the increasingly popular field of *regional science.

historical materialism The analysis of history, most closely associated with Marx, which stresses the material basis of society; pointing out that economic systems underlie the development of history and ideas: 'It is not life which determines consciousness, but consciousness which determines life'. A very simplified, and, as I believe, erroneous, example is the link between feudal systems and subsistence production. The idea is important to geographers who try to explain spatial patterns of activity and environmental change in terms of the social relations of prevailing economic systems, notably *capitalism.

histosol *See* *US soil classification.

Hjulström diagram A diagram showing the relationship in a channel between particle size and the mean fluid velocity required for *entrainment. It shows that an entrained particle can be transported in suspension at a lower velocity than that required to lift the particle initially. When the stream velocity slows to a critical speed, the particle is deposited.

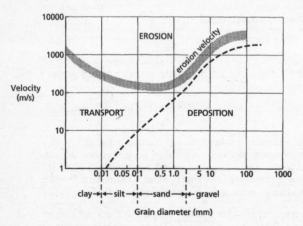

FIGURE 28: *Hjulström diagram*

Note that higher velocities are needed for the entrainment of clay-sized particles because of the electrostatic forces which bind them together.

hoar frost *See* *frost.

hogback A nearly symmetrical ridge with *dip and *scarp slopes of the same value. Hogbacks form where the dip of the beds has been tilted such that the dip is almost vertical. Gaishörndl, Austria, is a European example.

holism The view that the whole is more than its parts. In earlier geographies, the region has been seen as having a distinct identity which does not come entirely from its separate parts. A **holist** looks at the workings of concepts like 'culture' or 'society' rather than the workings of individuals.

hollow frontier A situation in which the agricultural frontier moves forward leaving behind it a tract of worked-over farmland with a shrinking population. Such hollows may then be used for different forms of cultivation which may support higher population densities. The term has been used in connection with agriculture in New England.

Holocene The most recent geological *epoch, stretching from 12 000 years ago to the present day. This epoch has seen the development of early man.

homeostasis In ecology, the process whereby constancy is achieved in an organism or community. **Homeostatic theory** is the contention that a population level remains constant in a pre-industrial society. When there is an imbalance between population growth and resources, there is a corrective response. *Malthus was one exponent of this theory.

homocline One of a regular series of hills from a large area of rock *strata of uniform thickness and *dip.

homogeneous nucleation The spontaneous freezing of water droplets at around −40 °C as clusters of water molecules within a droplet settle by chance into the lattice formation of ice, causing the entire droplet to freeze. *See also* *heterogeneous nucleation.

homoiotherm An animal which maintains an almost constant body temperature; a warm-blooded animal. Homoiothermy is the process whereby such a constant is maintained, despite variations in the ambient temperature.

honeycomb weathering The breaking away of weathered material from sandstone cliffs, while the cement is preserved, leaving a lace-like net of holes. *See* *tafoni.

honey-pot A location, such as Shakespeare's birthplace, which is particularly appealing to tourists. Planners often develop a number of honey-pot sites in National Parks; at these points there is large-scale provision of car parks, shops, restaurants and cafes, picnic sites, and toilets, so that other parts of the Parks will remain unspoilt. This works surprisingly well—few visitors walk more than a quarter of a mile from Land's End, for example.

horizon, soil horizon A distinctive layer within a soil which differs chemically or physically from the layers below or above. The **A horizon** or topsoil contains *humus. Often soil minerals are washed downwards from this layer. This material then tends to accumulate in the **B horizon** or subsoil. The **C horizon** is the unconsolidated rock below the soil. These three basic horizons may be further subdivided. Thus, **Ah horizons** are found under uncultivated land, **Ahp horizons** are under cultivated land, and **Apg horizons** are on *gleyed land. The B horizons are also subdivided by means of suffixes: **Bf horizons** have a thin iron pan, **Bg horizons** are gleyed, **Bh horizons** have humic accumulations, **Box horizons** have a residual accumulation of *sesquioxides and **Bs horizons** are areas of sesquioxide accumulation. **Bt horizons** contain clay minerals and **Bw horizons** do not qualify as any of the above. **Bx horizons**, or fragipans contain a dense but brittle layer caused by compaction. C horizons are also subdivided: **Cu horizons** show little evidence of gleying, salt accumulation, or fragipan; **Cr horizons** are too dense for root penetration; and **Cg horizons** are gleyed. Additional suffixes may be used. Some soil scientists use the term **D horizon** for the consolidated parent rock.

In addition to these soil horizons, other layers are distinguished. Thus, the layer of plant material on the soil surface is classified as: the **L horizon** (fresh litter); the **F horizon** (decomposing litter); the **H horizon** (well decomposed litter); and the **O horizon** (peaty). A *leached A horizon is termed an **E horizon** or *eluviated horizon.

horn A mountain peak formed when three or four *cirques have cut into it, back to back, leaving a *pyramidal peak. Examples include the Matterhorn, and Cir Mhòr, on the Isle of Arran.

horse latitude Those latitudes stretching from 30 to 35° North and South of the equator where winds are light and weather is stable and dry. The origin of this term is uncertain.

horst A block of high ground which stands out because it is flanked by normal faults on each side. It may be that the block has been elevated or that the land on either side of the horst has sunk.

horticulture Originally garden cultivation, this now refers to the intensive production of fruit, vegetables, and ornamental plants.

Hortonian overland flow An overland flow of water occurring more or less simultaneously over a *drainage basin when rainfall exceeds the *infiltration capacity of the basin. R. E. Horton (*Trs. Amer. Geo. Union*, 1933) maintained that such overland flow was a major contribution to the rapid rise of river flow levels, and was the prime cause of soil erosion. Hortonian flow is distinct from *return flow since it involves no movement of underground water back to the surface. Recent research indicates that the Hortonian model is not widely applicable.

hot desert Located on the west coasts of tropical and subtropical climes, these have average temperatures of over 20 °C and rainfall of less than 250 mm. Deserts are too dry for most plant species except for xerophytes. Xerophytic strategies for survival include the development of succulents to store water, the growth of ephemeral plants after rains, and the development of spines to ward off animal attack. Desert insects, reptiles, mammals, and birds are all adapted to drought. This is an extremely fragile *biome.

hot spot Also known as a *plume, this is an area of localized swelling and cracking of the earth's crust due to an upward welling of *magma. Volcanoes form above hot spots: the Hawaiian islands are cited as an example. One theory suggests that, as plates move across the hot spot, a line of volcanoes is formed. The cause of hot spots is not known; indeed, some writers deny their existence.

hot towers Immensely tall, tropical *cumulo-nimbus, at the rising branch of the *Hadley circulation, which siphon both sensible and latent heat up to the equatorial *tropopause.

Hotelling model A model, proposed by H. Hotelling (*Econ. J.*, 1929) of the effect of competition on locational decisions. The model is usually based on two ice-cream salesmen, A and B, on a mile of beach. The cost and choice of ice-cream is the same for each distributor. Buyers are evenly distributed along the beach. The first pattern of market share has the two salesmen positioned so that each is at the centre of his half of the beach and the market is split up evenly. If A now moves nearer to the middle of the beach, he will increase his market share. The logical outcome of this will have both salesmen back to back at the centre of the beach, as long as some customers are willing to walk nearly half a mile for an ice-cream, i.e. that the consumer provides the transport. This analogy indicates that locational decisions are not made independently but are influenced by the actions of others.

housing class A classification of urban social groups in terms of their access to suburban housing. J. Rex and R. Moore (1967), the originators of the concept, saw ethnicity as the key determinant in the struggle for this scarce resource, since immigrants have difficulties in securing loans.
    The concept has been criticized; one objection is that access to housing depends on status in the labour market, another that not everyone aspires to life in the suburbs.

human ecology Numerous, and rather differing definitions of this term are current. Some human ecologists stress the ecological disorder created by human societies; others use it to describe the approach to *urban social geography developed by the 'Chicago School' of the 1920s which applies *ecological concepts to human behaviour.
    In this second definition, the city is seen as a social organism, where human communities emerge through 'natural' processes such as impersonal competition, *segregation, dominance, *invasion and succession. Impersonal competition is a central concept, as individuals compete for favourable locations throughout the city; through the market mechanism a pattern of land rents emerges which brings about the segregation of different types of people according to their ability to meet these rents; this in turn leads to the development of *natural areas, or *communities, within the city. The dominance of a group within a natural

1.

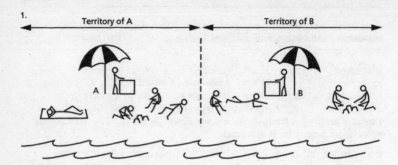

First pattern of market share

2.

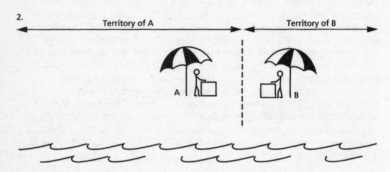

Second pattern of market share

3.

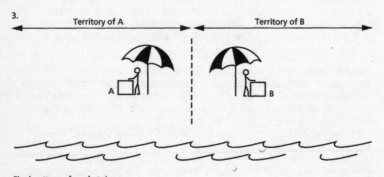

Final pattern of market share

FIGURE 29: *Hotelling model*

area is thus related to its relative competitive power. (This thinking is expressed in the *Concentric model.) Other concerns of human ecology are descriptions and delineations of 'natural areas' and the investigation of 'ecologies' associated with deviant behaviour. Although the effects of this school of thought have been far-reaching, it has also been criticized for over-emphasizing competition and neglecting the importance of cultural and motivational factors in explaining residential behaviour.

human ecological triangle The relationships between person, society, and environment.

human geography A generalized term for those areas of geography not dealing exclusively with the physical landscape or with technical matters such as *remote sensing. It is concerned with the relationships between man's activities and the physical environment, with *spatial analysis, and with those processes which lead to *areal differentiation. The term covers a number of fields; *see also* *cultural geography, *behavioural geography, *economic geography, *agricultural geography, *industrial geography, *political geography, *regional geography, *social geography, *urban geography.

humanistic Concerned with human interests and with the human race as opposed to the purely physical world. It is an approach which stresses distinctly human traits such as meaning, feeling, and emotion.

humanistic geography A view of human geography centred on human perception, capability, creativity, experience, and values. It maintains that any investigation will be subjective inasmuch as it reflects the attitudes and perceptions of the researcher who may also be an influence on the very field of his study. Two main strands may be distinguished. The first focuses on human experience and human expression and is concerned with the unique and the particular. The second takes constructions, like *existentialism, from the social sciences and explores the relationship between these and the time and space settings of ordinary life.

humic acid A complex acid formed when water passes slowly through humus. Humic acid is an example of an organic acid in that it is formed from carbon-based compounds. It is significant in chemical weathering and in the formation of soil. *See* *mor, *moder, *mull.

humidity The amount of water vapour in the atmosphere. It is more exactly defined as the mass of water vapour per unit volume of air, usually expressed in $kg\,m^{-3}$. This is absolute humidity. Relative humidity is the moisture content of air expressed as the percentage of the maximum possible moisture content of that air at the same temperature and pressure.

humidity mixing ratio The ratio of the mass of water vapour in a sample of air to the mass of dry air associated with that water vapour.

humilis *See* *cloud classification.

humus Material of vegetable or animal origin found in the soil. More exactly, humus is fully decomposed and finely divided organic matter. This decomposition is humification; the process whereby the simple mineral compounds released by weathering combine with the organic residues to form large, stable organic molecules which act as bonding agents in the structure of the soil. Humus is also important in its great ability to absorb *cations.

hundred An Anglo-Saxon term for a portion of a *shire or county, perhaps indicating an area of 100 *hides.

hunting and gathering An early form of society with no settled agriculture, or domestication of animals, and which has little impact on the environment. The hunting of animals and the collection of edible plants depends on the environment rather than changing it.

hurricane Also known as a typhoon, or tropical storm with winds over 140 km per hour, this is a disturbance about 650 km across spinning about a central area of very low pressure. The violent winds are accompanied by towering clouds, some 4000 m high, and by torrential rain; 150 mm (6 inches) frequently fall within the space of a few hours. There is, as yet, no complete understanding of how these storms develop; they can begin when air spreads out at high level above a newly formed disturbance at low levels. The upper level outflow acts rather like a suction pump, drawing away the rising air at height and causing low-level air to be pulled in. The winds spiral in to the centre because they are affected by the earth's rotation. The intense energy of these storms comes from the warmth of the tropical seas over which they develop. Thus, an extensive ocean area with surface temperatures of over 27 °C is necessary for hurricane formation.

The source regions must be far enough away from the equator—5° at least—for the *Coriolis force to have an effect. The removal of air at height may be along the eastern limb of an upper air trough. Moisture-laden air spirals into the centre and rises, condensing to form a ring-like tower of *cumulo-nimbus clouds. With this condensation, *latent heat is released which causes the air to rise further and faster. The condensation also causes torrential rain. In the upper *troposphere water droplets freeze and form *cirrus clouds which are thrown outwards by the spin of the storm.

At ground level, the temperature at the centre, or eye, of the storm is only slightly warmer than that at the margins, but, at heights of around 5000 m, the centre can be 18 °C warmer than the margins. This warm core maintains the low pressure which drags in the winds.

Hurricane modification research began with the first experimental cloud seeding of a typhoon in 1947, and in 1962 project STORMFURY was set up to investigate the effect of introducing freezing nuclei into the ring of clouds around the eye, in an effort to lower the horizontal temperature gradients within the storm, thereby reducing pressure gradients and hence wind speeds. In spite of some success with hurricane 'Debbie' in August 1969, project STORMFURY was cancelled in 1983.

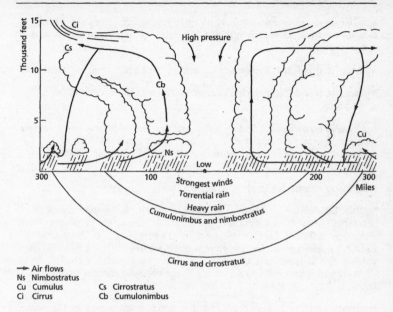

FIGURE 30: *The structure of a tropical hurricane*

husbandry In geography, the farming of animals.

hydration The incorporation of water by minerals. Hydration often causes swelling and is believed to be a major cause of the crumbling of coarse-grained igneous rocks which are disrupted by the expansion of their hydrated minerals. Compare with *hydrolysis.

hydraulic action In geomorphology, the force of the water within a stream or river. Hydraulic action is one component of fluvial erosion.

hydraulic conductivity The ability of a soil or rock to conduct water. The conductivity of dry soil or rock is low (dry hydraulic conductivity); little water is conducted since water entering a soil must form a film of water surrounding the soil particles. Until these films are formed, little conduction occurs. **Saturated hydraulic conductivity** refers to the maximum rate of water movement in a soil.

hydraulic force The force of water, including *cavitation and fluvial *plucking.

hydraulic geometry The study of the interrelationships exhibited along the course of a river. *Discharge is linked with the mean width of the channel, the mean depth and slope of the channel, the suspended *load, and the mean water velocity. Further links are thought to exist within meanders where the wavelength of the meander is related to the *radius of curvature.

hydraulic gradient The rate of change in *hydraulic head with distance.

hydraulic head The pressure exerted by the weight of water above a given point.

hydraulic hypothesis The view that the practice of large-scale irrigation stimulated urban development as the need for organized labour and supervisory authorities arose. Equally well, it might be that urban settlement stimulated irrigation.

hydraulic mean *See* *hydraulic radius.

hydraulic radius Also known as **hydraulic mean**, this is the ratio of the cross-sectional area of a stream to the length of the wetted perimeter. The wetted perimeter is the cross-sectional length of a river bed. The hydraulic radius is a measure of the efficiency of the river in conveying water. If the value of the hydraulic radius is large, a large area of water in the cross-section is affected by each metre of the bed, and there is thus little friction.

hydroelectricity Energy produced as generators are turned by the power of running water. The necessary conditions are a constant supply of water from rivers and lakes, steep slopes to aid the fall of water, and stable geological conditions for the construction of dams. By 1976, 27% of Western Europe's electricity supply was from hydroelectricity, with Norway generating 100% of its needs, and West Germany only 6%.

However, recent research indicates that the construction of dams may

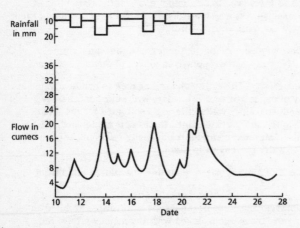

FIGURE 31: *Hydrograph*

trigger off earth movements. The energy generated is a function of the height of falling water as well as of the mass of water concerned. A high proportion of the energy is converted into electricity.

hydro-fracturing A form of weathering whereby water enters minute fractures in a rock. If the water freezes and expands at the open end of the fracture, the rest of the water may be pushed downward. The pressure thus exerted may then deepen the crack.

hydrograph A graph of *discharge, or of the level of water in a river throughout a period of time. The latter, known as a **stage hydrograph** can be converted into a discharge hydrograph by the use of a stage-discharge rating curve. Hydrographs can be plotted for hours, days, or even months. A storm hydrograph is plotted after a rainstorm to record the effect on the river of the storm event.

hydrological cycle Also known as the water cycle, this is the movement of water and its transformation between the gaseous (vapour), liquid, and solid forms. The major processes are *condensation by which *precipitation is formed, movement and storage of water overland or underground, *evaporation, and the horizontal transport of moisture. The length of time any water stays in the *atmosphere is about 11 days.

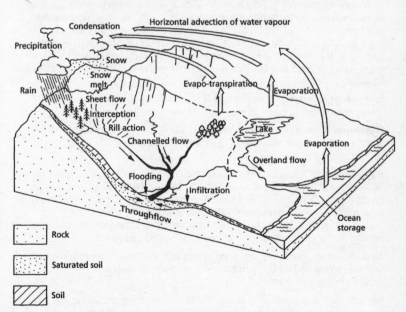

FIGURE 32: *Hydrological cycle*

hydrology The study of the earth's water, particularly of water on and under the ground before it reaches the ocean or before it evaporates into the air.

This science has many important applications such as flood control, irrigation, domestic and industrial water supply, and the generation of hydroelectric power.

hydrolysis The chemical reaction of a compound with water. Hydrolysis is an important component of soil formation, and of *chemical weathering—for example, as feldspars in granite decompose to make china clay.

hydromorphic Denoting areas with waterlogged soils. *Gley soils form in such conditions.

hydrosere A successional sequence of plants originating in water.

hydrosphere All the water on, or close to, the surface of the earth. Some 97% of this water is in the earth's seas and oceans; of the rest, about 75% is in *ice-caps and -sheets, about 25% in surface *drainage and *groundwater, and about 0.03% in the atmosphere.

hygrometer A meteorological instrument used to measure *relative humidity. A hair hygrometer uses a strand of hair, which responds uniformly to changes in relative humidity. A wet-bulb hygrometer has two thermometers; one covered with saturated lint, and one dry. Evaporation from the wetted wick cools the bulb below the temperature of the dry bulb. As evaporation rates depend, among other factors, on relative humidity, the disparity between the wet- and dry-bulb temperatures can be used, with the help of a book of hygrometric tables to determine relative humidity.

hygroscopic nuclei *Condensation nuclei which are hygroscopic, i.e. which tend to attract and condense ambient water vapour.

hypabyssal rock An igneous intrusion which has consolidated near the earth's surface above the base of the crust. Examples include dolerite and quartz porphyry.

hypermarket A huge complex with generous free parking offering a very wide range of goods. Out of town hypermarkets divert traffic from the city centre, simplify shopping, and can lower prices as costs of land are lower than at the city centre. However, they are designed for the more affluent and are difficult of access for those who don't own cars. They represent a serious threat to city-centre stores and neighbourhood shops.

hypolimnion The lower layers in a body of water which are marked by low temperatures and insufficient light for photosynthesis. Levels of dissolved oxygen are low.

hypothesis A general supposition made as a basis for reasoning but not held to be true until proven by reference to empirical evidence.

hythergraph A plot of monthly rainfall against monthly temperature over a year. *See* *climograph.

ice The rate at which snow is converted to ice depends on the temperature. Wet snow, falling with temperatures very near to freezing point is converted to an icy mixture; pressure between grains induces thawing, followed by re-freezing as the water penetrates voids between grains—dry, powdery, colder snow turns to ice much less slowly because pressure is less. *See* *firn, *névé.

ice age A length of time during which ice sheets are found on the continents. Thus, an ice age is occurring at the present day, as a part of the *Pleistocene glaciation, which began about 2 million years ago. Within an ice age there may be *interglacial periods of milder climate. Ice ages last for some tens of millions of years with intervals of about 150 million years between them. The term is used more loosely to identify the last time that ice sheets covered much of Europe and North America.

ice-cap A flattened, dome-shaped mass of ice, similar to an *ice sheet, but under 50 000 km² in area, such as the Barnes Ice Cap of Baffin Island, Canada. An ice cap does not necessarily obliterate relief. An **ice-cap climate** is a climatic regime where the average yearly temperature is below 0 °C. Ice and snow are permanent and precipitation is very light.

ice contact feature Any landform developed in contact with a glacier. An **ice contact terrace** is a synonym for a *kame terrace.

ice fall An area of *extending flow where the gradient of a glacier steepens, perhaps at a corrie lip, or over a rock step. Here the ice, marked by crevasses, begins to split up. The speed of flow is very rapid, so that the ice thins at this point.

ice floe A flat section of ice which is floating in water and not attached to ground ice. *See also* *iceberg.

ice front The floating, vertical ice cliff at the seaward end of an *ice shelf, or of a glacier extending over the sea.

ice lens An area of ice, often having convex upper and lower surfaces, which exists underground in *periglacial environments.

ice mound A *periglacial landform; a swelling in the ground due to the expansion of a lens of ice below the surface. *Solifluction displaces the material at the top of the mound, so that when the ice lens melts a depression forms, often water-filled, and surrounded by a rampart. Large ice mounds are called *pingos, and may be 50 m in height.

ice polygon, ice wedge polygon A 3 to 6 sided polygon of *ice wedges with straight to gently curving sides, formed by *ice segregation and the drying and shrinking of sediments. The initially random shape of the polygon becomes more regular with age. In north Baffin Island polygons up to 50 m in diameter are common.

ice segregation As a verb, the formation of discrete bodies of ground ice in *periglacial conditions. Segregation depends on the rate of freezing—it will not occur if freezing is very rapid—and the lithology of the material, developing well in silts, which appear to have the optimum size of pore space, but less well in clays and sands. Segregations tend to develop under large stones, because of their high thermal conductivity. The term is used as a noun to denote a discrete body of ice so formed.

ice sheet An area of ice spreading over more than 50 000 km². The snow line is low, and the ice creeps towards the edges with a slow, massive movement. The ice sheets in Greenland and Antarctica are the only two currently in existence, but during the last *ice age, ice sheets covered large areas of North America and north-west Europe. Ice sheets generally have the effect of grinding down relief, but *see* *nunataks.

ice shelf A sheet of ice extending over the sea from its land base. It is fed by snow falling on it or from glaciers on the land surface.

ice stream Within a glacier or ice sheet, a stream of ice moving more quickly than, and not necessarily in the same direction as, most of the ice.

ice wedge In a *periglacial environment, a near-vertical sheet of ice tapering downwards, and up to 12 m deep. When soils are cooled below −15 °C, the ice contracts, causing the ground to split into vertical, polygonal cracks. When the *active layer melts in spring, these cracks fill with water. As this water refreezes and expands, the cracks widen. This process is repeated many times. Wedges grow less than 10 mm a year. These wedges create a system of polygons with raised margins. *See* *ice wedge polygon. Fossil ice wedges, infilled with sediment are known as **ice wedge pseudomorphs**, or ice wedge casts.

iceberg A huge mass of ice, floating in the sea and usually broken off from a glacier. The depth of an iceberg is often far greater than that of an *ice floe.

ice-pushed ridges Ridges of ground, 0.6–4.5 m high, common in *periglacial environments, and depending on permafrost for their formation.

iconography In geography, the study of the way in which images of the landscape reveal symbolic meaning. A human landscape is not only shaped by a society and its culture, it also helps to shape that society; think for example, of the image of the thatched English village, or the whitewashed crofter's cottage in Scotland, both of which are often used in British party political broadcasts to promote patriotic feeling. The meanings of landscapes such as these are not fixed—to eighteenth-century Britons, the Lake District appeared as a bleak and desolate area, to be avoided—and they may also be highly political—the sight of pithead buildings can be highly charged politically.

idealism The view that human activity may only be explained in terms of the thought processes that bring them about; the social world consists of ideas originating from some root, and society, for example, only exists insofar as people think it does. Reality is based on, or evolved by, the mind; this is **metaphysical idealism** which claims that no material things exist independently of the mind. **Epistemological idealism** maintains that human understanding is limited to perception of external objects. The concept is used in geography in any study of how the *cultural landscape depends on the way in which people perceive their environment. The logic is that if human societies are structured by thought, then the only way to understand them is to investigate the way in which people think.

ideology A system of ideas and beliefs, especially of political ideas and cultures. From this term is derived the concept of *idealism. It is a set of beliefs and values often forming the basis of an economic or political theory or system. Thus, the ideology of capitalism, based as it is on *accumulation and on *market forces, is very different from a communist ideology.

idiographic Concerned with establishing the uniqueness of a phenomenon: an individual, a place, or a region, for example. The idiographic approach has been the underlying basis of *regional geography which is concerned with establishing and explaining the differences between places. *See* *chorography. This contrasts with the *nomothetic approach, which tries to find similarities between phenomena and to formulate 'laws' about social behaviour.

IGES International Graphics Exchange System.

igneous rock A rock which originated as molten *magma from beneath the earth's surface and subsequently came to the surface as an *extrusion, or remained below ground as an *intrusion. The nature of the rock depends in part on the rate at which it cooled; as intrusions of magma slowly solidify, enough time elapses for large crystals to form whereas extrusions cool quickly, leaving little time for crystal growth. Thus, a coarse-grained, intrusive igneous rock has a fine-grained, extrusive counterpart; granite is coarse rhyolite and gabbro is coarse basalt. Igneous rocks are also classified as *acid or *basic, according to whether their silica content is high (e.g. granite), or low (e.g. basalt).

illuviation The downward *translocation from the A-*horizon and subsequent precipitation in the B-horizon of clay-sized particles in a soil. Hence **illuvial horizon**; the B-*horizon, in which there is redeposition or entrapment of matter brought down from above.

image A picture built up by an individual from information arising from the social and physical *milieus experienced from birth. The image of a city, for example, is made up of meeting places, *paths, landmarks, limits, and areas, and this image fosters a sense of belonging. This image is also a way of organizing knowledge and is a source of ready reference for movement around the city. It is argued that similar individuals in similar milieus are likely to have similar mental images and hence exhibit similar forms of behaviour. *See* *mental map.

imageability is the extent to which an object or set of objects makes a strong impression on individuals.

image data In *Geographic Information Systems, data in the form of points, lines, *polygons, or a mixture of these.

imbricated Of deposits, laid down in overlapping sheets, as in the orientation of tabular blocks lying parallel to the slope in *periglacial environments.

IMF *See* *International Monetary Fund.

immigration The movement of a person as a permanent resident into another area, usually into a foreign country. Official figures for 1985 yield the following totals for immigration into selected European countries:

WORKERS (THOUSANDS):

West Germany	33.4
France	5.5
Belgium	1.9
Switzerland	33.4
Austria	60.3

GROSS IMMIGRATION (THOUSANDS)

Netherlands	46.3
Sweden	27.9

The term is also used in ecology for the movements of plant seeds and animals.

imperfect competition In economics, a state of affairs in which the necessary conditions for *perfect competition are not met. In such a situation, a major component of demand is the influence of advertising and product branding. In the real world, competition is generally imperfect, perhaps because competition is limited by the operation of restrictive practices and/or because price competition is limited.

imperialism The control of one or a number of countries by a dominant nation. This control may be political, economic, or both, and indicates a degree of *dependence in the subordinate nation. Many writers take the word as a synonym for colonialism, but imperialism can exist without the creation of formal colonies, which usually require military force and the institution of a colonial administration.

Imperialism is promoted by monopolizing the external trade of the subordinate nation. The imperial power takes raw materials from the colony and sells it finished goods in return, discouraging the development of any manufacturing industry which might compete with its own. There now exist few relics of political empires but **economic imperialism** is alive and well. *See* *neo-colonialism.

impermeable Not allowing the passage of a fluid; in *hydrology, the

fluid is water. Impermeability in a rock may be due to an absence of
*pores, *joints, or *bedding planes, or because pores are so small that the
water within them is 'locked in' by surface tension.

impervious *See* *impermeable.

import substitution A strategy for economic development which
encourages industrial growth within a nation in order to reduce imports
of manufactures, save foreign exchange, provide jobs, and reduce
dependency. The United Nations Commission for Latin America promoted
import-substitution policies in the 1960s, but they were not successful, and
such policies have been replaced by strategies grounded on export-led
industrialization.

imports Goods which originate from a foreign country and are bought by
a nation in trade. The 1990 breakdown of imports into the EC was:

EC 12	USA	Japan	Rest of World
58.2%	7.8%	4.3%	29.7%

Invisible imports are services, like insurance, bought from outside a
country. **Import penetration** indicates the extent to which the country is
dependent upon its imports; in 1982, for example, 49% of electrical goods
sold in UK were imported.

inceptisol *See* *US soil classification.

incidence matrix A square or rectangular table used to indicate
relationships between two sets. The rows and columns display the
elements of these two sets. The intersections of each row or column are the
cells of the matrix. Where a relationship exists between the two sets, the
cell is marked with a 1; where no relationship exists, it is marked with a 0.

incised meander A *meander formed when a *rejuvenated river cuts
deeper into the original meander. An **intrenched meander** is an incised
meander with a symmetrical *cross-valley profile; an **ingrown meander** has
an asymmetrical cross-section.

independent variable In any study of cause and effect, the independent
variable is the *causal factor which shapes or determines the *dependent
variable. For example, an investigation of sediment transport in rivers
would identify stream velocity as the independent variable which
determines the size of particle that can be transported. In *human
geography, it is not always as easy to decide which is the independent
variable; to some extent, it depends on the theoretical basis of the study,
but, as a rule of thumb, the independent variable should predate the
dependent variable, and should have a causal effect. In any graph of
related variables, the independent variable is plotted on the x-axis.

index numbers Figures which show the relative change in one or more
variables over time. The value of the variable for one particular year is
chosen to be the base value, expressed as 100. The figures for the other
years are then expressed as a percentage of the figure for the base year. For

the economic geographer, for example, index numbers provide an easy way of comparing figures such as food production per capita, relating all production to one base year.

index of centrality A measure of the importance of a settlement in terms of the goods and services it provides for the surrounding region. W. Christaller (1933) devised an index of centrality based on the number of telephones inside and outside the central place. *See* *centrality. R. E. Preston's (*Prog. Hum. Geog.*, 1985) index measures the importance of a town in the following terms:

$$\text{Centrality} \left(C \right) = R + S - \alpha M_t F_t$$

where R is total retail sales, S is total sales in selected service establishments, α is the mean percentage of median family income spent on goods and selected services, M_t is median family income in town t, and F_t is total number of families in town t.

index of concentration *See* *coefficient of localization.

index of circulation *See* *Rossby waves.

index of decentralization An index of the degree to which an activity, such as manufacturing industry, is centrally located within an area: region, conurbation, city, or town. An index of 0 represents maximum concentration at the centre while a value of 100 indicates maximum location at the periphery.

index of dispersion An index of the degree to which the values in a data set are grouped around a central point: the mean centre. This index is the mean of the distances of all the points to the centre, and indicates the extent to which values, such as monthly rainfall figures, vary from the mean value. The lower the index, the less the dispersal. The indices of two dispersions can be compared with each other in an **index of relative dispersion**.

index of dissimilarity This is often used in the study of residential differentiation in urban areas. For each district, the percentage of those working in each occupational group is calculated. The index of dissimilarity between two occupational groups is half the sum of the absolute differences between the respective distributions taken district by district. The values range from 0 (complete similarity) to 100 (complete segregation).

index of level of living An assessment of living standards using indicators such as access to health care, standard of education, house ownership, car ownership, take-home pay, employment rates, access to *amenity, and so forth. Fifty-three indicators may be used, and are analysed so that areas of high living standards have a low composite index and vice versa. The spatial analysis of these data may be set out for a region or for a nation. In UK, standards of living are best in South-East England and worst in the North and West.

index of primacy An index of the importance of the largest town in a country:

$$\text{Index of primacy} = \frac{P_1}{P_2}$$

where P_1 is the population of the largest town and P_2 is the population of the second largest town. The higher the index, the higher the degree of *primacy.

index of segregation A measurement of the degree of residential segregation between two sub-groups inside a larger population. For example, we might want to see if Jamaicans are more residentially segregated than Barbadians within a British city. One simple method uses the *Lorenz curve, with the cumulative percentage of each ethnic group from each sub-area of the city on one axis, and the cumulative percentage of the remaining groups for each sub-area on the other. If the line is diagonal, there is no segregation, and the percentages within each sub-area are the same as the percentages over the city as a whole. Alternatively, a *location quotient may be used, which compares the percentage of an ethnic group living within the sub-area with the percentage living within the city as a whole.

index of variability The *quartile deviation of a data set expressed as a percentage of the *arithmetic mean. It is useful for comparing two apparently similar data sets, for example, of two rivers with the same mean annual flow.

index of vitality An index to indicate the growth potential of a population:

$$I_v = \frac{\text{fertility rate} \times \text{\% aged 20-40}}{\text{crude death rate} \times \text{old-age index}}$$

where I_v = index of vitality. *See* *old-age index.

indifference curves In economics, a graph of the various levels of *utility achieved at different prices through buying two commodities, for example, magazines and paperback books. It is possible to imagine, at a given price level, various combinations of the two which would yield the same amount of utility; for example, someone might get the same utility from five paperbacks and two magazines as from four paperbacks and four magazines, or three paperbacks and six magazines. (The graph would show magazines on one axis and paperbacks on the other, and would have a negative slope, moving downwards to the right.) Each combination at a given price level gives the same utility, hence the term 'indifference'. The consumer then selects one of these combinations, within the limitations of his or her income.

The concept is important in the *Alonso model, where the combination is of money spent on housing and money spent on commuting.

individual data Items which are listed separately for depiction or analysis, as opposed to *grouped data.

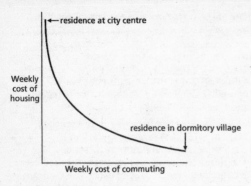

FIGURE 33: *Indifference curve*

indivisibility The difficulty of using only part of a plant for profitable production. For example, if a plant has a set of machines with different capacities, they will only be used economically if they are used to the full. This is an important concept in understanding not only *economies of scale, but also the diseconomies imposed by serving a market which does not require industrial plant to be used to the full capacity of the plant.

induction Using the observation of particular initial cases in order to infer a general law from them. The researcher devises a general law to fit the observations—such as 'what goes up must come down'—and then searches for examples which disprove that law; if any are found, the law is reformulated until it fits these exceptions. When no new discrepancies can be found (although this is a rather subjective decision), the generalization is accepted. Compare with *deduction.

induration In geology, the hardening of a rock, usually sedimentary, by drying, pressure, or *cementation.

industrial complex A large concentration of manufacturing industry in a relatively small area. Such a complex gains from *agglomeration economies and is usually well served by transport and financial provisions. **Industrial complex analysis** is a technique, developed by Isard, of studying the linkages between the industries in an industrial complex. It is a technique combining elements of *input–output analysis and *comparative cost analysis.

industrial diversification The spreading of employment and investment over a wide range of industrial activities. Diversification is not always easy to recognize; a firm may be described as a food manufacturer (one product), or as a manufacturer of many types of food (several products).

industrial estate A district of purpose-built workshops with supporting services, often located in suburbs or at the edge of a town or city.

industrial geography The study of the spatial arrangement of manufacturing industry. Manufacturing industry is specifically chosen for this definition since, it is argued, it is the basis upon which regional economies are built. Explanations of industrial patterns may be based on *location theory and on models with costs as the predominant locational factor. Other approaches are concerned with the nature of decision-making, and an understanding of change; where and why some regions grow while others, like inner cities and *depressed areas, decline. Strategies may then be suggested to aid ailing industrial areas and underdeveloped countries.

industrial inertia The survival of an industry in an area even though the factors which led to its location there no longer apply. It is often advisable to update and expand a factory rather than to relocate because of existing *agglomeration economies and *external economies and because of the difficulty of moving a skilled labour force.

industrial location policy There are different views of the role the state should play in economic activities. One is that the state should set a minimum number of rules to ensure that the market economy functions successfully. Another sees government intervening, especially for those areas that are in industrial decline. The latter type of government may attempt to attract industry by providing any or all of the following: land, buildings, financial incentives, and advice. Industrial development officers may be appointed to attract industry. Such policies may be carried out by local authorities as well as by central government. Industrial location policy may be adopted by less developed countries in an effort to industrialize. *See* *assisted area, *development area, *enterprise zone, *urban development corporation.

industrial location theory Theories of the forces leading to the location of industrial activity. One choice might be the least-cost location. Another is the *locational interdependence approach which stresses the influence of other enterprises especially under conditions of imperfect competition. A third is the profit maximization approach, although it is by no means certain that firms do maximize profits. *See* *least-cost location, *Lösch, *Weber.

industrial organization The make-up of an industrial unit, especially as it concerns decision-making. This is a major component of *industrial location theory. Organizations range from the owner-operated small firm to the multinational corporations. In the latter, the low-level functions may take place in the Third World but decisions are taken in countries with advanced economies.

industrial overspill The movement of industry from *conurbations to new locations outside the built-up area because of constricted sites.

industrial retention policy A policy which is developed to help maintain industrial activity by taxation incentives, subsidies, and government contracts.

industrial revolution Although there is some discussion about its timing, the industrial revolution is generally accepted as occurring in Britain in the late eighteenth and early nineteenth centuries. The revolution was in technology—new techniques involving new machinery and new processes—but was accompanied by social and political changes. These changes, beginning in Britain, took place over a long period of time but their effects transformed society.

industrial specialization The domination in a region of a limited range of industries. Such a region will have a high *location quotient in a few industries. Specialization may be advantageous as *functional linkages are facilitated and external and *agglomeration economies can be made, but an overdependence on a narrow range of industries can be dangerous and industrial diversification may be necessary.

industrialization The process by which manufacturing industries develop from within a predominantly agrarian society. Characteristic features of industrialization include the application of scientific methods to solving problems, mechanization and a factory system, the division of labour, the growth of the money economy, and the increased mobility of the labour force—both geographically and socially. One problem is that these are features of *capitalism, and capitalism is not the same thing as industrialization, although it was the first instrument of industrialization. Industrialization is generally accompanied by social and economic changes, such as a fall in the birth rate and a rise in per capita *GNP. Urbanization is encouraged and groups of manufacturing towns may form. Within the developed world, the growth of the factory system led to the separation of home and workplace with major repercussions for *urban social geography. Initially there is an emphasis on primary and secondary *industry, but as industrialization continues, there is a shift to tertiary industry.

Some writers argue that the term may also be used to describe the methods used to increase productivity in areas other than manufacturing, such as agriculture or administration. Although industrialization is often seen as a solution to problems of poverty in the Third World, its effects may well not benefit any but a small sector of society. Furthermore, the pollution associated with industrial activity may cause serious difficulties.

industry This term is now often used to cover any form of economic activity, such as 'the music industry', and hence covers a very broad sweep. More specifically, industry is divided into: primary industry—the acquisition of naturally occurring resources like coal and fish; secondary industry—the manufacture of goods; tertiary industry which serves the public as well as primary and secondary industry and includes distribution, transport, warehousing, and retailing. Some writers suggest the term quaternary industry to cover administration, finance, research, and the processing and transfer of information.

inequality Disproportionate opportunities or rewards for different individuals or groups within society. Geographers are concerned with the *spatial expression of inequality, whether it relates to gender, social class, or *ethnicity, and study its causes (see e.g. *cumulative causation), its consequences, such as migration from poorer to richer regions (*Mezzogiorno* to northern Italy, *'South' to *'North'), and the remedial measures taken by governments to redress inequality. *See* *territorial justice.

infant mortality The number of deaths in the first year of life per 1000 children born. *See also* *mortality.

inferential statistics To **infer** is to draw a conclusion from only partial evidence, such as a sample, so inferential statistics are techniques which may be used to evaluate the value of such conclusions; they tell us the *probability of a statement based on a sample being valid for the whole *population. This probability is expressed as a percentage, and a 90% probability rating tells us that that deduction would be made from 90 out of every 100 samples. This calculation of validity should be included in any geographical enquiry based on sampling. *See also* *descriptive statistics.

infield–outfield farming A type of farming, now generally superseded, whereby the infield—the land nearest to the farmhouse—was cropped continuously and manured heavily while the outfields were less heavily cropped and left fallow for long periods of time in order to recover their fertility.

infiltration The process of water entering rocks or soil.

infiltration capacity is the rate at which water can infiltrate the soil. The basic mechanism is that the upper soil surface receives precipitation so that existing soil moisture is displaced downwards by newly infiltrated water. Infiltration may be controlled by factors including cracks, cultivation, freezing, the intensity and type of precipitation, and the *porosity of the soil. The last factor is probably the most important. Infiltration may not occur if the speed of the water is too great or if the rock or soil is saturated.

informal sector, informal economy Employment which is not formally recognized; workers in the informal economy generally have no contracts, no fixed hours, and no employment benefits such as sick pay or maternity leave. There are no official figures on the informal economy, and it is untaxed. It is also called the *black economy* or *moonlighting*, although some sociologists would include perfectly legal activities such as subsistence occupations, skills which are bartered, and unpaid domestic work, for example. The informal economy can be a source of cheaper labour as no allowance need be made for tax. In less developed countries much of the work done by women is in the informal sector; this includes such activities as petty trading, small-scale agriculture, and crafts. Compare with *formal sector.

information city A city with a high proportion of its workers in services based on the management of information, as in finance, insurance, and law. Through their access to high-technology media, information cities are major sites for decision-making.

information theory One, mathematical, view of the problem of conveying a message from one point to another. The meaning of the information does not signify; what is important is the capability of the technology to code, transmit, and decode the message. Its success depends on the carrying capacity of the technology. In geography, this approach has been used to describe the distributions of populations.

infrastructure 1. The framework of communication networks, health centres, administration, and power supply necessary for economic development. Geographers and economists do not agree over the extent to which this underlying structure, also known as *social overhead capital*, should be provided before development takes place, and politicians argue over whether the state, the private sector, or both, should provide the infrastructure.
2. In Marxist theory, the structures used in the production of the material things of life. *See* *historical materialism.

ingrown meander A *meander which cuts sideways into the bank so that there is a slight overhang above the stream. *See also* *incised meander.

IKBS *See* *intelligent knowledge-based systems.

inner city An area at or near the city centre with dilapidated housing, derelict land, and declining industry. *See* *filtering down. The inner city is often home to those with low wages, living in multi-occupied housing. *Squatting is common. European examples range from Christiania, Copenhagen, to Toxteth, Liverpool, or, before *urban renewal, Kreuzberg, Berlin or Jordaan, Amsterdam. The existence of these impoverished zones is an example of *uneven development and has been attributed to *counter-urbanization and *decentralization. The British *urban development corporations were established to ease the problems of the inner cities. Also known as *twilight zone, zone of downward transition*.

innovation The introduction of a new feature or the new feature itself. An **innovation wave** is the diffusion of an *innovation from its point of origin. Initially a very few, near to the point of origin, accept the innovation so that the wave peaks close to the focus. At this stage, there is a strong contrast between the area of innovation and the rest of the country. The peak of the wave moves away from the source through time as locations at a greater distance from the point of origin adopt the innovation. Next there is a period of consolidation across the whole of the region concerned, and, finally, as the innovation saturates the region, diffusion slows down until the stage of maximum acceptance, when it ceases. The spread of 'pick-your-own' farm shops in the UK illustrates this phenomenon very neatly. *See also* *diffusion.

input–output analysis A view of the economy which stresses the interdependence of different sectors. The output of one sector is often the input of another. Primary inputs, like land or labour, come from outside the system while intermediate inputs originate within the system. Final outputs pass out of the system. The quantities, expressed in money values, are displayed in a matrix; the rows record the destination of the outputs while the columns show the origin of the inputs. The coefficients indicate the linkages between inputs and outputs. It is possible to predict the amount each sector must produce for a given requirement but the calculation is very large and very time-consuming because of the high number of sectors found in real life.

inselberg A steep, isolated peak rising abruptly from a *pediment; Ayers Rock, Australia, is perhaps the most famous example. There is some debate about the origin of inselbergs. Some writers attribute their formation to *parallel slope retreat; others believe that they are the revealed remnants of the deeply weathered rock typical of tropical climates. *See* *bornhardt.

insolation From *incoming solar radiation*, this is the *solar radiation received at the earth's surface. The amount of insolation varies with latitude, since the angle of the sun's rays and the duration of daylight change with latitude and season. Other contributory factors include the *solar constant, the slope and aspect of the surface, and the amount of cloud in the *atmosphere. Global variations in insolation are a prime factor in the *general circulation of the atmosphere.

insolation weathering *See* *thermal expansion.

instability The condition of a parcel of air which has positive *buoyancy, and thus a tendency to rise through the *atmosphere. It is the temperature of the parcel relative to the *ambient air which is critical, since this affects densities. A parcel remains unstable if it cools more slowly than the surrounding, stationary air. **Absolute instability** occurs when an air parcel, displaced vertically, is hastened in the direction of the displacement, and such a move will be checked when the temperature of the parcel is at one with its surroundings. The **potential instability** of a parcel of air can be calculated using a *tephigram.
   Conditional instability occurs when a parcel of air would become unstable if lifted by some other force. This can take place when a lower, stable parcel of air is overlain by an unstable layer. If any agency, such as a mountain, or the moistening of the lower-level air, can 'nudge' it upwards, it will become unstable. **Convective instability** is a tendency of an air parcel towards instability when it has been lifted bodily until completely saturated. This will often occur when it responds to the inevitable stirrings in the atmosphere.
   The atmosphere is said to be unstable if the environmental *lapse rate is greater than both the dry and saturated *adiabatic lapse rates; that is, when the fall of temperature with height of the environmental air is more rapid than that experienced by a rising air parcel, so that the air parcel continues to be less dense than the ambient air.

instrumentalism A philosophy of science which judges the worth of a theory by its fit with empirical evidence but requires no understanding of causal correlation. Thus, for example, the *gravity model works reasonably well, but has no theoretical underpinning.

integration 1. Social integration is the process whereby a minority group, particularly an ethnic minority, adapts to the host society and where it is accorded equal rights with the rest of the community. *Assimilation is integration such that the immigrants' culture is lost.

2. Economic integration can be the breaking down of trade barriers between nations in order to set up a *common market. The term is also applied to a firm which takes control of all the stages of production. A major example is an integrated iron and steel works comprising coke ovens, blast furnaces, steel forges, and a strip mill. **Horizontal integration** is the central organization of all units at the same stage of production, while **vertical integration** is the integration of units at all stages of production.

3. In *Geographic Information Systems, the fusion of data from different systems and sources to supply new information.

intelligent knowledge-based systems, IKBS Interactive computer systems using artificial intelligence for solving problems. *See* *Geographic Information Systems.

intensity Of an *earthquake, a term for the severity of ground movement at any location. *See* *Mercalli scale.

intensive agriculture Agriculture with a high level of inputs—capital and labour—and high yields. Outputs are valuable and often perishable. Examples include market gardening and veal production, and both outputs and inputs are measured in terms of cost per hectare. Intensive agriculture is usually found in regions of dense population and high land values.

interaction Also known as **spatial interaction**, this is the action between two points, upon one another. An interaction model describes the reactions of two or more processes or systems as they affect each other.

interception The holding of raindrops by plants as the water falls onto leaves, stems, and branches. When the plant can hold no more, the water will drip from the plant (throughfall) or run down the stem (stemflow).

interdependence The interlocking of parts within a system. Within human geography, it is a view of a system as a whole, stressing the role of each part of the system. For example, an advanced economy may depend on the raw materials of a less advanced economy just as much as the latter depends on the finished goods and technology of the former.

interface The zone of interaction between two systems or processes. Estuaries might be seen as the interface between *fluvial and *marine systems.

interflow Movement of water through soil, but at a greater depth than *throughflow. It is difficult to separate the two processes in the field.

interfluve The land between two adjacent rivers.

interglacial A long, distinct period of warmer conditions between *glacials when the earth's glaciers have shrunk to a smaller area.

interlocking spur One of a series of tapering ridges which alternately project into a river valley, and around which the river winds its course.

intermediate technology Agricultural or industrial processes using basic skills that are available in developing countries and that require a simple, easily learned and maintained technology. An example might be the use of a knapsack crop sprayer rather than an aircraft-borne spray, or a pedal-powered maize sheller rather than a commercial mill. Industries can be developed to suit the technology of the workers, to serve local needs, and to use plentiful supplies of labour. Where available capital is limited, it may be more effective to spread it over a number of projects rather than to concentrate on one high-technology industrial development. *See also* *appropriate technology.

internal deformation Inside the glacier, ice is subject to stress as a result of the pressure of overlying ice. Individual ice crystals react to additional stress either by elongating or by melting and re-crystallizing. Ice may also shear along separate shear planes. *See* *extending flow, compressive flow.

internal migration The temporary or permanent relocation of population inside the boundaries of a *nation-state.

International Date Line Any place just west of 180° is twelve hours ahead of Greenwich Mean Time; points just east of it are twelve hours behind. To reconcile these facts, an imaginary line—the International Date Line—has been established. The line follows 180° longitude except where it crosses land so there are some departures from the meridian. As the traveller moves from east to west over the International Date Line he or she 'skips' a day; the date is put forward one day. The traveller moving in an opposite direction keeps gaining time, so that a day must be repeated in order to reconcile the gains; thus, the date is put back one day.

international division of labour The allocation of various parts of the production process to different places in the world. In theory, different regions specialize in different activities, and everyone benefits, but some studies indicate that, in practice, low-skilled, poorly paid, and ecologically damaging work is switched to newly industrializing and developing countries, where the work is often done by non-unionized, female labour, working in poor conditions.

International Monetary Fund A fund established in 1944 with the aims of encouraging exchange stability and eliminating exchange controls, promoting international monetary co-operation, and expanding world trade. The IMF is not a bank (compare with *World Bank, to which it is a sister organization), but it facilitates access to funds, although often under very stringent conditions. *See* *structural adjustment.

international region Also known as a *geostrategic region*, this is an association of all, or parts of, nations with a common interest. The *heartland as defined by Mackinder is an example.

interpersonal space The linear distance separating one individual from others—the average distance by which one person separates herself or himself from the next person. Not surprisingly, the distance is least between couples, followed by family and friends, and greatest between strangers. It has also been observed that the extent of interpersonal space varies culturally, and that Anglo-Saxons tend to be more distant from their fellows than, say, Latins.

interpluvial A time of increased aridity in deserts; the time between *pluvials.

interpolation Forming an estimate of a value with reference to known values either side of it. This method is used for contour lines or other *isopleths.

interquartile range If the number of values of ranked data is divided into four equal parts, then the lines marking each division are quartiles. The interquartile range is the difference between the values of the upper and lower quartiles. The closer the clustering of values around the *median, the smaller the interquartile range. The value of the interquartile range is important when two sets of similar data are compared.

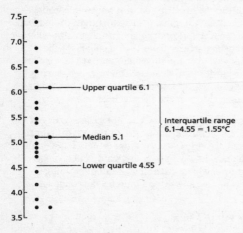

FIGURE 34: *Interquartile range*

interstadial A warmer phase within a *glacial which is too short and insufficiently distinct to be classed as an *interglacial.

inter-tropical convergence zone, ITCZ That part of the tropics where the opposing north-east and south-east *trade winds converge. It is not a continuous belt, more like a necklace with groups of clouds as the 'beads';

in places there may be two or more 'strings'. The zone is narrower over the oceans, and broader over the continents, where other wind systems may be involved; in West Africa the ITCZ is the convergence of the *harmattan and Guinea *monsoon. Here, pressure is low, humidity is high, and spasmodic rain is associated with shallow depressions.

The ITCZ moves north and south with the seasons; moving more over land, and arriving in the summer in each hemisphere. Its position is affected by the apparent movement of the overhead sun, the relative strengths of the trade winds, and the changing locations of maximum sea-surface temperatures. This means that the movements of the ITCZ are highly unpredictable. If it moves well away from the equator it brings unusually heavy rainfall; this caused floods in Khartoum in August 1988. If it stays close to the equator, droughts occur beyond, as in Ethiopia in 1984. The ITCZ can draw in moist air from the sea, bringing rain, and is more active in mountainous regions. However, in the dry interiors of continents it may not even bring cloud. Over the oceans, the ITCZ is broad, and often loses its identity. Winds are then absent, and such windless regions are known as *doldrums.

interval data Data which show both the order of magnitude (10^2, 10^3, and so on) and the degree of magnitude. Similarly, an **interval scale** shows the exact quantities of the variables, or the frequency of their occurrences, and **interval level measurements** are measurements made such that it is possible to assess the size of the differences between them. For example, if there are three towns with populations of 18000, 12000 and 9000, the difference between the first two may be reckoned and that could be compared with the interval between the second and third values. Such comparisons are not possible with *categorical data (in groups), *nominal data (by types, e.g. blonde, brunette) and *ordinal data, which are ranked by importance but given no absolute value. *Parametric tests should be used on interval data.

intervening opportunities theory This theory, advanced by S. A. Stouffer (*Am. Soc. Rev.*, 1940), states that the number of people travelling a given distance is directly proportional to the number of opportunities at that distance and inversely proportional to the number of intervening opportunities, that is, the number of chances of finding satisfaction in work or residence, for example, which may be encountered along the journey. As an illustration, a number of Jewish nineteenth-century migrants from Russia, bound for the New World, actually settled in the East End of London. The theory is also used to study patterns of consumer behaviour; for shoppers living west of Poole, Bournemouth has more retail outlets, but Poole lies between them and Bournemouth, and thus gets more of their trade. The concept indicates that opportunities nearby are more attractive than slightly better opportunities further away. One drawback of this theory is the difficulty of measuring 'opportunities'.

E. L. Ullman (1954) believed intervening opportunity to be one of the three fundamental principles underlying *spatial interaction. The other two are *transferability and *complementarity.

intervention price A guaranteed minimum price, set by a government, for agricultural produce. Should prices fall below this minimum, the government must buy the produce at this price. The intervention price is usually a percentage of the target price: the price hoped for in the open market. The use of intervention prices is a major part of the *Common Agricultural Policy of the EU.

intrazonal soil A soil affected more by local factors than by climate, unlike a *zonal soil. Thus, waterlogging gives rise to *gley soils and a parent rock of pure calcium carbonate will produce a *rendzina.

intrenched meander A deeper watercourse cut into an original *meander such that the banks steepen very suddenly above the stream. *See* also *incised meander.

intrusion A mass of igneous rock which has forced its way, as *magma, through pre-existing rocks and then solidified below the surface of the ground; hence **intrusive rock**. The crystals in intrusive rocks are large since the subterranean magma cools slowly giving time for crystal growth. Intrusions can occur along the *bedding planes as **concordant intrusions** (*see* *sill) or across them as **discordant intrusions** (*see* *dike). Some major forms of igneous intrusion are shown below.

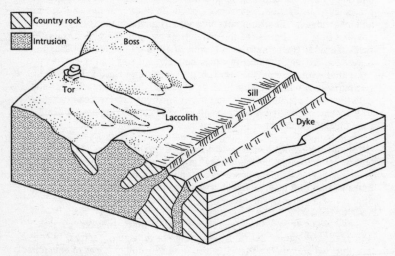

FIGURE 35: *Intrusion*

invasion and succession A model of change used in *urban ecology to represent changing land use within a neighbourhood. For example, a few in-migrants who are content with multiple dwelling invade a neighbourhood to the discontent of the original residents who will eventually leave. Succession is the end of the process when the area has changed completely. The concept is also used in plant *ecology.

inverse distance law Formulated by Zipf (1949), this states that the movement of people between two towns is inversely proportional to the distance between them.

inversion The increase of air temperatures with height. (This is the reverse of the more common situation in which air cools with height.) Inversions occur: when strong, nocturnal, *terrestrial radiation cools the earth's surface and therefore chills the air which is in contact with the ground; when cold air flows into valley floors, displacing warmer air (see also *frost pockets); where a stream of warm air crosses the cool air over a cold *ocean current; where warm air rises over a cold front; when air from the upper *troposphere, subsiding in a warm *anticyclone, is compressed and *adiabatically warmed. A **subsidence inversion** is a stable layer in the low troposphere of an anticyclone, caused by the subsidence of warm, dry air. It forms where the descending air meets small-scale upward-rising convection currents. In spite of the name, an inversion is not always present.
   The boundary between the top of the cold air and the beginning of the inversion is an **inversion lid**. Inversions are very *stable and damp or polluted air is often trapped below them. *See also* *trade wind inversion.

invisible exports Services like shipping and insurance which can earn foreign exchange without the transfer of goods from one country to another.

involution 1. The refolding of two *nappes differing in age so that parts of the younger nappe lie below older rocks.
   2. The convolution of layers of ground under *periglacial conditions. This may occur when the active layer is trapped at the start of winter between the frozen surface and the *permafrost below it. The resulting pressure distorts the trapped strata. Alternatively, involutions may result from the pressure exerted by expanding *ice segregations.
   A synonym for *cryoturbation.

ion An atom or group of atoms that has either lost one or more electrons, making it positively charged (*cation), or has gained one or more electrons thereby becoming negatively charged (anion). *See* *ionosphere.

ionosphere A layer of the *atmosphere containing *ions and free electrons. The ionosphere is warmed as it absorbs solar radiation, and it will reflect radio waves. Three bands are recognized: E, F, and F2, at about 110, 160, and 300 km above the earth. The ionosphere is in a state of constant motion, and is affected by tidal forces and by the earth's magnetic field.

iron band *See* *podzol.

irradiance The rate of flow of radiant energy through unit area perpendicular to a solar beam.

irrigation The supply of water to the land by means of channels, streams, and sprinklers in order to permit the growth of crops. Without irrigation *arable farming is not possible where annual rainfall is 250 mm or less and it is advisable in areas of up to 500 mm annual rainfall. To some extent, irrigation can free farmers from the vagaries of rainfall and, to that end, may be used in areas of seemingly sufficient rainfall because irrigation can supply the right amount of water at the right time.

Within Europe, it is in the Mediterranean regions where irrigation is vital; in 1990, Greece and Italy had 24% of crop land irrigated, Portugal 18%, and Spain 15%. However, 26% of Dutch farmland is irrigated—a reminder that intensive farming systems will use irrigation in seemingly well-watered regions. The UK figure was 2%.

Large-scale irrigation schemes may encounter difficulties if they cross national boundaries; the Punjab irrigation scheme of north-west India and Pakistan is a source of conflict between the two nations since the original scheme was set up before partition. Even within a nation there may be disputes about water supply; Arizona and California both use the water of the Colorado which acts as a frontier between the two states.

In its simplest form, irrigation is achieved by devices such as the *sakia and the *shaduf to lift water but, increasingly, modern pumps are used. Irrigation is not suited to saline soils since the salt will move to the surface and be so concentrated there as to inhibit the growth of most plants. Similarly, the use of *brackish water for irrigation is unwise since the salts remain in the soil after the water has been lost through *evapotranspiration.

Irrigated lands show regular and intricate systems of intensively cultivated fields dependent on water through canals, cuts, and irrigation channels.

isentropic, isentropic surface In meteorology a surface of constant potential temperature. Isentropic surfaces slope very gently upwards towards the cold air in the presence of a horizontal temperature gradient, and winds flow along these surfaces.

Isentropic analysis is the analysis of *radiosonde data to pick out winds and other meteorological observations on each of several such surfaces.

island A body of land completely surrounded by water.

island arc An island chain, mostly of volcanic origin, in the form of an arc. According to *plate tectonic theory, the arc is formed when *oceanic crust plunges into the *mantle where it undergoes *subduction. The *magma thus formed creates a chain of submarine volcanoes which are eventually built up into islands.

island biogeography The number of species living in an isolated space, such as an island, can be seen as a balance between the immigration of new species and the extinction of established ones. While the population is low, the balance will be non-interactive, i.e. different species multiply without interference. However, when populations are large enough, they interact and immigration and extinction are affected.

Distant islands will receive immigrants at a slower rate than the islands near the mainland, but extinction rates will be the same for both, so that distant islands will hold fewer species. On large islands, immigration is high since the 'target' is large. Extinction is also lower because there is more cover. Thus, large islands will have more species than small. The concepts of island biogeography may be extended to any community in an isolated habitat—even to an enclosed lake, which is an island of water in a sea of land.

isobar A line on a map or chart that joins places with the same atmospheric pressure. By reading the isobars over a large area it is possible to gain a visual impression of any anticyclones or depressions that may be present. **Isobaric areas** are parts of the atmosphere having uniform pressure, while an **isobaric surface** is a surface on which any point experiences the same atmospheric pressure.

isochrone A line connecting places of an equal journey time to the same location.

isodapane A line joining up places of equal total transport costs for industrial production and delivery between the points where the raw materials are located and the markets. *See* *Weber's theory.

isogloss In geolinguistics, a line marking the limit of use of a word, or other linguistic feature; a classic example is the boundary between the use of the terms 'pail' and 'bucket', which runs through southern New Jersey, USA. Isoglosses are not necessarily linguistic boundaries; they are usually highly simplified representations and do not depict an abrupt transition, although contrasting forms do exist on either side of these notional divides. Sometimes, a **bundle of isoglosses** may occur, where a number of isoglosses lie close enough together to indicate a true *dialect boundary. Very seldom, however, do even two isoglosses coincide along their whole length.

isohyet A line connecting points of equal rainfall.

isoline Any line on a map joining places where equal values, e.g. rainfall, temperature, atmospheric pressure, are recorded.

isophene A line connecting places with the same timing of similar biological events such as the flowering of a crop.

isopleth A line connecting places of equal value. These values may be physical such as height above sea level, precipitation, and so on, or of human values such as distributions of population, wealth, or transport costs. The word *contour* may be used as a synonym.

isostasy The continental crust of the earth has a visible part above the surface and a lower, invisible one. The balance between these two is isostasy. If part of the upper surface is removed by erosion, the continental crust will rise to offset this erosion, at least in part. Sections of the continental crust have been pushed down by the weight of glacial ice, the extent of the depression varying with the thickness of ice, and the density of the material below. It is believed that the critical size for an ice cap to depress a land mass is a 500 km diameter. Land masses will rise again if the ice melts. Evidence for isostatic readjustment includes the 30 m beach off the west coast of Scotland. Recovery tends to be slow, and uneven. An adjustment as a result of glaciation is *glacio-isostasy.

isosteric Of that part of the atmosphere having uniform density.

isotherm A line on a map or chart joining places of equal temperature.

isotim A line drawn about a source of raw materials or a market where transport costs are equal.

isotope One of two or more alternative forms of an element that have the same number of protons in their nucleus, but have different numbers of neutrons. An isotope of carbon is used in *carbon dating.

isotropic Having the same physical properties in all directions. The featureless isotropic plain is the basis of many location theories such as those of *Weber, *von Thünen, and Christaller.

jet stream The name given to any narrow belt of strong, upper-atmosphere winds, blowing at speeds of over 45 m s⁻¹, between 7.5 and 14 km above the earth's surface. Jet streams are several hundred kilometres wide and 2–4 km deep, owing their existence to the *conservation of angular momentum, and appearing as a fast-moving track inside lighter winds. Such is their strength that aircraft routes which run counter to jet movements are generally avoided. Jets are coincident with major breaks in the *tropopause.

The **polar-front jet stream** is a *frontal wind, located just below the *tropopause, blowing parallel to the surface fronts, moving with them, and draining the air rising from the fronts. It is strongest at the 200–300 mb level, and swings between latitudes 40 and 60° N, since it is located along the *Rossby troughs, so that speed and location vary from day to day with the Rossby waves. It is not necessarily continuous. This jet is coincident with strong horizontal shifts in temperature and pressure (see *baroclinic) since it marks the polar *front; the boundary between cold polar air and warm tropical air, where the steepness of the isotherms is at a maximum. It has important effects on *convergence and *divergence in the upper air. For example, at the 'jet entrance', the pressure gradient steepens, and the wind becomes super-*geostrophic, leading to high-level convergence. A strong polar-front jet is associated with rapidly moving *depressions; a weak jet with a *blocking pattern where northerly and southerly air streams dominate.

The **westerly subtropical jet** is at the poleward limit of the *Hadley cell, around 30° N and S; the northern subtropical jet is strongest at the 200 mb level, and above the Indian *subcontinent. This is one of the most powerful wind systems on earth, at times reaching speeds of 135 m s⁻¹, and it follows a more fixed pattern than the polar-front jet. It results from the poleward drift of air in the Hadley circulation and the *conservation of angular momentum. Some *anticyclones develop beneath the westerly subtropical jet, through high-level convergence and subsidence, but the subtropical, westerly jets do not seem to affect surface weather as much as the polar-front jets do.

The **tropical, easterly jet** develops during the summer months at 15° N, and is strongest at the time of the summer *monsoon.

The **stratospheric, subpolar jet stream** blows at a height of 30 000 metres, being westerly in winter and easterly in summer. *See* *conservation of angular momentum.

joint A crack in a rock without any clear sign of movement either side of the joint.

jökulhlaup A sudden flood of glacial meltwater released when volcanic activity heats the ice.

Jurassic The middle *period of *Mesozoic time stretching approximately from 190 to 136 million years BP.

just-in-time system, JIT A system of production which aims to deliver all the necessary inputs, such as raw materials, components and labour, just in time for the appropriate stage of production. This is essentially a flexible system of manufacturing, linked with small production runs and flexible methods of production. It is largely credited with being a Japanese innovation. *See* *post-Fordism.

juvenile water Water contained within *magma and which is emitted during volcanic eruptions.

Kamchatka current A cold *ocean current off the peninsula of the same name in Siberian Russia.

kame An isolated hill or mound of stratified sands and gravels which have been deposited by glacial meltwater. Some kame deposits show slumping on a side which previously had been held in position by a wall of ice. Many kames seem to be old deltas of sub-glacial streams.

 Kame terraces are flat-topped, steep sided-ridges of similar *fluvio-glacial origin, running along the valley side. They are *ice contact features, formed between the side of a decaying glacier and a valley wall. Moulin kames form below *moulins.

karre, karren (pl.) A collective name for the shallow channels formed by solution on exposed limestone, such as the Jurassic Quintener limestone of the Swiss Bernese Alps. They are also known as lapiés. **Kluftkarren** are the enlarged joints also known as *grikes. **Rillenkarren** are very closely spaced small runnels; small radiating rillenkarren are the overflow of surface solution pans. **Rinnenkarren** are both longer and deeper, as much as 10 m in length and 0.5 m in depth. They may develop as a result of coalescence of small channels. Their walls are sharp, in contrast with the large, rounded hollows known as **rundkarren**. See *limestone pavement.

karst The name comes from the Karst region of Slovenia, but is now applied to any area of limestone which is dominated by underground streams, and hollows and pits usually caused by subsidence into underground channels. See *doline, *polje and *uvala for surface, solutional karst features.

 Karst is the most strongly developed in humid uplands where very thick, strongly jointed limestones occur. Other typical karst features include *blind valleys, *sink holes, caves, *karren, and springs. Classic areas of karst scenery in Britain occur around Malham and in the Brecon Beacons. The term **karstic** is used for karst scenery; **karstification** is the formation of karst scenery, especially the formation of dry valleys and underground drainage systems with the associated development of passages and caves, and with protracted surface solution. See karren. **Karstland** is an area of karst scenery. See also *cockpit karst, *labyrinth karst, *tower karst.

karst lake A periodic lake formed in a *polje or *doline. The Lake District in Florida contains a number of karst lakes which result from rising water levels.

kata- From the Greek, *kata*—down—sinking, as in a *katabatic wind. The term is also used to describe the sinking of air in the warm sector of a *depression at a cold or a warm **kata-front**, bringing about a large-scale inversion of temperature at the fronts, which are fairly inactive. At a kata-warm front, cloud development is limited to cirrus and high stratus, and precipitation is restricted to light rain; at a kata-cold front strato-cumulus is common, and precipitation is similarly moderate.

katabatic Referring to downslope winds. Descending, *adiabatically warmed katabatic winds are *föhn winds. Cold katabatic winds result from the slumping down of very cold, and hence dry, air. Coastal Antarctica is dominated by katabatic gales; the gentler katabatic flows of hill slopes produce *frost hollows. Compare with *anabatic.

kettle hole Large masses of ice can become incorporated in glacial *till and may be preserved after the glacier has retreated. When one of these bodies of ice finally melts, it leaves a depression in the landscape; a kettle hole, also known as a *dead ice hollow*. These are particularly common in Mecklenburg, northern Germany.

key village A village designated to be developed in terms of goods and services. Key villages are to be expanded while other centres are run down, with their residents encouraged to leave. Key villages have been developed successfully in rural counties such as Norfolk and Devon but little has been done in the running down of small villages.

Khamsin *See* *local winds.

kibbutz, kibbutzim (pl.) A type of agricultural system, first established by Jewish settlers in Palestine and then supported by the Israeli state. All the agricultural land and resources are communally owned and all farming strategies are decided jointly. The domestic arrangements, including childcare, are shared by all. Later kibbutzim began to use waged labour. Although the collectivist ideology and economic viability of the kibbutzim have been challenged in recent years, nearly 3% of the population of Israel have been part of the movement.

kinematic wave A high-velocity wave in a glacier, recording velocities of over 100 m per day, generated by a major accumulation of snow above the firn line, and causing a *glacial surge when the wave reaches the snout.

kinetic energy The energy of motion; in geomorphology, the energy used by wind, water, waves, and ice. Kinetic energy for channelled flow is defined as:

$$\frac{MV^2}{2}$$

where M is the mass of water, and V is the mean velocity.

knick point, nick point A point at which there is a sudden *break of slope in the *long profile of a river. In areas of uniform geology, the presence of a knick point may be evidence of *rejuvenation; the river is forming a new, lower profile cutting first from the mouth of the river and working upstream as *headward erosion takes place.

knock and lochan topography A glacially scoured, lowland landscape, as in Lewis, north-western Scotland, south Sweden and Finland, or the

Barren Grounds of the Hudson Bay region. Low, rounded hills (**knocks**) and
*roches moutonnées alternate with *striated and eroded hollows, which
may be tens of kilometres long, often containing lakes (**lochans**). Drainage
is generally chaotic. While the overall pattern of relief is controlled by the
direction of ice movement, local rock type and structures such as joints
and faults dictate the details.

knock-on effect A *multiplier effect that operates in reverse, to the
detriment of a region. Thus, unemployment in a key industry leads to
unemployment in associated industries and therefore to unemployment in
service industries. *See* *cumulative causation.

kolkhoz A large-scale farming unit in the former USSR where the
state-owned land is leased to a collective. Decisions are made by an elected
committee and profits are shared between the members.

Kondratieff cycles A series of long waves of economic activity. Each
cycle lasts 50–60 years and goes through development and boom to
recession. The first cycle was based on steam power, the second on
railways, the third on electricity and the motor car, and the fourth on
electronics and synthetic materials. Kondratieff argued that one of the
forces which initiates long waves is the large number of important
discoveries and inventions that occur during a depression and are usually
applied on a large scale at the beginning of the next upswing. Each cycle
leaves its mark on the industrial landscape.

König number Also known as the **associated number**, this is the number
of *edges from any node in a *network to the furthest node from it. This is
a *topological measure of distance, in edges rather than in kilometres. A
low associated number indicates a high degree of *connectivity; the lower
the König number, the greater the *centrality of that node.

kopje A small, rocky hill, originally in South Africa, but now within the
whole of Africa. They are thought to be remnants of *inselbergs.

Kuro Shio A warm *ocean current which, fed by the North Equatorial
current, runs from the Philippines to Japan, thence feeding into the North
Pacific current.

kurtosis This applies to the degree to which the *frequency distribution
is concentrated around a peak, that is, it describes the sharpness of the
central peak of the curve.

k-value *See* *central place theory.

labour The manual or intellectual work which is one of the *factors of production. The quantity of labour is the amount of work done in terms of production or time whilst the quality indicates the degree of skill and intelligence required. A **labour intensive** enterprise is one, such as hairdressing, in which the input of labour is high (compare with *capital intensive). Labour costs are a major part of the *cost structure of many firms; in such cases they should focus on labour as the key to locational comparative advantage. Note, in this context, that in the late 1980s West German wage rates for car workers were over twice those in Japan. In times of employment decline, high-cost labour can bring about failure in small firms and a closure of some plants in multi-site enterprises. Some employers see labour as being unreliable where trade unionism is strong, or where *absenteeism is high. The **labour market** is the mechanism whereby labour is exchanged for material reward.

labour market The exchange of work for *capital. In *neoclassical economics, market forces acting on *economic man are held to bring about an equilibrium between the supply of capital and the supply of labour. This takes no account of differences caused by gender, race, or location, and state intervention is not considered.

labour theory of value The Marxist contention that the value of a product reflects the amount of labour-time needed to make it. If the capitalist pays low wages which do not reflect the labour expended, he will obtain surplus capital. This may be seen as exploitation, which can lead to class conflict.

Labrador current A cold *ocean current, running south from Greenland, off the Labrador coast, and bringing both nutrients and fog to the fishing grounds of the Grand Banks.

labyrinth karst The deep canyons of limestone formed by *carbonation. Initially the limestone shows *bogaz; these widen and deepen into long gorges known as karst streets with other, cross-cutting, lines of erosion. The remnant of this carbonation is *tower karst.

laccolith An *intrusion of igneous rock which spreads along *bedding planes and forces the overlying *strata into a *dome. Classic examples are found in the Henry Mts. of Utah.

lacustrine Of lakes, especially in connection with sedimentary deposition. **Lacustrine plains** result from the in-filling of a lake, while **lacustrine terraces** result from the formation of beaches along the shoreline of a former lake.

lagged time The interval between an event and the time when its effects are apparent.

lagoon A bay totally or partially enclosed by a *spit or reef running across the entrance, known in the Baltic as a *haff*.

lahar A downslope flow of volcanic debris, either dry or mixed with water as a mud flow. Lahars most commonly occur when a crater lake or an ice-dammed lake suddenly overflows; perhaps because of an eruption, the collapse of a dam, heavy rain, snow melt, or the mixing of a *nuée ardente with lake water. Velocities of flow may be up to 90 km hour. Around 5500 people were killed in a mud flow following the eruption of the Kelut volcano, Java, in 1919. The 5000-year-old Osceloa mud flow of Puget Sound, Washington, USA, is up to 150 m thick, and extends over 320 km^2.

laissez-faire economics The view that a *market economy will perform most efficiently if it is free from government intervention, and is subject only to *market forces. This view is criticized because it takes no account of considerations such as environmental degradation—except in so far as it might increase costs—or social justice. Furthermore, the models of the workings of the free market, on which *laissez-faire* economics are based, bear very little relation to reality; in the real world, markets are distorted by monopolies, lack of choice through cultural constraints, imperfect information, and so on.

Lamarckism The doctrine, devised by the French naturalist Jean Baptiste de Lamarck (1744–1829), that acquired characteristics are inheritable. This has been applied to the development of theories of social evolution and to varieties of *environmental determinism. However, the theory has now been discredited.

laminar flow A type of non-*turbulent flow where the movement of each part of the fluid (gaseous, liquid, or plastic) has the same velocity, with no mixing between adjacent 'layers' of the fluid. It may be seen at low velocities in a smooth, straight river channel, and at some glacier snouts. Compare with *turbulent flow.

land breeze A wind blowing from land to sea (an offshore wind) which develops in coastal districts towards nightfall. Pressure is relatively higher above the land than above the sea as the land cools more rapidly in the evening, and air therefore moves seawards in order to even up the pressure difference. *See also* *sea breeze.

land bridge A dry-land connection between continents.

land capability The potential of land for agriculture and forestry depending on its physical and environmental qualities. The main factor investigated is soil type, but climate, gradient, and aspect are also considered. Present land use is not taken into account.

land classification The division of land into categories according to the potential agricultural output. Soil quality is a major factor but any assessment should also take into account drainage, elevation, gradient, susceptibility to soil erosion, temperatures, and rainfall. Economic classifications may also be used, concerned with the layout of the farm, its workings, and prices and markets. Economic factors may change and so might the physical evaluation; chalklands are more extensively cropped than they were 40 years ago. Land classification maps have been produced but, given their intricacy, it is difficult to make useful generalizations. Furthermore, classification of land into a particular category tends to be somewhat subjective.

land consolidation A type of *land reform which aims to give each farmer one relatively large plot of land rather than scattered, small parcels of land. *See* *remembrement.

land economics The study of land use and of the factors which influence and shape it. Land values are a central part of this study as the patterns of land use and land values are interlinked.

Land Information System A system for the capture, storage, manipulation, analysis, and display of land-use data.

land reform A sweeping change in land tenure. It usually involves the breaking-up of large estates and the widespread redistribution of the land into smallholdings, but may also be *land consolidation. Another variation is the policy of certain revolutionary regimes of collectivizing the land, taking it out of private ownership.

land slide Used very loosely, the term land slide covers most forms of *mass movement.

land tenure The nature of access to land use. Common forms of land tenure are owner-occupied farms which range from large farms using hired labour to peasant plots and tenancies which vary very widely but basically involve payment, in one form or another, to the landlord from the tenant. This payment can be in the form of labour, in the form of a portion of the crop (*share-cropping), or in the form of cash. A plantation is owned by an institution and uses paid labour. Collectives may own land together and work together to an agreed strategy, sharing any profits which may accrue. There may even be land which is owned by nobody but is used by an individual or group. This last is typical of *shifting cultivation.

land use classification The analysis of land according to its use: agricultural, industrial, recreational, and residential. Comparisons are very difficult to make between different countries which may have different classifications. On occasion, the land may have more than one use, as in upland areas used for sheep farming and for recreation.

land use survey In the UK, the First Land Utilization Survey was carried out between 1931 and 1939. Showing seven categories of land, the original maps are kept at the London School of Economics where they may still be studied. The maps were also published at a scale of six inches to the mile.

The Second Land Utilization Survey was carried out between 1961 and 1969 using base maps of the same scale but having 256 categories. These maps can be consulted at the Land Use Research Unit at King's College,

London. Some of these maps were published at 1:25000, using 70 categories, but cover only 15% of the area of England and Wales.

land use zoning The segregation of land use into different areas for each type of use: agricultural, industrial, recreational, and residential. The 1947 Town and Country Planning Act required local authorities to zone land use in the future.

land-locked state A nation with no access to the sea. There are twenty-six land-locked states ranging in size from the tiny Vatican City to Mongolia. One important preoccupation of a land-locked state is access to the sea. This may be achieved along a river like the Danube, by the creation of a *corridor, or simply by negotiating a right of passage through a maritime neighbour.

landlord capital In an agricultural tenancy, the landlord supplies certain assets such as land, roads, drains, and buildings. *See also* *tenant capital.

landscape An area, the appearance of an area, or the gathering of objects which produce that appearance. Carl Sauer first used the term in geography in 1925, stressing the concept of the landscape as the expression of interaction between humans and their environment. *See* *iconography.

landscape architecture Originally the design of gardens, this term now also covers the planning and management of a landscape in order to meet aesthetic standards while also fulfilling some functions. An adventure playground may be laid out in a pleasing fashion or a motorway may be designed to clash as little as possible with the landscape.

landscape evaluation An attempt to assess the landscape in objective terms. Sometimes a consensus of views on the landscape is sought so that particular landscapes may be chosen as being outstandingly beautiful. Landscape description studies try to identify important items such as topography or buildings. Some kind of ranking method may be attempted to compare one landscape with another. Using this idea, Leopold attempted to calculate how close a landscape was to being unique. Personal preferences may also be used.

landscape preference It is argued that most cultures have a preferred landscape: the Dutch are said to be attracted by order and neatness while Americans are said to value a *wilderness where landscape elements are very large. A knowledge of the preferred landscape might enable planners to modify the landscape with a minimum of protest.

Landschaft A German concept of landscape which attempted to classify landscapes, usually distinguishing between the natural and the *cultural landscape.

landslide A form of *mass movement where the displaced material retains its form as it moves. Landslides are prompted by an increase in *pore pressure through snow melt, through precipitation, and through spring action. These all reduce the friction which binds the mass to the slope. Probably the world's largest landslide occurred in south-west Iran in 1937, when a segment of the Kabir Kuh ridge, about 15 km long, 5 km wide and 300 m thick, slid off the mountain, with enough momentum to travel 20 km.

lapié(s) The French term for *karre(n).

lapili *See* *pyroclast.

Laplacian determinism Laplace held that all present conditions, and all the laws of behaviour, would determine all future conditions; in other words, that an application of physical laws would enable people to forecast future events. But *see* *chaos theory.

lapse rate 1. In meteorology, the rate at which stationary or moving air changes temperature with a change in height. *See also* *adiabatic and *environmental lapse rate.
  2. In human geography, the decline of interactions between a central place and its surroundings with distance from the central point. Towns gradually shade into countryside and industrial areas can merge into areas of rural land use.

large nuclei *See* *nucleus.

latent heat The quantity of heat absorbed or released when a substance changes its physical state at constant temperature, e.g. from a solid to a liquid at its melting point, or from a liquid to a gas at its boiling point. The release of **latent heat of condensation** in the rising air of a *hurricane is the chief force fuelling that meteorological phenomenon.

lateral accretion In geomorphology, the build-up of sediments at the *slip-off slope of a river or during *braiding.

lateral erosion Usually of rivers; erosion of the banks rather than the bed. In a stream or river, it results in undercutting of the banks; in deserts, lateral erosion by *sheet flow may be responsible for the formation of desert scenery.

lateral fluvial migration Both *braided and meandering rivers change their course over time and move from side to side of the valley. Lateral movement within a braided river occurs when the water inundates the flood plain and a new course is established. Meanders migrate when the outside of a bend is undercut.

lateral moraine *See* *moraine.

laterite Thick, red, and greatly weathered and altered strata of tropical ground. Laterites are red because silicates have been leached out, and iron and aluminium salts now predominate. *Horizons are unclear and the nutrient status of the soil is low. Laterite is soft but hardens rapidly when exposed to the air until it has a brick-like hardness.

lateritic soils, latosols Soils of humid tropical or equatorial zones characterized by a deep weathered layer from which silica has been *leached, a lack of *humus, and an accumulation or layer of aluminium and iron *sesquioxides. The reddish colour of these soils is imparted by the iron compounds. *See* *laterization.

laterization The formation of *lateritic soils. Laterization takes place in warm climates where bacterial activity takes place throughout the year. Consequently, little or no humus is found in the soil. In the absence of *humic acids, iron and aluminium compounds are insoluble and accumulate in layers in the soil. Silica is *leached out.

latifundium, latifundia (pl.) A large farm or an estate, particularly in Latin America, originally set up as an imperial grant of land to New World settlers. The estate is farmed with the use of labourers who sometimes lease very small holdings from the landowner.

Latin America Free Trade Area, LAFTA A regional grouping of certain South American states formed in 1961 in order to foster trade between them, and to lessen their dependence on more advanced economies. Members are: Argentina, Bolivia, Brazil, Colombia, Ecuador, Mexico, Paraguay, Uruguay, and Venezuela. It was replaced in 1981 by the Latin American Integration Association.

latitude Parallels of latitude are imaginary circles drawn round the earth parallel to the equator. The parallels are numbered according to the angle formed between a line from the line of latitude to the centre of the earth and a line from the centre of the earth to the equator.

Those regions lying within the Arctic and Antarctic circles, having values of 66.5° to 90° are termed **high latitudes**. **Low latitudes** lie between 23.5° north and south of the equator, i.e. within the tropics. **Mid-latitudes**, also known as temperate latitudes lie between the two.

latosol A major soil type of the humid tropics with a shallow A *horizon but a thick B horizon comprising clay, sand, and sesquioxides of iron and aluminium which, respectively, endow it with a red or yellow colour. Much of the silica has been leached from latosols, and they tend to be of low fertility.

Laurasia One of the two original continents which broke from the supercontinent, Pangaea by *continental drift.

lava *Magma which has flowed over the earth's surface. *See* *extrusion. The *viscosity of lava depends on its silica content, pressure, and temperature. Temperature is the most important factor. **Basic lavas** have a low silica content and flow freely; **acid lavas** are more viscous. Water in lava also makes it more fluid. When the water in an underground magma chamber vaporizes, it expands instantly causing explosive eruptions.

Few deaths are caused by lava flows, due to their relatively slow speed, but nearly 22% of the population of Iceland died of famine in 1783, following the burial of an area of 560 km^2 by lava from a five-month-long *fissure eruption.

law A theory or hypothesis which has been confirmed by empirical evidence. It generally indicates a relationship between cause and effect. *See*, for example, *Reilly's law.

law of retail trade gravitation *See* *Reilly's law.

law of the sea A framework, agreed to by the majority of maritime nations, for administering the seas. It recognizes seven administrative zones: internal waters, the *territorial sea, the *contiguous zone, the *continental shelf, exclusive fishing zones of up to 200 miles from a nation's coastline, exclusive economic zones of the same extent as the fishing zones, and the high seas.

laws of drainage basins *See* *drainage basin geometry.

layer In *Geographic Information Systems, a subset of non-*spatial, digital map data, such as particulars of buildings.

leaching The movement of water down the *soil profile. This results in the movement of *cations, *sesquioxides, *clay colloids, and *humus to the lower soil *horizons. Specific types of leaching include: lixiviation—the removal of the soluble salts containing metallic cations; the removal of *chelates; *lessivage; and, in tropical soils, *desilication.

lead–lag model A statistical model which identifies differences of timing in fluctuations through regions and city systems.

league A more or less informal association of states for a particular purpose, such as *OPEC (the Organization of Petroleum Exporting Countries). OPEC is a particularly strong league in comparison with, for example, the British Commonwealth which is more a cultural than a political league.

least-cost location A site chosen for industrial development where total costs are theoretically at their lowest, as opposed to location at the point of maximum revenue. *See* *Weber's theory of industrial location.

least developed country A country with the poorest development indicators; *life expectancy, perhaps the most striking indicator, is below 45 years, indicating that the physical quality of life is at a very low ebb. According to a United Nations' classification, it is a country where manufactured goods account for no more than 10% of *GDP. The number of least developed countries grew from 24 in 1971 to 42 in 1989. In the mid-1990s, the West African state of Guinea was amongst the least developed of nations in terms of literacy, per capita *GDP, and life expectancy.

Lebensraum Literally, 'living space', the room needed for a nation's expansion. The concept was used by the Germans to justify their territorial growth. The term was introduced after 1870 and became a central concept in the propaganda literature of the Nazis. *See* *geopolitik.

lee The side (lee side) sheltered from the wind. Somewhat confusingly, a

lee shore is the shore towards which the wind is blowing.

lee depressions These occur when *lee troughs develop into *lows. They are frequent in winter where mountains block low-level air streams, as in areas east of the Rockies or south of the Alps.

lee trough As a *stable air column rises to cross over a ridge, it shrinks in the vertical, therefore *diverging in the horizontal. This gives it a negative relative *vorticity. Accordingly, in the Northern Hemisphere, the flow is deflected anticyclonically: to the right. Since low pressure always lies to the left of an airstream in this hemisphere (*Buys Ballot's law), the pressure is lower on the lee side of the mountain than on the windward side. The lee trough to the east of the Rockies is a major factor precipitating the north–south meanders of the Rossby waves.

lee wave Also known as a standing, rotor, hill, or mountain wave, this is a wave motion in a current of air as it descends below an upper layer of stable air, after its forced rise over a mountain barrier, which sets up vertical oscillations. The typical wavelength is 5–15 km with an amplitude of some 500 m. Lee waves are often disclosed by enhanced wind speeds and by the presence of *lenticular clouds, forming at the crest of the wave if air reaches *condensation level.

legend Another term for the key on a map.

lenticular cloud A lens-shaped cloud formed above a peak as air rises over mountain barriers and condensation occurs. *See* *lee wave.

lenticularis *See* *cloud classification.

leisure That time left over after time taken for work and other obligations. The term indicates that this time is spent on activities which are worthwhile in themselves to the individual. The **geography of leisure** studies the spatial patterns of people's behaviour in their free time. Throughout Western Europe, with the increases in the number of people over 65, the number of young people in further education, both of which groups might be argued to have more free time than the full-time employed, together with the explosive growth in car ownership, demands for, and the use of, leisure facilities have mushroomed, with an attendant impact on the landscape.

less developed country, LDC A country with low levels of economic development. Indicators of lack of development include high birth, death, and infant mortality rates (characteristically over 20; over 30; and over 50 per thousand, respectively); more than 50% of the workforce in agriculture; and with low levels of nutrition, secondary schooling, literacy, electricity consumption per head, and GDP per capita—generally below US$1000 per capita. *See also* *least developed country, *more developed country.

lessivage The *translocation of *clay colloids in a soil with no change in their chemical composition.

levée A raised bank of *alluvium flanking a river. The bank is built up when the river dumps much of its *load during flooding. Natural levée development along the course of the Hwang Ho, China, has been furthered by the heavy load of easily eroded *loess carried by the river. The breaching of these levées has caused periodic, catastrophic flooding. Man-made levées have been built along many rivers, such as the Mississippi, as a flood control measure.

level In *Geographic Information Systems, a set of digital map information, such as topography, roads, or buildings, which can be accessed and displayed separately, or in combination with other levels.

ley Within a rotation, the seeding of a field to grass or clover either for a few years (**short ley**) or for up to 20 years (**long ley**) before ploughing.

lexical diffusion In geolinguistics, phonological changes proceeding through the lexicon, word by word.

liana A creeper of the equatorial rain forest which winds and climbs around trees for support.

lichenometry A method of establishing the relative age of a deposit, especially one of glacial origin. These deposits are free from lichen when first formed so that the diameter of the largest rosette of lichen on such a deposit is assumed to be an indication of the time the deposit was formed. Lichenometry can only supply relative dates of formation.

life expectancy The average number of years which an individual can expect to live in a given society, normally derived from a national *life table. Life expectancy is usually given from birth but may apply at any age, and because, in all societies, *mortality rates tend to be rather high in the first year of life, life expectancy at birth is usually significantly lower than at one year old. Women consistently have a longer life expectancy than men, especially in *more developed countries where the risks of childbirth are less than those in *less developed countries.

Mainly because of very high infant mortality rates, life expectancy is much lower in less developed countries than in developed nations; Guinea, for example, had a figure of 42 in 1994, while the figure for Japan was 79. By the age of 70, the years of life remaining to an individual are similar in both types of society. Thus, the strong correlation between *GDP per capita and life expectancy becomes weaker as the age of an individual increases.

life space The limited time and space which an individual has in which to pursue a necessarily limited range of opportunities. Life space is the interaction of the individual with her or his behaviour setting.

life table A summary of the likelihood of living from one age to any other. In a life table, a hypothetical *cohort of 100 000 births is set up and then the loss by deaths is shown for each year of life. Averages of losses are calculated for a given year, and from this the actual diminution of the cohort is shown.

life world, lifeworld The day-to-day world in which the individual lives

out his or her life, and which is generally taken for granted. Some *human
ecologists suggest that, for human beings, there are three different levels
of encountering the world: the life world, philosophy, and science, and
that these three ought to be linked.

lift force The upward force produced when fluid rises over a particle. In
rivers and streams, the particle moves up from the bed into the flow when
the lift force exceeds the gravitational force provided by the mass of the
particle.

light industry The manufacture of relatively small articles (toasters as
opposed to girders), using small amounts of raw materials. In consequence,
the *material index is low, and such industries are more *footloose than
*heavy industries.

lightning An emission of electricity from cloud to cloud, cloud to
ground, or ground to cloud, accompanied by a flash of light. It is the result
of variations of electrical charge on droplets within the cloud and on the
earth's surface. This variation may be due to the break-up of raindrops, to
the splintering of ice crystals or to differing conditions between the
splintered ice crystals and pellets of soft hail. As a *cumulus cloud
develops, the frozen upper layer becomes positively charged. Most of the
cloud base is negatively charged, with patches that are positively charged.
These negative charges are attracted to the earth, which has a positive
charge. It is suggested that when the electrical field strength gets to about
1 MV m^{-1}, the electrical insulation of the air breaks down, and the spark
forms. The result is a leading stroke or stepped leader from cloud to
ground, which creates a conductive path between the two. The return
stroke, from earth to cloud, follows the same path. This return stroke may
be as much as 10 000 amps, carried through a pathway of air only
millimetres across. The intense heat of the stroke engenders light, and a
violent expansion of the air, making shock waves, heard as *thunder. Not
all the negative charge may be released; there may be several return
strokes, each prefaced by a downward dart leader which reactivates the
channel.

Where the path between ground and cloud is clearly visible, forked
lightning is seen. The illumination of other clouds by a concealed fork is
sheet lightning. Ball lightning has been described as a sphere of glowing
light meandering through the lower air. Little is known about it.

lightning conductor A sturdy electrical conductor, running from a high
point to ground, providing lightning with a route to earth which generates
little of the heat which causes lightning damage. Humans in open spaces
can act as lightning conductors; you are advised to lie down if caught in
the open during a thunderstorm.

limestone A general term for a *sedimentary rock which consists mainly
of calcium carbonate. Limestones vary in texture; **oolitic limestone** consists
of tiny, rounded grains, **pesolitic** of larger grains, whereas other limestones
are of a crystalline texture. Limestones also vary in mineral content, as
with dolomite and magnesian limestone, and in modes of origin. Different

limestones are classified according to geological age, as in Carboniferous or Jurassic limestones. *Chalk is a soft limestone.

limestone pavement A more or less horizontal, bare limestone surface, cut into by *grikes (deep fissures) running at right angles to each other, leaving *clints (the slabs of limestone) between them. The classic English example is at Malham, in Yorkshire; an Irish example is the Burren, County Clare. Some geomorphologists class such features as fault *karren.

limiting factor The success of an organism is limited by the presence or absence of the factors necessary for survival. Often growth of a *population is limited by an apparently minor factor in the environment, such as the presence of *trace elements in the soil.

limnology The scientific study of fresh-water ponds and lakes. Limnology covers all biological, chemical, meteorological, and physical aspects of lakes.

line In *Geographic Information Systems, a line segment with common attributes. By comparison with a *link, a line may have intersections.

line-haul costs incurred in transporting goods over a route but not including costs of loading and unloading. Line-haul costs vary directly with distance.

line squall A linear, *severe storm event, with rapidly rising pressure and wind speed, a sudden temperature fall, low, dark clouds, and often thunder. At certain cold fronts in the central and eastern United States, the advancing wedge of cold air scoops up the warm air before it to form a nearly continuous line of squall-bearing cumulo-nimbus. Line *squalls may last for two days, and foster *tornadoes.

linear city A planned city developed along a single, high-speed line of transport. Industry is developed along one side of the link, while shops and offices are located on the other side, with housing beyond them. The 1965 plan for Paris is based on the concept of a linear city, as development was planned along two motorway routes.

linear village An elongated ribbon of settlement usually formed along a routeway such as a road or canal. Linear villages may reflect the pattern of *land tenure or may have developed as clearings were cut along the road through a forest as in the German *Strassendorf.

linguistic distance In geolinguistics, the degree of contrast exhibited between two *dialects, as measured in terms of the percentage of items which differ from a fixed set.

link 1. The route or line joining two *nodes.
  2. In *Geographic Information Systems, a line segment with common attributes and a *node at each end, but having no logical connection with another line segment, except at the end. Also known as an arc, or edge. Compare with *line.

linkages Flows of inputs and outputs to and from a manufacturing plant in association with other plants. Movements of matter are **material linkages** as opposed to **machinery** and **service linkages** such as information, advice, and maintenance. Individual plants are also tied together by **forward linkages**—supplying customers—and **backward linkages** with their suppliers. **Horizontal linkages** occur between plants which are engaged in similar stages of a manufacturing process.

listed building In the UK, any building of such architectural or historical quality that permission must be granted before it may be changed or demolished.

lithification Processes by which loose sediments are converted into hard rock. These processes include the expulsion of air from the sediments or the suffusion into the rock of *cementing agents in solution, like quartz.

lithology The character of a rock; its composition, structure, texture, and hardness.

lithosphere The crust and that upper layer of the *mantle which lies above the *asthenosphere.

litter Dead plant material which reaches the ground. In soil science, the litter layer is the layer of dead and dying vegetation found on the surface of the soil. Some soil nomenclatures assign the litter layer to the A-*horizon, shown as A00.

Little Climatic Optimum The time period, roughly between AD 750 and 1200, when warmer conditions obtained in Europe and North America.

Little Ice Age The phase between AD 1550 and 1850 when temperatures were generally lower in Europe and North America than they are at present, and glaciers advanced.

littoral drift *See* *longshore drift.

livestock Domesticated animals in an *agricultural system. The rearing of livestock solely as food is costly, because animals are positioned high in the *food chain, meaning that large amounts of energy have been lost *en route*; more simply, there are more calories in the fodder used to provide meat than in the meat itself. However, such is the demand for animal protein that livestock products account for two-thirds of agricultural output in the developed world.

lixiviation *See* *leaching.

load The matter transported by a river or stream. **Solution load** is dissolved in the water. **Suspension load** refers to undissolved particles which are held in the stream. On the river bed, the material of the *bed load jumps by *saltation, or rolls along the bed. The deposits forming a channel bed are known as **bed-material load**.

loam An easily worked, fertile soil, composed of *clay, *silt, and *sand, roughly in a ratio of 20:40:40. A **clay loam** has a clay content of 25–40%, a

silt loam has more than 70% silt, and a **sand loam** has between 50 and 70% sand. Loams heat up rapidly, drain neither too slowly nor too easily, and are well aerated.

local climate The climate of a small area such as a moorland or city—a *mesoclimate—falling between a *microclimate and a *macroclimate. At this scale, such variables as local winds, *albedo, relief, slope, and *aspect are of considerable significance. *See* microclimate, *urban climate.

local winds Local winds blow over a much smaller area than global winds and have a much shorter time span. **Hot winds** originate in vast *anticyclones over hot deserts and include the Santa Ana (California), the Brickfielder (south-east Australia), the Sirocco (Mediterranean), the Haboob (Sudan), the Khamsin (Egypt), and the Harmattan (West Africa).

   Cold winds originate over mountains or other snow-covered areas and include the Mistral, funnelled down the Rhone Valley, and the *Bora. Some local winds, such as the Southerly Burster of Australia are associated with cold fronts.

   Other local winds include *land breezes and *sea breezes. *See also* *mountain winds and *föhn winds.

localization economies Advantages arising from the localization together of a number of firms in the same type of industry. *See* *agglomeration economies.

location **Absolute location** is expressed with reference to an arbitrary grid system as it appears on a map. **Relative location** is concerned with a feature as it relates to other features. Cox has argued, however, that location is a social product, rather than a thing in itself.

location coefficient Also known as the location quotient, this expresses the relationship between an area's share of a particular industry and the national share. Thus, the locational coefficient for a given region equals:

$$\frac{\% \text{ employed in a field in a given region}}{\% \text{ employed nationally in that field}}$$

A location coefficient of 2.0, for example, indicates that twice the percentage of workers are employed in a specific industry than the percentage employed nationally for that industry. It should be noted that a high location quotient for an industry in a region does not necessarily indicate high employment levels.

location reference In *Geographic Information Systems, the means used to relate information to a precise point.

location theory A group of theories which seek to explain the siting of economic activities. Various factors which affect location are considered such as localized materials and *amenity, but most weight is placed on transport costs. Early location theory was concerned with *industrial location theory, and with agricultural land use, as modelled by *von Thünen. Modern location theory has been concerned with the real individual, rather than with rational *economic man, reflecting the influence of *behavioural geography. Attention has shifted from the single factory producing a single product to the interrelationships within an organization or agglomeration, usually as part of a *capitalist economy. *See* *industrial location.

locational interdependence The response of a plant to its competitors in a given location. Plants may be attracted or repelled by the presence of rival plants and plan their locational strategies with regard to their competitors. Irregular arrangements may be made by two firms to locate so as to split the market between them.

locational triangle A model devised by *Weber to establish a least-cost location. In this model, S_1 and S_2 are the sources of the necessary raw materials, located at two corners of the triangle, and M is the market, at the third corner. Assuming equal transport costs in all directions, the least-cost location, P, located somewhere within the triangle, is derived from the amounts of 'pull' from each of the three corners.

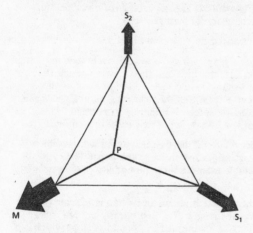

FIGURE 36: *Locational triangle*

location-allocation model A mathematical model used to establish the optimal location for larger, central facilities such as hospitals, factories, and schools. The model takes account of the location and demand of the customers, the capacity of the facilities, and operational and transport costs. These factors are used to calculate the number of facilities to be developed, together with their size and location.

loch, lough In Scotland and Ireland, respectively, a lake, or narrow arm of the sea.

locked zone An area along a *rift where *plates remain attached to each other. In such a zone, no new crust develops, so that the locked zone stretches, and the crust thins. Ultimately, the plates will separate, and the locked zone becomes a deformation at the edge of a continent.

lode A long, narrow vein of a mineral running through a rock.

lodgement The release and consolidation of debris from a glacier if the basal ice reaches its *pressure melting point as the ice moves. The moving ice aligns fragments of this debris, known as **lodgement** *till, in the same direction as the flow of the glacier.

loess, löss Originally referring to a loose, fine and sharp-grained soil occurring in the Rhine Valley, this term has been extended to refer to any unconsolidated, non-stratified soil composed primarily of silt-sized particles. It is a very fertile agricultural soil.

   The origin of loess is in dispute. Some writers believe the deposit to be wind-borne; others note the occurrence of the soil in *periglacial environments, and stress the importance of glacial grinding in the production of silt-sized particles. The loess may be derived from *outwash sands and gravels. A further school of thought points to the frequency of dust storms in deserts and postulates the importance of processes such as *salt weathering in the production of loess particles.

logical positivism *See* *positivism.

logistic curve A curve on a graph which shows relatively slow movement initially, becoming steeper through time and then slowing down. The curve has a characteristic S-shape; it starts with slow, linear growth, followed by *exponential growth, which then slows again to a stable rate. **Logistic growth** is characteristic of features like the increase in urbanization over time, or the innovation of new ideas in a community.

lognormal distribution A cumulative frequency distribution which appears as a straight line when plotted on log graph paper. For geographers, the most notable example is the plot of city rank against city size. *See* *rank-size rule.

long profile A section of the longitudinal course of a river from head to mouth, showing only vertical changes. The theoretically smooth curve shown by such a profile may be interrupted by *breaks of slope which can result from bands of resistant rock or from *rejuvenation.

long wave *See* *earthquake.

longitude The position of a point on the globe in terms of its *meridian east or west of the prime meridian, expressed in degrees. These degrees may be subdivided into minutes and seconds, although decimal parts of the degree are increasingly used.

longshore drift The movement of sand and shingle along the coast. Waves usually surge onto a beach at an oblique angle and their *swash takes sediment up and along the beach. The *backwash usually drains back down the beach at an angle more nearly perpendicular to the coast, taking sediment with it. Thus there is a zig-zag movement of sediment along the coast. **Longshore currents**, initiated by waves, also move beach material along the coast. The term *littoral drift* is synonymous.

loose-knit village A settlement where buildings are scattered at random but are not far enough apart to be deemed isolated. The village extends over a fairly large area without a clear nucleus.

lopolith A large *intrusion which sags downwards in the centre, forming a saucer-shaped mass.

Lorenz curve A *cumulative frequency curve showing the distribution of a variable such as population against an independent variable such as income or area settled. If the distribution of the dependent variable is equal, the plot will show as a straight, 45° line. Unequal distributions will yield a curve. The gap between this curve and the 45° line is the inequality gap. Such a gap exists everywhere, although the degree of inequality varies. *See* *Gini coefficient.

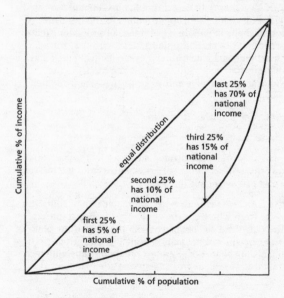

FIGURE 37: *Lorenz curve*

Lösch model A model of central places developed by A. Lösch (1954) which is less narrow than that of *Christaller, in that it treats the range, threshold, and hexagonal hinterland of each function separately. The resulting pattern of central places is much more complex than Christaller's, and yields a continuous, rather than a stepped, distribution of population sizes.

löss *See* *loess.

low A region of low atmospheric pressure. In Britain, the term low is generally applied to pressures of below 1000 *millibar.

low-order goods and services Goods and services with a low *range and a low *threshold population like daily newspapers, bread, and hairdressing. The goods are often *convenience goods. *See* *central place theory.

lowest bridging point The point on a river's course which is nearest to the sea and crossed by a bridge. Such a point is favoured since it is served by road, by river, and by sea.

Lowry model Developed by I. S. Lowry in 1964, this is a model of the evolution and distribution of urban land use—residential, industrial, and service—and the urban qualities of total population and *primary, *secondary, and *tertiary industry. It is based on the assumption that activities can be predicted from a given level of *basic employment. The model uses *economic base theory to determine the total population and employment in service industries. The population is assigned to zones of the city in proportion to their *population potential. Service employment is allocated in proportion to the market potential of each zone. These allocations have to meet land use constraints, notably housing densities, and threshold populations for the various services. The model is run several times while allocations are determined until the system reaches equilibrium.

lumbering The extraction of timber from forests.

lunette *See* *sand dune.

lynchet In earlier, possibly prehistoric times, small belts of uncultivated land were sometimes left between ploughed areas. These now can be seen in the landscape as part of a system of terraces.

lysimeter A block of soil, covered with vegetation, placed in a container and replaced in the site from which it came. The input of precipitation is measured with a rain gauge and the drainage from the base of the block is also recorded. The block of soil is repeatedly weighed. With the aid of these measurements, estimates of the loss of water by *evapotranspiration may be made.

maar A shallow, circular crater of volcanic origin, usually filled with water. The term is German; in the Eifel district of the Rhineland, many maars contain lakes, as in the Schalkenmehren Maar.

macro- Large-scale. Thus, a **macroclimate** is the general climate of a region extending across several hundred kilometres, such as the Great Plains of North America, **macrometeorology** is the study of large scale meteorological phenomena which can cover hundreds of kilometres or may encompass the whole globe, from *monsoons to the *general circulation of the atmosphere, and **macro-economics** is the study of an economy as a whole.

macrogeography The use of *centrographic methods in order to identify patterns in the spatial distribution of observed phenomena. At its centre lies the concept of *population potential which, it is alleged, is related to many patterns in social and economic geography.

magma The molten rock found below the earth's crust which can give rise to *igneous rocks. Molten magma may pick up pieces of existing rock—xenoliths—and is also charged with gases. It may dissolve and absorb the surrounding rocks in **magmatic stoping**.

magnetic pole reversal The earth's magnetic field resembles that of a bar magnet located at the earth's centre. The axis of the imaginary bar magnet emerges from the earth's surface at the magnetic poles. The north and south magnetic poles have repeatedly changed places while the axis has stayed in the same position. The timing of the intervals between such magnetic polarity reversals seems to be irregular. *See also* *geomagnetism.

magnetic stripes When *igneous rocks form, they take up the prevailing pattern of the earth's magnetic field at that time, but the north and south *magnetic poles have repeatedly changed places while the axis has stayed in the same position. As a result, areas of the ocean bed where *sea-floor spreading has taken place are characterized by parallel bands of igneous rock with differing magnetic polarity. The existence of these magnetic stripes did much to validate theories of *plate tectonics.

magnitude Of an *earthquake, an expression of the total energy released. *See* *Richter scale.

Magnox reactor Magnox is an alloy of magnesium which can be used to clad the uranium fuel in a nuclear reactor. It is used because it does not react with carbon dioxide, which can thus be used to carry the heat from the reactor to the generator.

malapportionment A device used by some politicians to improve their chances in an election by drawing up particular electoral units. The most successful ploy is the establishment of small constituencies for one's own party while creating large constituencies for the opposition. In this way, votes for one's own party will go further. *See also* *gerrymandering.

Malthusianism In 1798, Thomas Robert Malthus (1766–1834) published his *Essay on Population* in which he put forward the theory that the power of a population to increase is greater than that of the earth to provide food. He asserted that population would grow geometrically (1, 2, 4, 8, and so on) while food supply would grow arithmetically (1, 2, 3, 4, and so on). When population outstrips resources, **Malthusian checks** to population occur:

1. Misery: famine, disease, and war.
2. Vice: abortion, sexual perversion, and infanticide.
3. Moral restraint: late marriage and celibacy.

Malthus's predictions were not borne out in eighteenth-century Britain, perhaps because of the *agricultural revolution, with its associated increases in output, and with the opening up of the New World, which provided an outlet for excess population in the form of emigration and agricultural production. Other theorists pointed out that the capacity of a population to feed itself depended on the prevailing economic system; Marx, for example, believed that capitalism, rather than excess population, was responsible for low living standards.

More recently, however, the *Club of Rome has put forward Malthusian-type predictions of disaster due to population increase.

mammatus Breast-shaped lobes of cloud, hanging from the undersurface of the *anvil of a *cumulo-nimbus cloud, manifesting pockets of negative *buoyancy.

mammilated Smooth and rounded in appearance. The term can be used for landforms of different sizes from a rock to a landscape.

mandate A territory, once part of the German or Ottoman Empires, governed by a member of the League of Nations between the First and Second World Wars—Britain had a mandate in Palestine. The territory was held in a mandate until it was deemed to be capable of, or, indeed, demanded, independence.

mangrove swamp The term mangrove is applied to a number of types of low trees and shrubs which grow on mud flats in tropical coastal areas where the tidal range is slight. The roots extend out into the mud flats, and trap silt which accumulates to form the swamp. Mangroves are significant agents of *progradation along tropical coasts, and are especially well developed in south and east Asia.

Manning's roughness coefficient, Manning's *n* The resistance of the bed of a channel to the flow of water in it. Representative values of the coefficient are 0.010 for a glassy surface to 0.020 for alluvial channels with large dunes. The coefficient is expressed as n in **Manning's equation**:

$$Q = \frac{A\left(R^{\frac{2}{3}} \times S^{\frac{2}{3}}\right)}{n}$$

where Q = discharge, A = cross-sectional area, R = hydraulic radius, and S = slope, measured as a fraction.

If the Manning coefficient is known from tables, Q may be calculated. This is useful during times of flood.

Mann's model This model of British urban development, proposed by P. Mann (1965) combines the *sector theory with the *concentric zone model. Four basic sectors are postulated: middle class, lower middle class, working class, and lower working class. Each sector displays four zones. In each case, there is the *CBD, the *transitional zone, a zone of smaller houses, and the outermost zone made up of post-1918 housing.

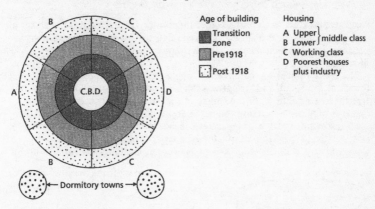

FIGURE 38: *Mann's model*

manor The smallest area of land held in the Middle Ages by a feudal lord. It usually covered one single village, and consisted of the lord's holding (the *demesne*) together with *open fields farmed on the *three-field system. It had its own court for dealing with minor offences.

mantle The middle layer of the earth, with a density of up to 3.3, and a thickness of some 2800 km. The mantle lies between the *crust and the *core of the earth. The upper layer, immediately below the *Mohorovičić discontinuity, is rigid, and forms the lower part of the *lithosphere. The lower layer of the mantle—the *asthenosphere—is floored by the Gutenberg channel.

manual digitizing *See* *digitizing.

manufacturing industry Also known as *secondary industry*, this is the mechanized, and usually large scale, processing of materials into partly finished or finished products.

map A cartographic representation of specifically chosen spatial information. The information is transmitted through images constructed from symbols. We tend to restrict the term to visual maps, but spatial information may be represented on a computer screen, through braille, or verbally through spoken description, and these categories of spatial representation may also be described as maps. **Map reading** is the process whereby people interpret and analyse map images, and an understanding of the physical and psychological means used in map reading helps *cartographers to improve the maps they produce. *See also* *mental map.

map generalization Decreasing the detail on a map when reducing its scale.

map overlay In *Geographic Information Systems, a *layer, not part of the base map, which corresponds to a specific theme.

map projection Both a method of mapping a large area and the result of doing so. The earth is a sphere; a map is flat, so that it is impossible to produce a map which combines true shape, true bearing, and true distance.

The term 'projection' comes from Mercator's projection, which was drawn as the shape and size of the shadows which opaque landmasses and transparent seas, lit from the centre of the earth, would throw onto a cylinder of paper encircling the earth and touching it at the equator.

Mercator's projection exaggerates the size of the northern continents and, relatively speaking, diminishes the size of tropical areas. Consequently, it has been criticized as over-emphasizing the importance of

FIGURE 39: *Map projection*

Europe and North America, although such was not Mercator's intention. No projection is perfect: for example, Mollweide's and Peters' are **equal area projections** (correct in area), but distort shapes. **Azimuthal projections** show true direction; **gnomic projections** show the shortest straight-line distance between two points; **orthographic projections** convey the effect of a globe. **Interrupted projections** show the earth as a series of segments joined only along the equator. Details of the projection used are given below each map in a good atlas.

maquis The evergreen brushwood and thickets of Mediterranean France.

march A frontier zone, often debated, between two nations. The region along the border between England and Wales is still referred to as the Welsh Marches.

margin of cultivation The distance from a market where the revenue received for a product exactly equals total cost. Given that production costs are the same whatever the distance from the market, transport costs rise with distance, and are, therefore, the determining factor of the location of the margin of cultivation.

margin of transference In von Thünen's model, the point at which the *economic rent gained from one form of agricultural production is bettered by that from another. Thus, the margin of transference is the boundary between one form of production and the next.

marginal analysis In economics, a concentration on the boundaries, or margins, of an activity rather than looking at it in its entirety. It may be concerned with, for example, the *utility or costs of an extra unit of production.

marginal costing The expenditure incurred by producing a further unit of a product or service, or the expenditure saved by not producing it. Marginal cost pricing is the fixing of the price of all units at the cost of producing the last unit.

marginal land Land, such as upland, or desert border, which is difficult to cultivate, and which yields little profit.

marginal propensity to save, MPS The proportion of an increment of income that is saved. An MPS of 0.5 indicates that the worker will save half of the increase paid.

marina A harbour with moorings for pleasure yachts.

marine Of, or concerned with, the sea.

maritime Of climates near the sea coast. Such climates have less extremes of temperature, both diurnally and seasonally, than their continental counterparts.

market The places where goods and services are bought and sold; this includes any convenient arrangement whereby people can buy and sell goods, services, and factors of production and is therefore not a particular site.

market area analysis The analysis of the way in which the market area of a firm is established. A. Lösch (1954) postulated an *isotropic plain with settlements regularly spaced. As one settlement develops manufacturing, its trade area can be represented by a demand cone. At its centre, the point of production, demand is high since the price need not include transport costs. With movement away from the point of production, the cost of the product will rise as transport costs are added. Transport costs may be incurred by movement of goods to the customer or by the customer in travelling to the point of production. Either way, buyers will be paying more than they would at the point of production. The market area will extend to the point where costs are enough to make the product prohibitively expensive. Beyond this point, a competitor may locate. In time, a hexagonal pattern of market areas will arise since the hexagon represents the most efficient shape of trade area. Thus, *locational interdependence determines the pattern of market areas. The market will be supplied by a system of regularly spaced plants, and the density of these plants increases as industry develops. In reality, the evolution of market areas is far more complex than the analysis given above.

market cycle A series of periodic—usually one day—markets such that a trader moves from one location to the next in a weekly cycle.

market economy An economy in which the major parts of production, distribution, and exchange are carried out by private individuals or companies rather than by the government, whose intervention in the economy is minimal. Although decisions about resource allocation are made by innumerable, independent producers and consumers, the whole thing is co-ordinated by the *market mechanism. Market economies are characterized by specialized production, the freedom to exchange commodities between individuals, and the use of the *market mechanism to determine prices. A market economy is characteristic of *capitalism.

market gardening The intensive production of fruit, vegetables, and flowers.

market orientation The tendency of an industry to locate close to its *market. Industries locate near the market if the cost of transport of the finished goods to the customers is a major part of the selling price. One example is the brewing industry where large, bulky quantities of water are used to make the finished product. Industries may also locate near the market in order to benefit from *agglomeration economies. An industry may be market orientated at regional scale but material orientated at sub-regional scale.

market mechanism The interaction of supply, demand, and prices. Here is a simple example: imagine that two producers of fizzy drinks are in competition. One produces orangeade, the other lemonade. If tastes swing away from orangeade to lemonade, demand falls for the former and rises for the latter. In response to falling demand, the orangeade producer lowers prices; in response to rising demand, the lemonade producer raises them. The consumers react to the higher prices by buying less lemonade, and to the lower prices by buying more orangeade, so that demand for the two returns to the original level.

market potential The intensity of possible contacts with markets. If a plant could be located in a number of locations, it is useful to be able to estimate the probable volume of sales which could be achieved at each possible site. This is the market potential, which is expressed as the sum of the ratios of the market to distances to each of the points under consideration. Thus:

$$P_i = \sum_{j=1}^{n} \frac{M_j}{d_{ij}} = \frac{M_1}{d_{i1}} + \frac{M_2}{d_{i2}} + \frac{M_3}{d_{i3}} \cdots + \frac{M_n}{d_{in}}$$

where the market potential (P_i) at point i is the summation (Σ) of n markets (j), accessible to the point i divided by their distance (d_{ij}) from that point. M is usually a measure of actual retail sales, and transport costs may be substituted for d.

Once P_i has been calculated for all the possible locations, the site with the highest market potential may be found, but in practice the prediction of likely sales rarely reflects the true situation.

market town Any town which has a trading market. In earlier times in Britain, permission was granted by the monarch to hold a market.

marketing geography The geographical application of the way that production of goods is linked with their marketing. Retail functions are at the heart of the study, especially as they affect the internal planning of the city, but all the marketing functions such as consumer behaviour, information flows, and the role of transport are also studied.

marketing principle The principle, as used by *Christaller, for an arrangement of settlements so that low-order places are as near as possible to the higher-order *central places. The numbers of settlements at progressively lower levels follows the sequence 1, 3, 9, 27 . . . This hierarchy is termed $k = 3$.

market-system firm A firm which operates as a response to changes in demand, not having a large enough part of the market to influence the price of its product. Most small manufacturing firms fall within this category, producing a limited range of products from a single plant.

Markhov chain An unfolding of events where each happening is partially determined by previous occurrences and partly by chance.

marl A mixture of clay with lime (calcium carbonate). Marling is the addition of marl to a light soil to increase its water-retaining capacity and to improve its texture.

marriage rate The number of marriages per thousand population or the number of persons married per thousand population. Both definitions are used.

Marxism A view of world events based on the work of Karl Marx (1818–83) and Friedrich Engels (1820–95). Marx saw man's history as a natural process rooted in his material needs. The historical evolution of mankind is seen as the outcome of the *modes of production which finally determine the nature of each historical epoch, the specific forms of property prevailing in it, and its class structure. The struggle between the classes is limited by the mode of production which determines the social struggle. This struggle provides the impetus for change. All history is the history of the class struggle. Marx believed that the outcome of this struggle was revolution.

Capitalism fosters large scale economic and social development but produces conditions which hamper its development (*see* *Marxist geography). Through systematic impoverishment of the masses, it creates a proletariat of exploited industrial workers who sell their labour as a marketable commodity. It is suggested that the proletariat will eventually rebel to emancipate mankind as a whole. This rebellion will put an end to all class distinction and all forms of exploitation. The sense of depersonalization and powerlessness felt by the working class will cease as the means of production become common property.

Marxist geography Marxist geography attempts to explain the world and also to change it. Marxism sees human beings gradually transforming themselves from stage to stage until they reach social perfection, and this transformation is seen as an aim towards which societies should be moving. This change is brought about by 'dialectical' processes—conflict between opposing forces—bringing forth a new synthesis which again is contradicted, and so on. The forces shaping society are seen as entities, which include capital, labour, capitalism and other *modes of production, the state, class, society, and the market.

Marxist geography highlights the dialectical relationships between social processes and the natural environment and spatial relationships. It is concerned with the modes of production which underlie the superstructure of society. It sees spatial and environmental problems, such as the destruction of habitats or uneven development, originating deep within the social formations of capitalism. It is aimed at changing the fundamental operations of social processes by changing the workings of production. Marxist geography is the study of the inherent contradictions of capitalism as they appear in the landscape and as they relate to each other.

Examples may be given of the changing structures and contradictions of capitalism. There is an inherent contradiction in a capitalist state which seeks to generate better conditions funded by taxation. Higher demand for raw materials generates higher costs. More output leads to more pollution and, in an environment where the authorities require strict controls on pollution, costs again rise. The result of these contradictions is the movement by multinational firms from established industrial regions in search of new environments to develop (despoil) and of new, politically virgin, labour to hire (exploit).

These views have been criticized. Many writers object to the passive role apportioned to individuals, who have been turned into non-decision-makers.

Marxist theory of rent The theory that part of surplus value is paid to landowners. Different rents may reflect the quality of the land or the amount of capital invested. Monopoly rent is the rent paid to the landowner when he leases the land to produce goods which are sold in a *monopoly. Absolute rent is extracted when the landlord can regulate the supply of land and force up prices.

mass movement, mass wasting The movement downslope of rock fragments and soil under the influence of gravity. The material concerned is not incorporated into water or ice, and moves of its own accord, but slides are often triggered by increase in water pressure on rocks and soil.

A widely used classification of mass movement uses the combination of types of movement (*falls, *topples, *slumps, *slides, and *flows) with the nature of the material (bedrock, *debris, and fine soil). In this classification, the term *creep is synonymous with *flow. Many cases of mass movement include more than one type of movement.

mass strength The strength of a rock in its resistance to erosion. Mass strength will vary according to the innate strength of the rock, but other factors are important, such as the jointing and bedding of the rock, and its state of weathering.

matched samples In two sets of measurements, where one measurement from the first set can be paired with one, and one only, measurement from the other set. An example would be the rates of evapotranspiration for different temperatures.

material index The ratio of the weight of localized materials used in the manufacture of a product to the weight of the finished product. A material index of much greater than 1 indicates that there is a considerable loss of weight during the manufacturing process (as in 'instant' mashed potato) so the factory should have a *material orientation, while a material index of less than 1 (where weight is gained during manufacturing) would suggest a *market orientation. An index of less than 1 could be achieved by an industry using largely ubiquitous materials, like water, as in the brewing industry. The use of material indices is central to *Weber's theory of industrial location.

material orientation The tendency of an industry to locate close to its raw materials. Industries with a high *material index locate near their raw materials, as do industries where the costs of raw materials are a major part of the selling price. Industries may be material orientated at sub-regional scale but market orientated at a regional scale.

mathematical geography The study of the earth's size and shape, of time zones, and of the motion of the earth.

matrix In statistics, an ordered array of numbers. The y-axis shows units of observation in columns, such as locations, while variables are shown across the x-axis. For example:

NATIONALITY	NUMBER OF CHILDREN IN FAMILY				
	1	2	3	4	5
A	20	40	10	10	10
B	10	15	25	30	20
C	20	25	35	20	5

Each row of the matrix gives an inventory of the variable for a given area whilst each column shows variations of one characteristic.

maximum sustainable yield The greatest yield of a renewable resource while keeping steady the stock of that resource.

mean *See* *arithmetic mean.

mean centre The 'centre of gravity' of a spatial distribution, such as a population or an industry over an area. It is determined by imposing an arbitrary grid on a map of the distribution. The co-ordinates for each point are recorded, and the means of the x- and the y-co-ordinates are calculated. Plotting these averaged co-ordinates gives the mean centre of the distribution.

mean deviation The mean of the sum of the deviations from the mean of all the values in a data set. The deviations are summed regardless of sign and the total is divided by the number of observations.

mean information field, MIF In *diffusion, the field in which contacts can occur. It generally takes the form of a square grid of 25 cells, with each cell being assigned a probability of being contacted. The possibility of contact is very high in the central cells from which the diffusion takes place, becoming markedly less so with distance from the centre, that is, there is a *distance decay effect. The probability values for the field may be based on observation, on a pre-existing theory, or arbitrarily. The model can then be used to simulate the diffusion of an innovation from a central point. To use the model, the MIF is placed with the centre over the source. A random number is then used to find the cell containing the destination of the innovation. From this receiving cell a random number is again used to find the receptor of the second generation of diffusion. This model can be run through a computer to foresee complicated diffusions, but the workings of the model are based on many assumptions which do not apply in the real world.

meander A winding curve in the course of a river. A *sinuosity of above 1.5 is regarded as distinguishing a meandering channel from a straight one. The dimensions of a meander are related to the square root of water discharge, Q:

$$\lambda = k_1 Q^{0.5};$$
$$A_m = k_2 Q^{0.5};$$
$$w_c = k_3 Q^{0.5}$$

where λ is meander wavelength, A_m is meander *amplitude, w_c is the channel width, and k_1, k_2, and k_3 are coefficients whose value varies with location. In addition, wavelength is about ten times the channel width, but the exact nature of these relationships is not understood. There may also be a relationship between meander development and *pool and riffle successions.

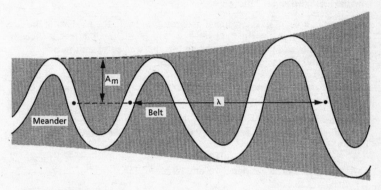

FIGURE 40: *Meander*

No satisfactory cause has been advanced for the formation of meanders, and possible explanations include: the rotation of the earth (although meandering is not restricted to those channels which are suitably orientated); deflection by obstacles (which may be relevant in some, but not all cases); secondary flow, which gives rise to *helicoidal flow; and that the meander form is the most efficient for the maintenance of stable discharge of water and sediment.

The **meander belt** is the total width across which the river meanders. *See also* *point bar, *ingrown meander, *intrenched meander.

meander core When a river cuts through a ridge between closely spaced meander loops, a meander core is left within the abandoned meander, which remains as an *ox-bow lake, or dries up. The Cirque de Navacelles, in the French Massif Central, is a good example.

means of production The ingredients necessary for the production of goods and services, including the social relations between workers, technology, and other resources used. Marx believed that *capitalism was characterized by the split between the capitalists, who owned the means of production, and the proletarians, who had only their labour services to sell.

mechanical erosion Erosion by physical means, such that the eroded material undergoes no chemical change, for example *abrasion, *freeze–thaw, *quarrying, *thermal erosion. The main agents of such erosion are wind, water, and ice. *Mass wasting is commonly held to be separate from erosion.

mechanical weathering The splitting up *in situ* of rock without chemical change. Processes include *exfoliation, *frost shattering, *granular disintegration, *salt weathering, spalling, and thermal weathering.

medial moraine *See* *moraine.

median In a set of data arranged according to rank, the median is the central value. Thus, for the data set: 1, 2, 3, 4, 5, the median is 3. If there is an even number of values, the median lies midway between the two central values. Thus, for the data set: 1, 2, 3, 4, 5, 6, the median is 3.5. For *grouped data, the median is found by plotting the data on a *cumulative frequency curve. The median value occurs halfway along the y-axis of cumulative frequency: at 50%.

medicinal geography The application of geographical methods to medical problems, such as disease, *morbidity, and *mortality. For example:
   1. The relationship of the distribution of disease due to geographical variables such as the incidence of bilharzia and the spread of irrigation schemes.
   2. A statistical analysis to discover whether spatial patterns of disease are due to chance or causal factors.
   3. The *spatial organization of health care.

Mediterranean climate A climate of hot, dry summers and warm, wet winters. This climate is not only characteristic of Mediterranean lands, but is also found in California, central Chile and the extreme south of Africa.
   In summer, the climate is dominated by *subtropical anticyclones, and *trade winds prevail. Daily weather is greatly influenced by *sea breezes and *land breezes. In winter, mid-latitude depressions bring rain. *Local winds, such as the Mistral of southern France and the Santa Ana of California are of great significance.

Mediterranean soils Soils formed in *Mediterranean climates. In the wet winters, there is *leaching of clays and carbonates and the release of iron which imparts a red colour to the soil. Leaching is slight during the dry summers so there is often a build-up of a carbonate *horizon in the soil.

megalith Any large stone which has been sited by humans and which may have been erected as a monument. Most megaliths date from between about 3000 to 2000 years BC. Megaliths may be arranged in rows or circles; the most famous British example is Stonehenge.

megalopolis Originally designating the seaboard of the USA from Boston to Washington, this is now any many-centred, multi-city, urban area of more than 10 million inhabitants, generally dominated by low-density settlement and complex networks of economic specialization. It is usually formed by the coalescence of *conurbations. Köln–Düsseldorf–Duisberg–Bochum–Dortmund furnish a European example.

melting pot A concept that a number of ethnic groups, cultures, and religions in a society will fuse together to produce new cultural and social forms. The idea was common in the USA in the first half of the twentieth century, and is exemplified in the motto on American coinage *e pluribus unum*: one out of many.

meltwater Water given out when snow or ice melts. Most snow and non-glacial ice melts in spring, often causing widespread flooding. Glacier ice may cause an increase in river levels later in the year.
    Glacial meltwater is produced by melting at the surface or by pressure and *geothermal heat at the base, such that *subglacial streams develop. Some surface meltwater may percolate through the ice to emerge at the base. *See* *fluvio-glacial.

meltwater erosion Meltwater derived from glaciers can be a powerful agent of erosion. Water flows on, within, and at the base of, decaying ice, often under pressure. Because of this pressure, meltwater can carry large quantities of debris; this load promotes *abrasion. For the same reason, meltwater can flow upslope.
    Four major types of meltwater can be distinguished. **Ice margin meltwater** follows the edge of the glacier when drainage routes have been cut off by ice. **Tunnel valleys** occur below the ice and can cut steep-sided, flat-floored valleys. **Spillways** are channels cut by streams overflowing from *proglacial lakes. **Coulees** are canyons which result from the sudden and violent release of water from ice-dammed lakes when the barriers which impound them are breached.

menhir A single standing *megalith.

mental map A map of the environment within the mind of an individual which reflects the knowledge and prejudices of that individual. Such a map reflects the individual's perceptions of, and preferences for, different places and is the result of the way in which an individual acquires, classifies, stores, retrieves, and decodes information about locations. Thus, for many of the residents of South-East England, the 'North' is compressed into Yorkshire and Lancashire. The North extends much further in reality than is generally recognized in the South. Mental maps may also include images of locations—the South may still be seen as an area of privilege, of white collar workers, and Lloyd's 'names'. Thus, different locations may be ranked in order of attractiveness as perceived by an individual or group and this may affect decisions about relocation or recreation, for example.
    Mental maps are conditioned by the way in which individuals organize the space available and, in turn, reflect an individual's perspective. Recent studies have shown that an industrialist's conception of the location of assisted areas may be wildly inaccurate.
    Sequential mapping focuses on links between places while **spatial mapping** concentrates on landmarks and areas rather than on paths.

Mercalli scale A measurement of the intensity of an *earthquake.

MODIFIED MERCALLI SCALE OF EARTHQUAKE INTENSITY

   I Felt by very few, except under special circumstances.
  II Felt by a few persons at rest, especially on the upper floors of buildings.
 III Felt noticeably indoors, although not always recognized as an earthquake. Vibration like passing lorry.
 IV Felt by many indoors during daytime, but by few outdoors. Some awakened at night. Vibration like lorry striking building.
  V Felt by nearly everyone; many awakened. Some breakages, disturbances of trees, telegraph poles.
 VI Felt by all; many run outside. Some heavy furniture moved.
VII Everyone runs outside. No damage in well-built buildings; moderate damage in ordinary structures; considerable damage in poorly constructed buildings.
VIII Considerable damage except in specially constructed buildings. Disturbs people driving cars.
 IX Damage even in specially designed structures. Buildings shifted from foundations; ground cracked; underground pipes broken.
  X Ground badly cracked. Railway lines bent. Landslides considerable.
 XI Few brick-built structures remain standing, if any. Bridges destroyed. Broad fissures in the ground.
XII Total damage. Waves observed on ground surface. Objects thrown upward into the air.

Adapted from US Geological Survey.

mercantilism The view, current in early modern Europe, that one nation's gain is only achieved by another nation's loss; that trade between states is a 'zero sum game'. According to this view, a trading nation can only prosper if it encourages the export of manufactures and the import of raw materials, but discourages the import of manufactures and the export of domestically produced raw materials, through the erection of *tariff barriers. Adam Smith (1723–90) countered this view, arguing that an increase in prosperity for all was possible by, as it were, increasing the size of the economic cake.

mercantilist model The view, propagated by Vance, that the most important urban function is wholesaling; cities in the New World developed because of trading contacts with the Old World and are more affected by their external markets than by internal markets. Vance would seem to be woefully ignorant of pre-colonial urban structures.

meridian An imaginary circle along the world's surface from geographic pole to geographic pole; a line of longitude. Meridians are described by the angle they form west or east of the **prime meridian**, which has a value of 0° and runs through Greenwich, England. It will be noon at the same time throughout the length of each individual meridian.

meridional circulation In meteorology, air flowing longitudinally, across the parallels of latitude. Meridional flow occurs in *atmospheric cells and results in part from changes in temperature along lines of longitude.

meridional temperature gradient The change in temperature experienced by movement north or south along a *meridian.

mesa A steep-sided plateau or upland which is formed by the erosion of nearly horizontal *strata. This landform resembles a *butte but is much more extensive. Mesas and buttes are evidence of *parallel slope retreat.

mesoclimate *See* *local climate.

Mesolithic The middle period of the *Stone Age, from about 10 000 BC to 4000 BC, but starting rather later in Britain: about 8000 BC.

mesometeorology The study of middle-scale meteorological phenomena; between small features, like *cumulus clouds, and large features, like anticyclones. Meso-scale features are up to 100 km across and last for less than a day, for example, the *polar low which, in a British winter, brings an hour or two of falling pressure followed by some three hours of snow.

mesophyte A plant which requires a moderate climate in terms of temperature and precipitation.

meso-scale In *meteorology, describing systems, or patterns of systems between small and *synoptic scale; dimensions of between about 10 and 100 km across, in the horizontal, have been suggested.

meso-scale precipitation event A band of intense precipitation, in or near a *front. They coincide with bands of *cumulo-nimbus, formed as differing air streams come together. The cumulo-nimbus greatly amplify the rate of precipitation, either directly, or through the provision of *freezing nuclei, and are more effective at cold, rather than warm, fronts. **Meso-scale precipitation areas (MPAs)** are small-scale clusters of *convection cells arranged into bands 50–100 km wide, in association with a low-level *jet stream.

Mesozoic The middle *era of earth's history stretching approximately from 225 to 190 million years BP.

metamorphic aureole That area of rock altered in composition, structure, or texture by contact with an igneous *intrusion. A metamorphic aureole surrounds the Dartmoor granite intrusion.

metamorphic rock Rocks which have been changed from their original form by heat or by pressure beneath the surface of the earth. Metamorphic transformations include limestone to marble, shale to slate, and slate to schist. When *magma forms an *intrusion, it heats and alters the surrounding rocks by **contact metamorphism**, which forms a ring of altered rocks—the *metamorphic aureole—around the intrusion.

  Dislocation metamorphism occurs through friction along fault planes or *thrust planes. **Regional metamorphism** (also known as dynamic metamorphism) occurs during an *orogeny.

metasomatism The change in *country rock brought about by the invasion of fluid into that rock. Granites are formed at great depths by invasion of granitizing fluids.

meteoric water Water precipitated from the atmosphere, as opposed to *juvenile water.

meteorology The study of the character of the atmosphere and the events and processes within it, together with the interaction between the atmosphere and the face of the earth.

metropoles d'équilibre Eight metropolitan areas chosen to be *growth poles in the Fifth French Plan between 1966 and 1970. The aim was to promote regional development and to shift economic activity away from Paris.

metropolis A very large urban settlement usually with accompanying suburbs. The term is used rather loosely as no precise parameters of size or population density have been established.

metropolitan city In the USA the equivalent of a *conurbation.

Metropolitan Statistical Area, MSA Formerly a *Standard Metropolitan Statistical Area, this is an economically and socially integrated, large urban unit. Urban areas may be made up of more than one MSA.

micelle A cluster of molecules in a colloid. *See* *clay micelle.

microclimates The climates of those parts of the lower *atmosphere directly and immediately affected by the features of the earth's surface. The height of this part of the lower atmosphere varies according to the size of the influencing feature, which is usually a type of vegetation, or building, but could be put at four times the height of the feature.

microclimatology The study of *microclimates such as those beneath standing crops, within forests, or in built-up areas. *See* *urban climates.

microfauna Any animals, such as bacteria, too small to be seen individually by the naked eye.

microflora Any plants, such as algae, too small to be seen individually by the naked eye.

microgenetic Concerned with human adaptation to environmental change.

micrometeorology The study of small-scale meteorological phenomena operating near the ground surface, such as the investigation of climates within a field of grain or in a forest.

mid-latitude depression An area of low atmospheric pressure occurring between 30° and 60°, shown on a weather map as a circular pattern of isobars with the lowest pressure at the centre. This low is some 1500–3000 km in diameter and is associated with the removal of air at height and the meeting of cold and warm *air masses in the lower atmosphere. At the fronts between the air masses, a horizontal wave of warm air is enclosed on either side by cold air. The approach of the warm front is indicated by high *cirrus cloud. With the approach of the front, the cloud thickens and lowers. Rain falls. As the warm sector passes over, skies clear and the temperature rises. The cold front is marked by heavier rain and a fall in temperature. *See* *polar front.

The air behind the cold front moves more rapidly than the warm sector air and eventually pinches out the warm air, lifting it bodily from the ground to form an *occlusion.

Mid-latitude depressions move at around 30 km per hour in summer and 50 km per hour in winter and last between four and seven days.

mid-oceanic ridge *See* *oceanic ridge.

migration The movement of people from one place to another. The terms **in-migration** and **out-migration** are used for **internal migration**, where no national boundaries are crossed, and the simplest classification separates this from **international migration**.

While **voluntary migration** refers to unforced movements, **compulsory migration** describes the expulsion of minorities from their country of birth by governments, or by warring factions. In the 1970s, Asians were expelled from Uganda and Kenya, and the 1990s have seen the introduction of the term 'ethnic cleansing' in relation to the former Yugoslavia.

Migrations may be temporary or permanent. In the case of *commuting, migration is a daily act, but, because there is no change of residence, a purist would not call commuting a migration, preferring the term **mobility**. **Temporary migrations** may be **seasonal**, as migrant workers move in search of work, or **periodic**, as when a worker, usually male, moves to an industrial, urbanized area and sends money back to the women and children, perhaps over a period of a year or two. A good example of periodic migration is the movement of males from their homes in Botswana and Lesotho to work in the gold and diamond mines of South Africa.

Other classifications are based on the nature of the points of origin and arrival, such as rural–rural or urban–rural. **Rural–rural migration** may be seen in the movement of *nomadic people while **urban–rural migration** might include the movement of elderly people when they retire or when richer people move from the city to suburbs. *Rural depopulation describes rural–urban migrations.

Further classifications are concerned with the motive for migration. Compulsory movements, such as the repatriation of Ghanaians from Nigeria are seen to be entirely due to *push factors. Other cases include *pull factors: in **innovative migration** people move to achieve something new, like settlement of new lands, in **economic migration** people move from a poor to a richer area, and in **betterment migration** people move to uphold a lifestyle which is being threatened.

See also *emigration, *gravity model, *immigration, *intervening opportunity, *Ravenstein's 'laws' of migration, *refugee.

migration chain *See* *chain migration.

Milankovitch cycles There are three interacting, astronomical cycles in the earth's orbit around the sun: in the shape of the elliptical orbit (about 95 000 years), in the axis of rotation (about 42 000 years), and in the date of perihelion—the time of year when the earth is closest to the sun—(about 21 000 years). Building on the theory of M. Milankovitch (1930), a link has been established between these variations and long-term climatic change. Glacial conditions, for example, are favoured by conditions of small axial tilt, and perihelion in the Northern winter.

milieu The sphere in which each individual lives and which he or she is affected by. This sphere includes the tangible objects and people, the social and cultural phenomena, and the *images which influence human behaviour. *See* *phenomenal environment.

millibar, mb A unit of atmospheric pressure, measured by a barometer. Each rise of one square centimetre of mercury in a barometric column represents a rise in air pressure of 1000 dynes.

million city An urban area with a population in excess of one million. By the 1990s there were more than 250 million cities, more or less evenly divided between the *more developed and *less developed worlds.

mineralization 1. In soil science, the breakdown of organic residues by oxidation to form soluble or gaseous chemical compounds which may then take part in further soil processes or be utilized by plant life. Mineralization is an essential process in the formation of humus.
   2. The fossilization of buried plant and animal matter as the organic parts are replaced by other minerals.

mineralized zone An enriched zone of mineral deposits around an *igneous intrusion.

minimum efficient scale (m.e.s.) of production The smallest possible size of a factory which is compatible with profitable production. Below this size, *economies of scale do not apply. This minimum can be defined in terms of production or, less often, in terms of employment. Minimum sizes vary with the nature of different industries and may be so large that there is insufficient capital for production.

minimum requirements method A technique for assessing the number of *non-basic workers within a city. Specific size groupings are established for settlements of over 10 000. The percentages of labour forces are established for each occupation in each size group. The lowest number of workers in each employment group is noted within each size grouping, and these lowest numbers are regarded as the minimum requirements for the relevant occupation and city group. The minimum requirements are compared with occupational groups in centres of similar size. Any workers in excess of the minimum requirement are assumed to be non-basic workers. The formula for establishing the number of non-basic workers is:

$$S = e_i \frac{e_t}{E_t} \times E_i$$

where S is the minimum requirement for that industry, e_i is the employment total for that industry, e_t is the total employment for that area, E_i is the national employment figure for that industry, and E_t is the total national employment.

mirage When air near ground level is heated strongly by contact with the hot earth, it becomes less dense. Incoming rays of light are bent when entering this layer, so that a patch of sky is mirrored in the hot air. This often gives the appearance of water.

Occasionally, when air at ground level is very cold, rays of light from regions beyond the horizon are bent downwards. As a result, features usually beyond the horizon become visible.

misfit stream A stream which appears to be too small to have made the valley in which it is flowing. This valley may have been cut by a former glacier, as was the Nant Ffrancon, Snowdonia, now occupied by the River Ogwen. Alternatively it may have been cut by a larger stream which either suffered a decrease in *discharge through climatic change, or has been *captured.

mist A suspension of water droplets in the air which restricts visibility to between 1 and 2 km.

Mistral *See* *local winds.

mixed economy An economy where there is more government intervention than in a free *market economy; many of the activities of production, distribution, and exchange are undertaken by central government, but where there is more economic freedom for the individual than in a *command economy. There is therefore a mixture of *socialism and *capitalism.

mixed farming A type of commercial agriculture concerned with the production of both crops and animals on one farm. Stock on a mixed farm used to be grazed on fallow land, but many modern mixed farms produce some, or all, of their fodder crops.

mobility A general term used to describe any kind of spatial movement; not solely *migration, which involves a permanent change of residence, but also, for example, tourism, commuting, or studying away from home. These latter forms of movement have been classified as *circulation. W. Zelinsky (*Geog. Rev.*, 1961) developed a **mobility transition model** which suggested that, as a region or nation develops, there are changes in mobility. Geographical mobility, initially restricted, increases greatly with development, until the point when sophisticated transport means that migration may be replaced by circulation (one might, for example, commute from Leeds to London, or take regular international flights instead of relocating), or when phones, faxes, and modems reduce the need to circulate.

modal centre *See* *centrality of population.

modal split The varying proportions of different transport modes which may be used at any one time. The choices of modes may be determined by the costs, destinations, capacities, and frequencies of the modes together with the nature of the goods carried and their destinations. Modes of transport may be seen as competing services, and particularly so in the rivalry between the private car and public transport systems. In many cases the travelling time and comfort of a car journey outweigh costs so that non-cost factors play an important part in determining the modal choice. By the mid-1990s it seemed that the shift to private motoring which characterized British government policy in the 1980s was waning, and that the later 1990s might see restrictions for motorists.

In the transport of freight, water transport and pipelines are most suited to high volumes of freight over a long haul while small-scale local movements are best served by road.

mode The figure in a set of data which occurs most often. If the mode is used to indicate the predominant class grouping, rather than an individual value, the term **modal class** is used. However, the modal class will change according to the class limits (the size of the value range for each class), so that a change of class limits may well give a different distribution.

mode of production The way in which society organizes production. Marxists claim that the system of ownership of the *means of production is the foundation of all social systems, and that the *superstructure—the politics and ideology of a society—takes its character from the mode of production. *See* *historical materialism. Furthermore, Marxism claims, the mode of production, especially the *capitalist mode, reinforces and reproduces the *status quo*. In the primitive, communal mode of production, land and raw materials are communally owned and labour requirements are shared. In the slave mode of production, the labourer is a chattel who may be bought and sold. Under a *feudal system the proprietor owns the land and most of the produce, but the peasant, who is tied to the land, may own some of the *factors of production. Under *capitalism, the worker may sell his labour for a wage but does not own the other factors of production. Marx saw socialism as a transitional stage; the point at which the state defeats the capitalist class and at which class distinctions break down, paving the way for communism. Marx did recognize that more than one mode of production might be present in a society.

This rather deterministic view of the critical influence of the mode of production has been challenged, for example, because of the Eurocentric nature of the argument. None the less, geographers recognize that each mode of production creates its own geography. Capitalism seems to be inextricably wedded to *uneven development, and the rise of corporate capitalism in the USA has transformed city form and structure.

model A representation of some phenomenon of the real world made in order to facilitate an understanding of its workings. A model is a simplified and generalized version of real events, from which the incidental detail, or 'noise' has been removed. An **iconic model** represents reality on a smaller scale, an **analogue model** shows reality in maps and diagrams, and a **symbolic model** uses mathematical expressions to portray reality. **Probabilistic models** take into account the fact that human behaviour cannot be predicted with absolute certainty, while **simulation models** use mathematical laws of probability to simulate the consequences of human behaviour. Finally, in an **economic model**, the variables are defined in cash terms.

In geography, models were at their most popular in the 1960s; since that time, few new models have been created, and many classic models, such as those of von Thünen or Hägerstrand have been reworked.

model township A planned settlement first conceived in late nineteenth-century Britain by philanthropic industrialists to house their workforce. Bourneville, built in suburban Birmingham by Cadbury in 1879, is one example.

moder A type of humus which is less acid than *mor but more acid than *mull. The degree of mixing of this humus with the mineral content of the soil is greater than that of mor, but it still shows as a stratified layer.

modernization The change in society towards a more efficient government and control, better provision for health and social security, increased educational opportunities, and, possibly, increased social mobility.

modifiable areal units Areas studied in geography; such as states, counties, census areas, or *enumeration districts, which may be arbitrary units unconnected with any spatial patterns which have developed. Thus, the boundaries of many African states, for example, cut across cultural and ethnic regions.

In many cases the investigator has to use the units for which data are available rather than the units which are more suited to the investigation. The use of large units, such as census areas, may increase correlation coefficients, while at the same time hiding smaller-scale patterns and masking changes within the area. Analyses using different areal units may produce different results.

Moho, Mohorovičić discontinuity The boundary in the earth's interior between the *crust and the upper *mantle. It occurs at about 35 km below the continents and at around 10 km beneath the oceans, and marks a change in rock density from 3.3 (crust) to 4.7 (mantle).

Moh's scale *See* *hardness.

moist coniferous forest Some writers include this area with *coniferous (boreal) forest but it is found only on the west coast of Canada and the USA where the climate is cool and moist throughout the year. Very large evergreens such as redwoods and Douglas fir grow. It is difficult to know why evergreens, rather than broad-leaved deciduous trees, abound. Undergrowth is sparse.

moisture index A measure of the water balance of an area in terms of gains from precipitation (*P*) and losses from *potential evapotranspiration (*PE*). The moisture index (*MI*) is calculated thus:

$$MI = \frac{100(P - PE)}{PE}$$

See also *lysimeter.

mollisol In *US soil classification, soils with a rich humus content, developed under grassland. *Chernozems, *rendzinas, and *chestnut soils fall into this category.

monadnock An isolated peak, the remnant left by long-term subaerial denudation, now standing above the level of a *peneplain. The term comes from the classic example of this type of landform; Mt. Monadnock, in New England. Mars Hill, Maine, is a further example. *See* *cycle of erosion.

monoculture A farming system given over exclusively to a single product. Its advantages are the increased efficiency of farming and a higher quality of output. Disadvantages include a greater susceptibility to price fluctuations, climatic hazards, and the spread of disease.

monopoly The exclusive ownership or control of a resource; in economics, the provision of a good or service by a single supplier who then has the power to set prices, since competition does not operate. In practice, a monopoly occurs where one firm controls most of the output of a particular industry, but it is also common to find a small number of firms dominating the market. Such firms may agree, formally or informally, to limit competition between themselves—in other words, to set up a *cartel. Legal prohibition of monopolies is common in capitalist economies, so that firms have to seek new products rather than establishing a monopoly if they wish to continue to grow in the domestic market. In fact, *industrial diversification of this type makes sense, since a single-product firm is vulnerable to a fall in the demand for its product.

monsoon Colloquially, a sudden wet season within the tropics, but, more explicitly, a seasonal shift of air flows, cloud, and precipitation systems. Monsoons have been described in West and East Africa, Northern Australia, Chile, Spain, and Texas, but the largest is the south-west monsoon. This Asian monsoon is, fundamentally, the atmospheric response, complicated by the presence of water vapour, to the shift of the overhead sun, and therefore zone of maximum heating, from the Tropic of Capricorn in late December to the Tropic of Cancer in late June. Associated with this response are major changes in *jet stream movements and a meridional shift of the rain-bringing inter-tropical convergence zone.

In winter, pressure is high over central Asia. Winds blow outwards. Some depressions, guided by upper-air westerlies, move from west to east, bringing rain. During the spring, a *thermal low develops over northern India. Rains enter Burma in April and May, and India in late May, or early June. These rains are related to a trough in the upper air. In early summer the direction of the upper air changes from westerly to easterly. With this change, the monsoon 'bursts', giving heavy rain across the southern half of the subcontinent. The tropical easterly jet stream is semi-permanent about 15° N for the rest of the summer. By late June there is a continuous southerly flow of warm, moist air into the monsoon trough lying across northern India. This flow is a continuation of the south-east *trades, altered to south-westerlies by the Coriolis force as they move north, across the equator, bringing huge quantities of water vapour, and reaching the west coast of India as the south-west monsoon. Here, the rainfall is *orographically enhanced. *Monsoon depressions are formed in association with these air flows, steered by the now easterly jet. Subsiding upper air prevents rainfall in the Thal and Thar deserts of the north-west of the subcontinent. By autumn, the easterly jet stream is replaced by a narrow band of the westerly subtropical jet stream, which follows the southern Himalayas. The south-west monsoon begins to retreat in September. Thereafter, the north-east trade winds dominate, and *hurricanes are common in the Bay of Bengal.

monsoon depression A low pressure system occurring in summer and affecting southern Asia. It is 1000–2500 km across with a *cyclone circulation up to 8 km. It lasts 2–5 days and occurs roughly twice a month, bringing 100–200 mm of rain per day.

monsoon forest A type of tropical forest found in regions showing a marked dry season followed by torrential rain; a monsoon. The vegetation is adapted to withstand the drought so that trees are semi-deciduous or evergreen. The forest is more open and has more undergrowth than *tropical rain forest.

Monte Carlo model A simulation model—a method of representing reality in an abstract form which incorporates an entirely random element and describes a sequence of events from one state to another in terms of probability. It is impossible to predict any outcome as one process may give rise to any number of events. Change is represented by random sampling from a number of probabilities and though each stage of the model is dependent on the stage before, it is also subject to chance factors.

mor An acid humus developed from the tough, acidic leaf litter of conifers and heathland vegetation. The activity of soil fauna, such as earthworms, is slight, and mor remains unmixed with the mineral content of the soil.

moraine Any landform directly deposited by a glacier or ice sheet. The material which makes up moraines is often partly stratified, since some may have been formed under water.

  Ablation moraine is a grainy, sandy *till which sometimes overlies ground moraine: coarse because meltwater has washed out finer particles as the glacier shrank. They are common on retreating glaciers, such as the Suldenferner of the Italian Tyrol. Certain moraines are deposited at the side of the glacier as **lateral moraines**. Very well-marked lateral moraines are found each side of the Tschierva Glacier in Switzerland. Where two lateral moraines combine, a central, **medial moraine** may be formed. Moraine beneath a glacier may exist as a blanket covering the ground. This is **ground moraine**, also known as a till sheet, which covers, for example, large parts of north Germany, west and north Russia, and the northern Prairie states of the USA.

Other moraines have been moulded by ice parallel to the direction of ice movement. These include **fluted moraines** which are long ridges, possibly formed in the shelter of an obstruction. *Drumlins are streamlined moraines.

End moraines, or **terminal moraines** mark the end of a glacier; several run in an arc through the North German Plain, and others are found around the Great Lakes. They are ridges of *till, not usually higher than 60 m. In plan, they often form a series of crescents, corresponding with the lobes of the glacier; a well-developed example indicating that the ice front was at that location for some time. Not all former ice fronts are marked by terminal moraines; some may have been destroyed by meltwater. **Recessional moraines** mark stages of stillstand during the retreat of the ice. A moraine running across a glacier is a recessional, rather than an end moraine, if the up-glacier surface shows streamlining. Such streamlined transverse moraines are **rogen moraines**. These are ridges, up to 30 m in height and crescentic in shape, with the horns of the crescent pointing in the direction of the ice flow. Other transverse moraines form where a glacier meets its *proglacial lake. These are **de Geer moraines** and consist of *till, layered sand, and lake deposits.

At the margin of a glacier is **an ice-dumped moraine. Push moraines** occur when a glacier is retreating in the melt period but re-establishing itself in the cold season when the advancing glacier pushes up last year's moraine.

morbidity Ill-health. **Morbidity rates** are of two types: the *prevalence rate*, which gives the numbers suffering from a specific condition at any one time, and the *incidence rate*, which gives the number of individuals suffering from a particular condition within a given period of time, usually one year. Regional variations in morbidity are of interest to *medicinal geographers, and may indicate regional differences in living standards and life-styles. At the time of writing, attention is focused on the fact that the incidence of bowel cancer is higher in Scotland than in the rest of the UK, and the incidence of breast cancer is higher in the UK than in the rest of Europe.

morphogenetic region A region, such as a *periglacial area, in which a distinctive complex of land-forming processes are determined by climate and give rise to a distinctive set of surface features. *See* *climatic geomorphology.

morphological mapping A technique of mapping landscape features using standardized symbols. For example, slope elements may be displayed: convex, free face, concave, break of slope. Recognition of such features is not objective, however, and some landforms lie more in the eye of the beholder.

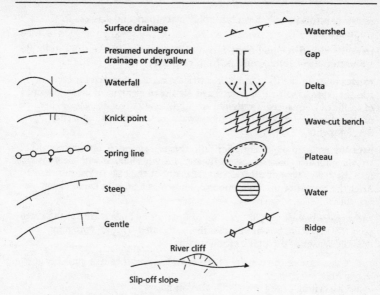

Surface drainage		Watershed	
Presumed underground drainage or dry valley		Gap	
Waterfall		Delta	
Knick point		Wave-cut bench	
Spring line		Plateau	
Steep		Water	
Gentle		Ridge	

River cliff

Slip-off slope

FIGURE 41: *Morphological mapping*

morphology 1. The study of form. However, the term is now used as a synonym for the form itself as in the morphology of the landscape or of the city.

2. In geolinguistics, the phonological shapes of words that adapt to special grammatical functions. Some words, for example, inflect to make a past tense, as in 'walk', 'walked'; others are the basis from which additional words may be derived as in 'hard', 'hardness'.

mortality Death; a low **mortality rate** indicates a longer life expectancy. Mortality rates are generally higher in the Third World than in developed nations, although some LDCs have surprisingly low rates, which reflect the youthful nature of their populations. Thus, in 1994, the death rate in Sri Lanka was lower than the UK rate.

Rates differ within countries; in the UK, mortality is higher in the 'old' industrialized regions of the north and north-west than in the more recently industrialized areas of the south. Regional variations hint that life expectancy is lower in soft-water regions than in areas of hard water. There are indications that country dwellers live longer than town dwellers. Mortality also varies between the socio-economic classes of the *Registrar General; mortality rates increase down the scale. This is thought to be a reflection of life-styles rather than affluence. *Infant mortality also increases down the social register and from developed to less developed countries.

moshav, moshavim (pl.) An Israeli agricultural co-operative where each family controls its own farm but holds machinery and services in common.

mother cloud A cloud from which another cloud develops; for example strato-cumulus cumulo-genitus (from cumulus).

motorway In the UK, a road of restricted access, reserved for certain specific types of vehicle and designed for the movement of heavy volumes of traffic at high speed. Gradients are slight and curves are gentle. Because of this ease of transport, many motorways have attracted industrial development.

moulin Also known as a glacier mill, this is a rounded, often vertical hole within stagnating glacier ice. Meltwater, heavily charged with debris swirls into the hole. Some of this debris settles out at the base of the moulin. After the retreat of the ice, a mound, known as a moulin kame, is left behind.

mountain building The creation of uplands by movements of the earth's crust This is often, but not necessarily, associated with an *orogeny. Possible causes of the uplift of mountains are:

1. A decrease of the density of the crust causing it to rise, possibly above a *hot spot.
2. Thickening of the crust at *collision margins.
3. *Subduction below continents, causing them to rise.
4. Overthrusting of sedimentary rocks during collision.
5. Compression at *converging or at *conservative plate margins.

mountain ecology A change in altitude up a mountain is equivalent to a latitudinal shift away from the equator, and vegetation types accordingly change with height. Thus, in the San Francisco mountains there is a change upslope from hot desert cactus at 200 m to oak scrub at 800 m. By 2000 m, the forest is coniferous shading into spruce and Douglas fir. At 4000 m, the tree line ends and Alpine tundra is encountered. Since each stage is of limited area, the full range of each *biome is not represented.

mountain meteorology Mountains give rise to distinctive climates. Upland areas are cooler than lowlands of the same latitude, since temperatures fall by 1 °C per 150 m. Heating of the valley floor to a higher temperature than the valley sides may lead to *anabatic mountain winds. Equally, colder and heavier air at the peaks may spill down into the valley floor as a *katabatic wind.

Warmer, moist air will be forced to rise over mountain barriers. As the air is chilled with height, *orographic rain falls. When such air has passed over the barrier, it descends and is *adiabatically warmed. This may bring warm winds, like the *chinook. Since the warmed air is now capable of holding the remaining moisture, a 'rain shadow' of drier air develops in the lee of mountains.

On a larger scale, mountain barriers affect low pressure systems as they pass over and *lee depressions may also develop. The location of major mountain barriers such as the Rockies and Himalayas affects the formations of air in the upper atmosphere.

mountain wind A wind which occurs when heavier, cold air flows downslope from mountain peaks or from a glacier. *See* *katabatic wind and *bora.

movement minimization theory The view that consumers choose the shortest journey to buy goods and services, whatever the cost. However, consumers may be more attracted by lower prices than by short journeys, and many consumers are motivated by quality, service, and ease of parking.

moving average A method of calculating *central tendency over time. The average is calculated, for example, over five years. For each year after this, the earliest value is dropped from the calculation and the most recent one is added in, again to make an average over five years; for example: 1988–93, 1989–94, 1990–5. A moving average is calculated in an attempt to even out short-term oscillations and thus identify long-term trends.

mud flat An accumulation of mud in very sheltered waters. Mud from the shore is carried into *estuaries and sheltered bays and settles at low water. *See* *salt marsh.

mud flow The *flow of liquified clay. Debris of all sizes, including large blocks of rock may be transported by mud flows. In 1891, at Ganderbach in the Pusertal, 475000 m^3 of mud was displaced in a mud flow.

mull 1. In soil science, a mild humus produced by the decomposition of grass or deciduous-forest litter. Earthworms and other soil fauna mix this humus thoroughly with the mineral content of the soil.
    2. In Scotland, a promontory or headland.

multicell storm A *severe local storm comprising a succession of convective cells. The first cell is formed by a powerful updraught of moist air. As the ensuing downdraught stifles this updraught, a new updraught develops to the right of the first. This sequence continues until the surface air cools.

multicultural society *See* *plural society.

multinational corporation, MNC A firm which owns or controls production facilities in more than one country through direct foreign investment. Although multinationals grew most rapidly in the 1960s, the foundations were laid in the inter-war period, notable examples being Ford, Vauxhall, and Phillips. In the mid-1980s, multinationals accounted for 14% of UK employment and 30% of UK exports. The corresponding figures for France were 24% and 32%.

Multinationals are made possible by improved international communications which provide rapid containerized transhipment and foreign travel, easy communication of information, and international mobility of capital. When one market is saturated, the multinational can rapidly develop others, since foreign investment cuts transport costs, and makes possible a rapid response to local markets. It also eases tariff barriers—the UK has been an attractive location for many Japanese manufacturers, for example, because it is within the European Union, but has opted out of the EU's social charter. Multinationals can compare costs at different locations, and can switch activities to different areas as appropriate.

MNCs are probably the major force affecting world-wide shifts in economic activity, since the largest have a turnover greater than the *GNP of many *less developed nations. Although a developing nation may benefit from the construction of a plant for an MNC in terms of jobs and markets, it has been argued that the price is a loss of local control. *See* *transnational corporation.

multinational state A state which contains one or more ethnic groups as identified by religion, language, or colour. European examples include Belgium (Walloon and Flemish) and Spain (Castilians, Catalans, and Basques). If the minority is concentrated in a particular area or is associated with a historic homeland, pressure may build up for the ethnic group to achieve either greater autonomy, or even to be independent.

multiple land use The use of land for more than one purpose as when an area of forestry contains nature trails, and picnic or camping facilities.

multiple nuclei model A model of town growth advanced by C. D. Harris and E. L. Ullman (*Ann. Am. Pol. Soc. Sci.*, 242) based on the fact that many towns and nearly all large cities grow about many nuclei rather than around a simple CBD. Some of these nuclei are pre-existing settlements, others arise from urbanization and *external economies. Distinctive land-use zones develop because some activities repel each other; high-quality housing does not generally arise next to industrial areas, and other activities cannot afford the high costs of the most desirable locations. New industrial areas develop in suburban locations since they require easy access, and outlying business districts may develop for the same reason. While the layout of the model is generally standard in most reference books, the location of the various sectors is infinitely variable, in contrast to the *concentric model. *See also* sector theory.

multiplier The economic consequence, intended or otherwise, of an action, whether in terms of the number of jobs created or the extra income. Basically, if a given amount of money is injected into a region, the income of that region increases not by the amount of cash injection but by some multiplier of it. In a closed system, the multiplier may be calculated if the *marginal propensity to save is known for the population.

Thus, if the government builds a factory giving each worker £1000, half is spent on, say, food and half is saved. Of that £500, the food supplier might spend £250 on clothes and save £250, and so on. The value of the multiplier for a closed system is shown as:

$$k = \frac{1}{MPS}$$

where k is the multiplier and MPS is the marginal propensity to save. In reality, much of any increment may well be spent outside the region. This draining away of money is known as leakage. Thus:

$$k = \frac{1}{MPS + P}$$

where P is the percentage of additional income spent on leakages. A low regional multiplier makes the possibility of rejuvenating a problem area very remote.

The multiplier can work in reverse if a source of income is cut.

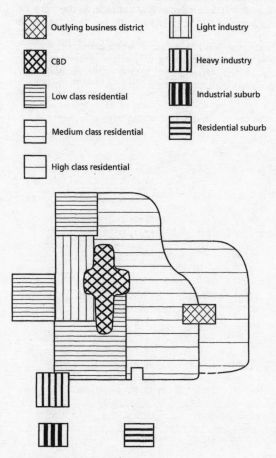

⊠ Outlying business district   ▥ Light industry

⊗ CBD   ▥ Heavy industry

☰ Low class residential   ▥ Industrial suburb

☐ Medium class residential   ☰ Residential suburb

☐ High class residential

FIGURE 42: *Multiple nuclei*

multivariate analysis, MVA Any statistical technique analysing the relationship between more than two variables; in other words an analysis which looks at the simultaneous and combined effects of a number of variables. Simplifications have to be made; for example, the interrelations between many variables have to be reduced to data on the correlations between each pair.

A frequently used technique in MVA is the use of the **multiple linear regression equation**:

$$Y_c = a + b_1X_1 + b_2X_2,$$

where Y_c is the estimated value of the dependent variable, a is the Y intercept, X_1, is the value of the first independent variable, X_2 the value of the second independent variable, b_1 the slope associated with X_1, and b_2 the slope associated with X_2.

MVA methods are commonly used to reduce a large number of intercorrelated variables to a much smaller group, while preserving as much as possible of the original variation, yet having properties such as independence. Such methods include *factor analysis and *principal components analysis.

muskeg A term used in Canada to describe spring bogs.

mutualism *See* *symbiosis.

nacreous cloud A rare, iridescent cloud formation, occurring in the *stratosphere at a mean height of 24 km. It develops above high latitudes, shortly after sunset or before dawn. Its stationary nature suggests that this is a mountain wave cloud. *See* *noctilucent cloud.

nappe A sheet of rocks, part of a broken recumbent fold which has been moved forward over the rock formations beneath and in front of it, finally covering them. Nappes are typical of most mountains of the Alpine type of structure; the western Alps display a complex series of nappes, formed when the African foreland was pushed northwards against the European foreland.

natalist Promoting population growth. The government of Romania, before the fall of Ceaușescu in 1989, banned contraception and abortion in an attempt to increase population.

natality The production of a living child; birth.

nation A large number of people of mainly common descent, language, culture, and history, usually associated with some specified territory. A nation need not be a *state—many *multinational states exist—but the rise of nationalism in the late twentieth century has brought about many struggles by nations within multinational states to achieve independence. Chechnya's attempt to secede from Russia in the 1990s is just one example.

nation-state Ideally, a *nation whose boundaries coincide with the boundaries of the state which governs the territory of that nation; a condition which rarely occurs in practice. Within its own territory, the nation-state claims the monopoly of the means of violence (army and police violence is the only type sanctioned), protects its inhabitants from external invasion, and maintains sovereignty over its space.

National Park An area less affected by human exploitation and occupation, with sites of particular scenic or scientific interest, which is protected by a national authority. This protection is limited in England and Wales by the need for farmland, forestry, and other commercial uses, but there are fairly strict controls on development. New roads can be hidden under cut-and-cover tunnels or within cuttings, camping and caravan sites can be hidden behind trees, and new buildings should be made of the same stone and to the same design as were traditionally used.

National Parks are generally sited in areas of low industrialization and the purpose of the park may conflict with local demands for employment. Furthermore, the attraction of an area to visitors may diminish as the park becomes overcrowded. One solution is to concentrate visitors in a few points served by car-parks, gift shops, and information centres, such that the open countryside remains thinly populated by tourists. *See* *honey-pot.

The first National Park was Yellowstone, designated in 1872 by the US Congress, and now all but the very smallest of West European states have National Parks, although the percentage of land area covered ranges from 1.44% of the Netherlands to 0.07% of Denmark. A more useful measure may be area of National Park land per capita. This ranges, in km^2, from 9.115×10^{-4} in Norway to 6.085×10^{-6} in Denmark (1989 figures).

The National Parks of England and Wales were first planned in 1949 with the National Parks and Access to the Countryside Act. As a result of this Act, some 20% of the total land area of Britain is officially protected.

National Transfer Format, NTF A UK standard, administered by the Association for Geographic Information (AGI), for the transfer of geographic data. The standard was published in 1987, but substantially revised in 1989. Four levels of data are identified: 0, *raster, or gridded; 1, simple *vector; 2, multiple attribute and quality data; and 5, a used-definable data description.

nationalism The feeling of belonging to a group linked by common descent, language, and history, and with a corresponding ideology which values the *nation-state above everything else, expecting supreme loyalty from its citizens. Associated with this is the belief that *nations which are not nation-states should have the right to determine their own futures, and the end of the twentieth century is witnessing a number of actual and potential struggles for *nations to be independent.

natural area In *human ecology, a spontaneously arising and individual area of an urban society with common social, economic, and cultural characteristics: a distinctive community. Most cities contain numbers of natural areas which are delimited by informal boundaries, such as canals, parks, railways, rivers, and main roads.

natural hazard A hazard is an unexpected threat to humans and/or their property. By this definition, the Indian monsoon is not a hazard, but its failure is. The most frequently occurring hazards are climatic: *drought, *hurricanes, *floods, ice, *snow, and *fog; *tectonic: *earthquakes, *volcanoes, and *tsunami; or due to *mass movement: *landslides and *avalanches. Some geographers distinguish between **environmental hazards** which are geophysical events, and **human-made hazards**, but others stress that human activity plays a part in the development of most hazards.

The impact of a hazard on the human environment is greater when it is not prepared for—small variations in rainfall distress Britain much more than the summer droughts which afflict the Mediterranean—or when a region cannot afford the protective measures necessary. Some hazards are unavoidable; Bangladesh is unable to forestall the catastrophic effects of a tropical cyclone because of poverty, and because of the lack of higher ground on which to evacuate the population at risk.

Human beings continue to inhabit hazard-prone areas, either because the benefits of doing so outweigh the risk—volcanic areas, for example, have rich soils—or because they are too poor to move, or because they do not take the risks seriously. *See* *cognitive dissonance.

natural increase The growth of a population brought about as births exceed deaths. The **natural increase rate** is calculated by subtracting deaths from births and then dividing the result by the number of the total population. Thus, for the UK in 1992, subtracting death rate, 11.2/1000, from birth rate, 13.9/1000, gives a natural increase rate of 2.7/1000.

natural region A region unified by its physical attributes. Three general factors have a great bearing on the character of any region: latitude, relief and structure, and location. These work together to form regions which may be characterized according to topography, climate, and vegetation.

natural resource Any property of the physical environment, such as minerals, or natural vegetation, which humans can use to satisfy their needs. Technically speaking, a property only becomes a resource when it is exploited by humans; by this definition, climate may be considered as a natural resource, especially for countries dependent on tourism. Natural resources may be classified as *renewable and *non-renewable.

natural selection The theory proposed by Charles Darwin which is popularly summarized as 'the survival of the fittest'. Organisms produce numbers of offspring, some of which are successful. The characteristics of these 'fittest' offspring—those which fit their habitat best—are then reproduced in the next generation.

natural vegetation In theory, the grouping of plants which has developed in an area without human interference. Most landscapes have been changed by humans through forest clearance, agriculture, and industry. It is argued that there are relatively few areas of truly natural vegetation left.

nature conservation order An order designed to protect certain plant or animal species or to conserve features of national importance. If development of a site is banned by an order, the landowner may be compensated.

nautical mile Theoretically, the length of one minute of arc on a *great circle drawn on a sphere of equal area to the earth. In Britain it is taken to be 1853.18 m.

nearest neighbour analysis The study of settlements in order to discern any regularity in spacing by comparing the actual pattern of settlement with a theoretical random pattern. The straight line distance from each settlement to its nearest neighbour is measured and this is divided by the total number of settlements to give the observed mean distance between nearest neighbours. The density of points is calculated as:

$$\frac{\text{number of points in the study}}{\text{size of area studied}}$$

The expected mean in a random distribution is calculated as:

$$\frac{1}{2\sqrt{\text{density}}}$$

The expected mean is compared with the observed mean.

$$R_n = \frac{observed\ mean}{expected\ mean}$$

where R_n is the nearest neighbour index. An index of 0 indicates a completely clustered situation. 1 shows a random pattern, 2 a uniform grid, and 2.5 a uniform triangular pattern. The interpretation of index values can be difficult since these values are not part of a continuum.

nebkha *See* *sand dunes.

neck A mass of lava which has solidified in the *pipe of a volcano. Erosion of the material surrounding the neck may reveal a steep tower or an erosional neck. Castle Rock, Edinburgh is a British example.

needle ice *See* *pipkrake.

nehrung A synonym for *barrier beach.

neighbourhood A district forming a community within a town or city, where inhabitants recognize each other by sight. It has been claimed that neighbourhoods evolve their own distinctive characteristics, or subculture. This is the basis of the *neighbourhood effect. Others deny the existence of identifiable subcultural areas, and some believe that, with the increasing mobility of the population, both in changing place of residence or in travelling outside one's immediate home area for socializing or for work, neighbourhoods do not really exist. *See* *balanced neighbourhood.

neighbourhood business district, neighbourhood centre A rank in the hierarchy of business districts below the *CBD and regional business centre. Neighbourhood centres offer frequently needed *low-order goods and services and serve populations of 5000–10 000.

neighbourhood effect It is claimed that individuals are affected by the principles and standards of the neighbourhood in which they live, and that the social environment of the neighbourhood conditions people's behaviour. If the neighbourhood is very run-down, inhabitants may not treat it with respect while the peer pressure in a high-class residential district may encourage the residents to maintain their property. The concept is attacked by those who would argue that the neighbourhood is less important than other social and economic factors, and by those who believe that city populations are too mobile for true neighbourhoods to develop. *See* *contextual effect.

neighbourhood unit The concept of the neighbourhood as a distinctive residential area was advanced by Clarence Perry in 1929 and has been used in the planning of *new towns. The town is planned to contain units of between 5000 and 10 000 people, each unit having its own low-order centre supplying convenience goods, medical facilities, and primary education, all within walking distance. Through traffic is discouraged. Most of the early British *new towns were designed on this principle, with the aim of fostering a sense of community in each neighbourhood unit. The development of neighbourhood units has not always proved to be successful; not every resident prefers to live entirely within a restricted area, while others argue that this form of planning discourages the integration of all the neighbourhoods into the new town.

nematodes Roundworms, sometimes called threadworms or eelworms. They are important in the breakdown of soil *microflora into humus.

neoclassical economics Classical economists, notably Adam Smith and David Ricardo, argued that the price of goods was determined by the 'invisible hand' of competition in the market and not by the producing firms, that the workings of the market alone should regulate the economy, and that the best government policy was *laissez-faire*.

Neoclassical economists argue that firms buy or rent the *factors of production which they operate in at the highest possible level of efficiency in order to maximize profits, but that firms have no control over the costs of these factors or of the price at which their finished goods are sold. In the same way, households may sell their factors of production: possibly land or capital, but often simply their labour. Neoclassical economists argue that consumers seek to maximize utility, subject to the constraints of their incomes and prevailing price levels, and stress the concept of marginal utility—the added satisfaction (utility) which comes from acquiring extra (marginal) goods or services. Consumers maximize utility when the ratio of marginal utility to purchase price is the same for all the goods and services consumed; in other words, if the marginal utility per expenditure is lower for one good than for another, it will not be bought. The whole process is governed by the forces of demand and supply.

When applied to *uneven development, these principles of neoclassical economics would suggest that surplus labour and *capital will move to areas where labour and capital are in short supply, so that inequality will be eradicated, but this argument ignores the frequent lack of mobility of these factors. Other important ideas include a belief in trickle-down economics (where the profits from growth in one sector or region trickle down to other sectors and regions) and the development of *growth poles.

Until recently, all the models created within neoclassical economics were based on the assumption of rational decision-making, *optimizing behaviour, and perfect information, assumptions also inherent in the neoclassical approach to industrial geography, which sees all firms or individuals acting in a manner similar enough to make generalizations possible.

neo-colonialism, neocolonialism The control of the economic and political systems of one state by a more powerful state, usually the control of a developing country (*LDC) by a developed one (MDC). Neo-colonialism is marked by the export of *capital from the LDCs on the periphery to the controlling MDCs at the core, adverse *terms of trade for the periphery or satellite nations, a reliance by the LDCs on imported manufactures from the MDCs, and a pervading process of Westernization. The means of control are usually economic, including trade agreements, investment, and the operations of *transnational corporations, who are often seen as the primary instruments of neo-colonialism. It is also argued that the imposition of Western business methods on a developing country creates a new, alien, class structure which divides societies. *See* *dependency theory.

neoglacial A time of increased glacial activity and extent during the *Holocene. Glaciers expanded and retreated during this phase; the *Little Ice Age may represent one such advance.

Neolithic In Britain, the period approximately between 3000 and 2000 BC during which more sophisticated techniques such as grinding and polishing were applied to the making of stone implements. This era also saw the beginnings of the domestication of animals and the planting of crops.

neotectonics The study of the causes and effects of the movement of the earth's crust in the neogene, i.e. the late *Cenozoic era. Some restrict the term to studies since the Miocene; others use it to refer to *Quaternary movements alone.

neritic Pertaining to shallow waters.

nesting The way in which one network fits into a larger one, for example, in *central place theory, where a smaller market area fits into a larger one.

net migration balance The figure derived when natural increase is subtracted from the total change in population.

net primary productivity In ecology, the amount of energy which primary producers can pass on to the second *trophic level. This represents the amount of carbon dioxide taken in by a plant minus the carbon dioxide it emits during respiration. Respiration rates can be measured if the plant's carbon dioxide output in the dark is recorded. Gross primary production may be calculated since:

gross primary production = net primary production + respiration

net radiation Also known as net radiative balance, this is the balance of incoming solar radiation and outgoing terrestrial radiation, which varies with latitude and season. Net radiation is generally positive by day and negative by night.

network A system of interconnecting routes which allows movement from one centre to the others. Most networks are made up of nodes (or vertices), which are the junctions and terminals, and links (or edges) which are the routes or services which connect them. The form a network takes will reflect not only relief, population distribution, and level of economic development, but will also be influenced by historical and political factors. **Network connectivity** is the extent to which movement is possible between points on a network.

Network analysis is a method of studying networks in terms of *graph theory, and **network density** is the total area covered by the network divided by the total length of the links between points. A planar network can be represented on a flat surface; a non-planar network exists in three dimensions. *See* *accessibility, *alpha index, *beta index, *connectivity.

network chain In *Geographic Information Systems, a *chain without left and right identifiers, but with node identifiers.

neutrality Some states, such as Sweden and Switzerland, have chosen to be permanently neutral except when their territory is at risk, while others, like Eire in World War II, have been neutral in particular conflicts. Austria is bound to neutrality under the terms of the Allied peace treaty of 1955. Neutral zones are often established between rival states, and are usually policed by United Nations forces.

névé An alternative term for *firn.

new international division of labour, NIDL A global division of labour associated with the growth of transnational corporations (TNCs) and the *de-industrialization of the advanced economies. The most common pattern is for research and development to take place in *more developed countries, while the less skilled processes are carried out by cheap labour in *less developed countries.

There are three ways of looking at the reason for this international division of labour. The first argues that TNCs, benefiting from improved transport systems and global communication networks, have looked for cheaper locations for their plants. The second is that the TNCs have been forced out of the old-established industrial nations through falling profits, and the third that the NIDL is a response to capitalism, depending, as it does, on persistent *accumulation.

The result has been the closing down of certain types of manufacturing industry in the industrialized countries, and the introduction of the same manufacturing processes in foreign subsidiaries of the TNCs.

new town A newly created town, either on a green-field site or around a pre-existing settlement, planned to relieve overcrowding and congestion in the major conurbation by taking in the *overspill population. The aim was to create a town which would be economically viable, with light industry, services, and shops. The Greater London plan of 1944 laid down proposals for ten new towns, and after the New Towns Act of 1946, work started on the development of the first group of new towns such as Bracknell and Crawley. Housing densities were low—about five houses per hectare—and housing was arranged in *neighbourhood units of around 5000 people with their own facilities such as shops, schools, and medical centres.

The second generation of new towns, such as Redditch and Telford, designated between 1955 and 1965, also used the neighbourhood unit but less rigorously, and greater emphasis was placed on the development of centralized functions and on solving the problems caused by mass ownership of cars. To cut down the length of journey to work, industries were not as rigidly zoned and separated from residential areas.

The third generation of new towns such as Milton Keynes and Peterlee were more individual in their plan and were more often created around pre-existing settlements. All British new towns were planned to give a balance of social groups, and while the first new towns were planned to house up to 80 000 people, the third and last group may house up to 500 000.

The development of new towns began in Britain but they may be found in countries as different as Russia and the USA.

new village As villages expand, they may be swamped by new developments. It may be better to build new villages rather than to spoil existing ones. To this end, new villages have been built at Bar Hill in Cambridgeshire, Studland Park in Suffolk, and New Ash Green in Kent.

newly industrializing country, NIC A country, once designated as *less developed, but which has undergone recent, rapid industrialization. The first countries to have been identified in this category were Hong Kong, Singapore, Malaysia, and South Korea, who all achieved spectacular industrial growth in the 1970s and 1980s through early *protectionism, high levels of literacy, government subsidy, and devaluation. More recent candidates for the title include Brazil and Mexico.

NIC *See* *newly industrializing country.

niche A set of ecological conditions which provides a species with the energy and habitat which enable it to reproduce and colonize. A niche is usually identified by the needs of the organism.

niche glacier A small patch of glacier ice found on an upland slope. Niche glaciers differ from *cirque glaciers in that their ice has little effect upon the topography.

nimbo-, nimbus Referring to clouds bringing rain, as in nimbo-*stratus clouds.

nitrogen cycle The cycling of nitrogen and its compounds through the *ecosystem. Micro-organisms living in the root nodules of leguminous plants can 'fix' atmospheric nitrogen, that is, they can assimilate atmospheric nitrogen into organic compounds. Fixing can also occur via blue-green algae, from lightning strikes, and by industrial processes. The ammonia and nitrates resulting from fixing are picked up by plants and animals and changed into proteins and amino acids. These are returned to the soil as faeces or as dead tissue. 'Denitrifying' bacteria then act upon these wastes to release free nitrogen into the air.

nitrogen fixation The alteration of atmospheric, molecular nitrogen to nitrogen compounds. The fixation mechanisms responsible are: biological micro-organisms, such as those in the root nodules of leguminous plants, *lightning and other natural ionizing processes, and industrial means. These processes are responsible per year, respectively, for 1, 2, and 8 billionths (10^{-9}) of the mass of the reservoir of atmospheric nitrogen. The natural processes of denitrification cannot balance this rate of nitrification, so that the net amount of nitrogen in the atmosphere is, infinitely slowly, diminishing, and could be reduced by 50% in 250 million years.

nivation The effects of snow on a landscape. These include *abrasion and *freeze–thaw. Furthermore, melted snow triggers mass movements such as *solifluction and *slope wash. These processes may produce the shallow pits known as **nivation hollows**. In time, these hollows may trap more snow and may deepen further with more nivation so that *cirques or *thermocirques are formed. Nivation is 2–3 times as active on shaded, pole-facing slopes. *See* *aspect. Snow has the greatest effect on a landscape where it is thin and melting.

noctilucent cloud A rare, frail, glistening cloud, occurring at night-time over high latitudes at a height of some 80 km. These clouds may be formed of ice crystals or dust.

nodes 1. In *network analysis, destinations or intersection points which are part of a *network. These are also known as vertices.
   2. In behavioural urban geography, the strategic points in a built-up area around which the individual plans his or her movements.
   3. In *Geographic Information Systems, the start- or end-point of a *line or *link.

nomadism A form of social organization where people and animals move from place to place in search of pasture. The itinerary of movement may take the form of a routine pattern but, as rainfall varies, there may be movement away from this routine. True nomads have no fixed abode and no *sedentary agriculture. Semi-nomads like some Australian Aborigines wander for some of the year and grow crops for the rest of the year.
   As international boundaries are increasingly well defined, with border guards, nomadism will decline. Governments try to immobilize nomads as an attempt to bring them in line with more advanced societies for the purposes of taxation as well as to improve their health and literacy. However, many researchers now consider that nomadism represents the best use of fragile ecosystems.

nominal scale A scale with data classified by names, rather than by any quantitative description, such as a soil classification or a classification by various cultural groups. All the categories are different from each other and cannot be ranked by size; it is not possible, for example, to say that a *gley soil is bigger than a rendzina! *See* *categorical data.

nomothetic Concerned with finding similarities between places or phenomena. Thus, for example, models of urban morphology are derived by looking for resemblances in cities—this is a nomothetic approach, and contrasts with *areal differentiation.

non-basic worker A worker concerned with serving the city in which she or he lives and who is thus not bringing wealth into the city. *See* *basic worker.

nonconformity A series of sedimentary strata overlying an igneous or metamorphic rock.

non-ecumene The uninhabited or very sparsely populated regions of the world. It is not easy to draw boundaries between the *ecumene and the non-ecumene as regions of dense occupation merge into sparsely populated regions. If there is a boundary, it is not static.

non-parametric statistics Statistical tests which are not based on a *normal distribution of data or on any other assumption. They are also known as distribution-free tests and the data are generally ranked or grouped. Examples include the *chi-squared test and *Spearman's rank correlation coefficient.

non-renewable resource A finite mass of material which cannot be restored after use, such as natural gas. Non-renewable resources may be sustained by *recycling.

normal distribution The line graph showing the expected frequency of occurrences in each class of any set of data for a given variable. The normal distribution is shown as a bell-shaped curve which is symmetrical about the mean. The laws of probability state that between $+1\sigma$ and -1σ 68.27% of the items in the data set will be found, between $+2\sigma$ and -2σ 95.45% of all the items in the data set will be found, and between $+3\sigma$ and 3σ 99.97% of all the items in the data set will be found. In other words, a difference of more or less than 3 *standard deviations from the mean is only to be expected once in every 300 observations. So, if in a sample data set of 50 items, one value exceeds $+/-3$ standard deviations from the mean, the data may be suspect and should be checked.

normative theory Any theory which seeks to explain or predict what would happen under theoretical constraints; what *ought* to be, such as *perfect competition, rather than what *is*, or *will* be (*imperfect competition).

In geography, normative models, like the *von Thünen model of land use, generally rely heavily on assumptions and preconditions and most *spatial analysis is normative. Many normative models are wildly unrealistic since they ignore the complexity of the real world, concentrating on idealized concepts such as rational, *economic man, and *isotropic plains.

North Atlantic Drift A warm *ocean current, driven by the prevailing south-westerlies from Florida to north-west Europe, at velocities of 16–32 km per day. Onshore winds transfer heat from this current to coastal areas, thus bringing warmer conditions than would be expected at high latitudes in north-west Europe.

North Following the terminology of the Brandt report, 1979, a portmanteau term used to describe the advanced economies, or the more developed countries of the *First (non-communist) World.

North Pacific Current A cold *ocean current.

nuclear family The small family unit of parents and children. This is not the most frequently occurring household unit; in the UK, for example, more households contain one person than any other category, and there are increasing numbers of single-parent and step-parent families, while, in other parts of the world, the *extended family is a common household unit.

nuclear power A form of energy which uses nuclear reactions to produce steam to turn generators. Naturally occurring uranium is concentrated, enriched, and converted to uranium dioxide—the fuel used in the reactor. This fuel readily undergoes nuclear fission which produces large amounts of heat. Some of the highly radioactive spent fuel may be reprocessed while the bulk must be disposed of. Both are costly and hazardous undertakings. The main advantage of nuclear power is the relatively small amount of an abundant fuel which is required. The major disadvantages are very high construction and decommissioning costs, highly technical operations, the problem of waste disposal, and the major problems which may arise with any accident, such as the Chernobyl disaster of April 1986. Furthermore, nuclear power stations have a short lifespan.
    The locational requirements of a nuclear power station include very large amounts of water as a coolant, stable and firm geological conditions, and distance from large centres of population because of the radiation hazard. In 1990, 33.5% of Western Europe's electricity was produced from nuclear power stations, major producers being Belgium (61% of total electricity generated), and France (74.5%). The UK figure was 21.7%.

nuclear winter A series of nuclear explosions would produce large quantities of smoke and dust. These particles might intercept incoming solar radiation and reflect it back into space. If this were to occur, very much lower temperatures would obtain at the earth's surface, giving severe wintry conditions.

nucleated settlement A settlement clustered around a central point, such as a village green or church. Nucleation is fostered by defence considerations, localized water supply, the incidence of flooding, or rich soils so that farmers can easily get to their smaller, productive fields while continuing to live in the village.

nuclei, nucleus (sing.) Minuscule solid particles suspended in the *atmosphere. Three types of atmospheric nuclei are distinguished: **Aitken nuclei**, of radii less than 0.1 μm, **large nuclei**, radii 0.2–1 μm, and **giant nuclei**, with radii greater than 1 μm. These nuclei may be scraps of dust, from volcanic eruptions or dust storms, or salt crystals, or given off when bubbles burst at the surface of the sea. Atmospheric nuclei can scatter sunlight enough to lower temperatures, if enough are present, and play an important role in *cloud formation. *See* *condensation nuclei.

nuée ardente A glowing cloud of volcanic ash, *pumice, and larger *pyroclasts which moves rapidly downslope. The material from a nuée ardente consolidates to form ignimbrite, also known as welded tuff.

null hypothesis A hypothesis, to be tested statistically, that no difference is to be seen within the groups tested or that no correlation exists between the variables. If the null hypothesis can be rejected, the existence of a difference or a relationship can be proved. Thus, if a researcher wanted to prove that pedestrian flows fell with distance from the CBD, she or he would set up the null hypothesis that there is no correlation between pedestrian flows and distance from the CBD. Researchers prefer to set up null hypotheses, since it is more satisfactory to disprove a null hypothesis than to prove a hypothesis: *see* *induction.

nunatak A mountain peak which projects above an ice sheet. Nunataks are generally angular and jagged due to *freeze–thaw and, after the ice has retreated, contrast with the rounded contours of the glaciated landscape below.

nuptiality The frequency of marriage within a population, usually expressed as a *marriage rate. Nuptiality varies with the age of first marriage common in a society and with the age structure of the population; obviously it will be lower in an ageing population than in a very young one. It is a major factor in *fertility, although less so now with a large proportion of children born out of wedlock. The most basic rates express the number of marriages per thousand population, or the number of people per thousand marrying in a given year.

nutrient cycle The uptake, use, release, and storage of nutrients by plants and their environments. In temperate *ecosystems, far more nutrients are stored in the soil than can be used immediately by plants. The soil then acts as a reservoir for nutrients. In tropical ecosystems, almost all the nutrients are stored in the plants and the soil is impoverished. The living parts of the ecosystem fulfil a number of functions in the nutrient cycle such as regulation of nutrients in outgoing water, the storage of combined nitrogen, and the control of water loss through *transpiration.

oasis A fertile, watered spot in a desert. Oases result when the *water-table reaches the surface, perhaps because a *deflation hollow has formed.

object In *Geographic Information Systems, a collection of *entities which together form a higher-level entity within a specific data model. An **object-oriented database** is a *database based on objects.

occlusion A stage which may occur in a *mid-latitude depression where the cold front to the rear catches up with the leading warm front, lifts the wedge of warm air off the ground, and meets the cold air ahead of the warm front. If this overtaking air is colder than the cold air which is ahead of it, it will undercut it, forming a **cold occlusion**. If, on the other hand, it is warmer than the cold air which is ahead of it, it will ride over it forming a **warm occlusion**. After its formation, the still-deepening low pressure centre moves polewards and westwards. The centre of circulation is now cold—the cold core. Precipitation may persist for several days.

occupancy rate The number of people dwelling in a house per habitable room (kitchens and bathrooms are not counted). A rate of one person per room is taken as acceptable; more than one person per room represents overcrowding.

occupational mobility The ability of the individual to change jobs after the acquisition of a new skill.

ocean current A permanent or semi-permanent horizontal movement of unusually cold or warm surface water of the oceans, to a depth of about 100 m. The global system of winds is the most important cause of these currents which are also affected by variations in the temperature, and hence density, of the water, and by the *Coriolis force. These currents are an important factor in the redistribution of heat between the tropics and the polar regions.
   Cold currents originate in high latitudes and can greatly modify the temperatures of coastal areas as far inland as 100 km. As tropical air streams move over these currents, advection *fog forms over the sea. The air streams are thus stripped of most of their moisture; onshore winds are therefore dry. Cold currents thus contribute to desert conditions. Conversely, **warm currents** originate in tropical waters, and bring unusually warm conditions to the higher latitudes affected by them.

ocean trench A long, narrow, but very deep depression in the ocean floor where, at the junction of two *plates, one plate dives steeply beneath another and penetrates the *mantle.

oceanic Describing a moist, mild climate, with a small temperature range.

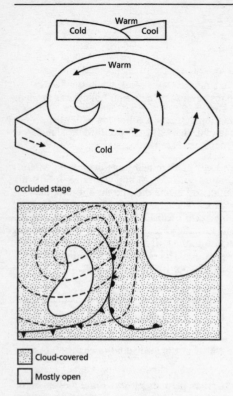

Cloud-covered

Mostly open

FIGURE 43: *Occlusion*

oceanic crust That portion of the outer, rigid part of the earth which underlies the oceans. The oceanic crust seems to be mostly a basalt layer, 5–6 km thick. Its average density is around 2.9, which is more than the *continental crust, but less than the *mantle.

oceanic ridge An underwater mountain range developed at a section of oceanic crust where *magma rises up through a cracking and widening ridge. Some magma cools below the crust, some of it forces into fractures, and much flows out to form new crust, which is then pushed away from the ridge. As new crust is created by the extrusion of *lava each side of the ridge, it takes up the prevailing *magnetic polarity of the earth, which reverses from time to time. As a result, symmetrical bands of crust, with alternating polarity, develop on either side of the ridge. These magnetic patterns are used to calculate the rate of the *sea-floor spreading resulting from the lava flow. The term **mid-oceanic ridge** properly refers to the ridge at the centre of the Atlantic Ocean, which comes to the surface at points such as Tristan da Cunha and Ascension Island. Other ridges, such as the

Pacific–Antarctic ridge, are not truly at the centre of the ocean. *See also* *plate tectonics.

oceanography The study of the oceans. This covers the shape, depth, and distribution of oceans, their composition, life forms, ecology, and water currents, and their legal status.

offshore **1.** In meteorology, moving seawards from the land, as in an offshore wind.
   2. In geomorphology, the zone to the seaward side of the breakers.

offshore bar *See* *bar.

ogive In the shape of a pointed arch. Bands of dark and light ice stretch across glaciers in this shape because the central ice of a glacier moves more rapidly than the sides. The term has been extended to refer to the bands themselves.

oil crisis, oil price shock In 1973, the *OPEC countries, headed by Libya, tried to take control of the market in oil. The *cartel quadrupled the price of oil, and many member states nationalized the oilfields in their own countries. At this time, no fewer than 12 European countries, including the UK, were more than 60% dependent on OPEC oil, with Portugal and Norway over 80% dependent.
   The results were many: oil companies began to develop alternative resources which had previously been uneconomic or where physical conditions were unpropitious as in Alaska and the North Sea; attention shifted to alternative sources of energy, energy conservation, and energy-efficient technology (kilometres/litre rose sharply for most motor vehicles); and many Third World countries, lacking alternative sources of supply or the technology to develop alternatives, found the cost of their oil imports rising painfully.

oil refining The processing of crude oil into petrol, paraffin, diesel oil, and lubricants. Oil refineries require stable, firm geological conditions, large sites, and easy access to oil supplies. Countries without sufficient domestic oil supplies will refine oil imported by tankers; a coastal location is generally used.

oil shale A *shale from which oil and natural gas may be distilled.

okta In meteorology, a measure of the extent of cloud cover. This runs from 1 okta (scant cloud cover) to 8 oktas (complete cloud cover). Oktas are shown on *synoptic charts by a circle which is progressively shaded in as cloud cover increases.

old age *See* *cycle of erosion.

old-age index The number of people over retirement age in the population, as a percentage of the total adult population.

old fold mountains *See* *fold mountains.

oligopoly The domination by a few firms of a particular industry. In order to maintain their share of the market, such firms are forced to imitate each other's behaviour. A classic example was the introduction of unleaded petrol by both Shell and Esso almost at the same time. Elsewhere, more dubious examples are price-fixing and the way in which the market is shared out between competing firms.

oligotrophic Poor in nutrients, usually of lakes, soils, and peat bogs.

omnivore An animal which eats both plant and animal matter.

onion weathering *See* *exfoliation.

ontogenetic Concerned with the development, notably intellectual development, of human beings, especially of the individual.

ontology The study of the nature of being, of what exists or what can be known.

ooze A deep, sea-floor deposit either of tiny organisms such as diatoms (a type of algae) or of fine inorganic sediments.

OPEC The Organization of Petroleum Exporting Countries, a central body which, at regular intervals, fixes the price of oil on the international markets. Although a supplier of oil, Britain is not one of the OPEC countries since they are all at odds with the old colonial powers who controlled the oil industry in its early stages. OPEC increased petroleum prices very dramatically in 1973 and 1974 to the great discomfort of most Western nations. *See* *oil crisis.

The founder members are Iran, Iraq, Kuwait, Saudi Arabia, and Venezuela. Later members include Algeria, Ecuador, Gabon, Indonesia, Libya, Qatar, Nigeria, and the United Arab Emirates.

opencast mining A system of mining which does not use shafts or tunnels and is, hence, cheaper. The layer above the mineral seam is removed and the exposed deposit is extracted using earth-moving machinery. The overburden may then be replaced. Most open-cast mining in Britain is permitted or licensed on the condition that this takes place.

open field A distribution of farm land associated with *feudalism in Europe. Each manor had two or three large open fields and each farmer was awarded a number of strips within each field. Holdings were scattered so that no one had all the good or all the poor land.

open system A system which is not separated from its environment but exchanges material or energy with it.

open systems interconnection, OSI The international standards whereby open electronic systems communicate with each other.

opportunist species Species which survive by the rapid colonization or recolonization of a *habitat. Such species have very good powers of dispersal in order to seek new habitats. Life is short, fecundity and reproduction are high, and these species are able to withstand difficult conditions. The term **fugitive species** is also used for opportunist species since they are seen to colonize those marginal habitats, such as deserts and saline areas, which are not used by other organisms.

opportunity cost A term used in *neoclassical economics to express cost in terms of sacrificing the alternatives. For example, if land is kept as *green belt, the sacrifice will be in terms of the increased value of that land if it were available for development. This is an important idea for geographers who hope to explain resource allocation between different activities, and the idea is also significant in the theory of *comparative advantage.

optimization model A model used to find the best possible choice out of a set of alternatives. It may use the mathematical expression of a problem to maximize or minimize some function. The alternatives are frequently restricted by constraints on the values of the variables. A simple example might be finding the most efficient transport pattern to carry commodities from the point of supply to the markets, given the volumes of production and demand, together with unit transport costs.

optimizer A decision-maker who seeks the best outcome, usually to maximize profits, especially those received from manufacturing industry. In choosing an industrial location, the profit maximizer is assumed to know all the relevant factors at given locations including the cost of assembling materials and distributing products, the price of labour, and *agglomeration and *external economies. Optimizing behaviour is central to *neoclassical economics and the concept of *economic man, as well as much locational analysis in economic geography. Compare with *satisficer.

optimum city size Some writers argue that there are benefits to be gained as a city grows—in terms, for example, of cheaper transport and the more economical provision of services—up to and until the optimum size is reached. A figure of around 500 000 has been suggested as an optimum size. Beyond this point *diseconomies operate: pollution and congestion set in, and the city is a less efficient and less pleasant place to live in.

optimum location The best location for a firm in order to maximize profits. It is argued that profit-maximizing firms will force less-successful firms out of business and that this will result in firms moving towards optimum-location patterns. In the absence of complete knowledge by the decision-maker, it may not be clear where to locate for maximum profits.

optimum population A theoretically perfect situation where the population of an area can develop its resources to the greatest extent, and achieve maximum output while enjoying the highest possible standards of living.

ordinal data, ordinal scale Data (or a scale) presented by their relative importance or order of magnitude—as in small, medium and large—but not their absolute values. *See also* *categorical data, *interval data, *nominal data.

ordinate The vertical or *y*-co-ordinate on a graph.

ordnance datum, OD Mean sea level at Newlyn, Cornwall, UK, from which all other spot heights on Ordnance Survey maps—and hence all *contours—are established.

Ordovician A *period of *Paleozoic time stretching approximately from 500 to 430 million years BP.

ore A naturally occurring deposit which contains a mineral, or minerals, in sufficient concentration to justify commercial exploitation. **Ore dressing** is the crushing of an ore to separate out the minerals it contains by chemical processing, sedimentation, and flotation.

organic acid Acid compounds of carbon, such as acetic acid, which are produced when plant or animal tissues decompose.

organic weathering The breakdown of rocks by plant or animal action or by chemicals formed from plants and animals. *Humic acids break down rock, and bacteria reduce iron compounds.

Organization of African Unity, OAU An association of independent African states constituted in 1963, and designed to encourage African unity, to discourage *neocolonialism, and to promote development.

Organization of American States, OAS An association of Latin American states with the USA. Constituted in 1948, it is designed to encourage American solidarity, to aid co-operation between members, to maintain present boundaries, and to arbitrate in disputes between members.

Organization for Economic Co-operation and Development, OECD Constituted in 1961, this is a group of nations, comprising most of Western Europe, together with Australia, Canada, Japan, New Zealand, and the USA and formed to develop strategies which will boost the economic and social welfare of the member states.

orogen The total mass of rock deformed during an *orogeny.

orogeny Movements of the earth which involve the folding of sediments, faulting, and *metamorphism. Two major types of orogeny are recognized: a cordilleran type in which *geosynclinal deposits are severely deformed, and one created by a continental collision when oceanic crust and sediment are trapped between two masses of continental crust.

orographic precipitation, orographic rainfall Also known as *relief rainfall*, this forms when moisture-laden air masses are forced to rise over high ground. The air is cooled, the water vapour condenses, and precipitation occurs. Some authors maintain that relief merely intensifies the precipitation caused by *convection or formed at fronts. The term **orographic intensification** is, therefore, used occasionally.

orthogonal At right angles. Orthogonals plotted through the crests of waves in plan illustrate the process of *wave refraction.

OSI *See* *open systems interconnection.

osmosis The passage of a weaker solution to a stronger solution through a semi-permeable membrane. In soils, the more dilute soil moisture passes by osmotic pressure into plant roots. In this way, soil moisture is taken up by plants.

outlet glacier A glacier streaming from the edge of a body of ice located on a plateau.

outport With the increasing size of modern ships, some old harbours are no longer deep enough to accommodate them. Consequently, a new port—an outport—is built seawards of the original port, where there is deeper water. Ouistréham is the outport for Caen, Normandy.

outwash, outwash sands and gravels Sorted deposits which have been dropped by *meltwater streams issuing from an ice front. Such streams tend to be heavily laden with debris, so that deposition occurs with only small decreases in stream velocity. The material deposited is coarse near the ice front, becoming progressively finer with distance from it. Outwash fabric tends to be well *bedded, and *current bedding is common. The finest elements of the outwash may deposited in *proglacial lakes to form *varves. While one stream may form a gently sloping outwash fan in a lowland area, a series of streams flowing over a plain may produce several fans which coalesce to form an outwash plain. This latter term is synonymous with *valley train.

overbound(ed) city A city where the administrative boundary includes areas which are not *urban. Oxford is an example.

overburden The layer of rock and soil overlying a particular rock *stratum. *See* *opencast mining.

overdeepening A phenomenon found in *cirques and in the steps of *glacial troughs, where the middle section(s) of the feature are lower than the mouth. One suggestion is that the overdeepened section is the zone of maximum ice thickness, another that overdeepening coincides with less resistant rocks. Overdeepening in glacial landscapes is possible because ice can move upslope within the general direction of ice flow. Overdeepening also occurs in well-jointed sections of the landscape, as evidenced by the Yosemite valley.

overflow channel A steep-sided and relatively narrow channel cut by meltwater from a *proglacial lake as the level of the lake rises and spills over the relief barrier which contained the water. It is suggested, however, that some features once identified as overflow channels were formed by subglacial meltwater channels.

overland flow Water flows overland because the soil or rock which it flows over has become saturated, that is, because the water-table has come to the surface. This is **saturated zone overland flow** which occurs in small valleys in humid climates, on land bordering streams, in hillside hollows with high water-tables, and where soil moisture levels are high. **Hortonian overland flow** occurs when rainfall exceeds *infiltration, but this is unusual in humid temperate regions where rainfall intensity is generally low and vegetation cover encourages infiltration.

overpopulation Too great a population for a given area to support. This may be because population growth has outstripped resources or because of the exhaustion of resources. The symptoms of such a situation include high unemployment, low incomes, low standards of living, and, possibly, malnutrition and famine. *Malthus was probably the first European to identify population problems but his views were discredited for some time. At present, some neo-Malthusians are supporting the same argument. Marxists, however, view overpopulation as the result of the maldistribution and underdevelopment of resources. In the developed world, some would suggest that pollution and the desecration of the countryside are indicators of overpopulation.

overspill The population which is dispersed from large cities to relieve congestion and overcrowding, and, possibly, unemployment. It occurs with redevelopment in the city where new building is at much lower densities so that some people—the overspill—cannot be housed in the city.

overthrust A nearly horizontal fold subjected to such stress that the strata override underlying rocks.

ox-bow lake A horseshoe-shaped lake once part of, and now lying alongside, a meandering river. The lake was once part of a meander, and erosion at the neck left only a short distance across it. When the river breaks through this narrow stretch of land, the old meander becomes a temporary lake. Ox-bow lakes quickly fill up and become hollows in the landscape.

oxidation The absorption by a mineral of one or more oxygen *ions. Oxidation is a major type of chemical weathering, particularly in rocks containing iron. Within soils, oxidation occurs when minerals take up some of the oxygen dissolved in the soil moisture.

oxisol A soil of the *US soil classification. *See* *ferrallitization.

ozone A form of oxygen, and an atmospheric *trace gas, made by natural photochemical reactions associated with solar ultraviolet radiation. Ozone has three atoms of oxygen combined in one molecule, rather than two atoms, as in free oxygen. The proportion of ozone in the atmosphere is very small, but it is of vital importance in absorbing solar ultraviolet radiation. The **ozone layer**, also known as the **ozonosphere**, is an ozone-rich band of the atmosphere, at 10–20 km above the earth, but is at its most concentrated between 20 and 25 km. When the ozone layer is breached (the 'hole' over Antarctica is an example) increased solar ultraviolet radiation reaches the surface of the earth, with consequent damage to human health.

Pacific-type coast Also known as a *concordant, or *Dalmatian coast this is a coastline where the trend of ridges and valleys is parallel to the coast. If the coastal lowlands are inundated by the sea, a coast of interconnected straits parallel to the shore may result. The term Pacific comes from the archetype: the coast of British Columbia.

pack ice All sea ice except that which is attached to terrestrial ice.

pahoehoe A type of lava flow which spreads in sheets, associated with highly fluid, basic lava, such as that ejected from *Hawaiian volcanoes. The surface is a glassy layer which has been dragged into ropy folds by the movement of the hot lava below it.

palaeoecology The reconstruction of past environments from the evidence of fossils. It is possible to use the fossil record to reconstruct the histories of communities. The origin, development, and extinction of species may also be studied in order to test ecological hypotheses. *See* *palynology.

Palaeolithic The Old Stone Age; a time when humans began to make simple stone tools, particularly of flint. This is the earliest period of human prehistory.

palaeomagnetism *See* *geomagnetism.

Paleozoic The *era of earth's history stretching approximately from 570 to 225 million years BP.

Palmer Drought Severity Index, PDSI A measure of water shortage devised by W. C. Palmer (1965), based on a method of soil moisture budgeting that considers precipitation and temperature for a given area over a period of months or years. The PDSI attempts to combine the impacts of precipitation and temperature on soil moisture, groundwater shortage, and low stream flow.

palsa In an area of permafrost, a peat mound, several metres in height, and up to 100 m in diameter, which obtrudes because it is better drained and thus more subject to *frost heaving than wetter areas. Palsas contain slim ice lenses, but the core is of silt, which distinguishes them from *pingos, which enclose a solid ice core.

palynology The study of pollen grains as an aid to the reconstruction of past plant environments. One weakness of this study is that most of the pollens found come from wind-pollinated species and animal-dispersed pollen is underrepresented.

pampas The natural grassland of southern Brazil, Argentina, and Uruguay. Trees were found in the river valleys, and shrubs occurred wherever hills rose above the level of the plains. Most of the pampas has now been altered by cultivation or grazing, and, probably, by fire.

pan A large, shallow, flat-floored depression found in arid and semi-arid regions. Pans may be flooded seasonally or permanently.

Pangaea A supercontinent consisting of the whole land area of the globe before it was split by *continental drift.

pannage The woodland diet of swine, and, from that, the right to feed swine in the woods of the manor.

parabolic dune *See* *sand dune.

paradigm The prevailing pattern of thought in a discipline or part of a discipline. The paradigm provides rules about the type of problem which faces investigators and the way they should go about solving them. For geographers, for example, the paradigm would be referred to when questions such as 'what is geography?'; 'what are the legitimate areas of investigation for geographers?'; 'how should geographers go about their investigations?' are asked. Perhaps the most powerful paradigm for Western thinkers has been the 'scientific method'.

Thomas Kuhn, who, in 1972, first used the term in this sense, argued that the evolution of a new paradigm marks a new stage in thinking. According to Kuhn, the paradigm, or shared view, persists for a while but then becomes obsolete because it becomes disturbed by too many 'anomalies', which do not fit into, and cannot be explained by, the existing paradigm. It is then replaced by a new paradigm, which is able to explain the anomalies. It should be noted that Kuhn's arguments have been contested, not least because the consensus in any one science is never complete or fixed; and while R. J. Chorley and P. Haggett (1967) argued that the *quantitative revolution had established a model-based paradigm for geographers, their arguments have been criticized as not fitting with those of Kuhn, who based his thinking on physical, rather than social sciences.

Thus, although it is not difficult to detect changes in the nature of geography; for example, the traditional 'paradigm' of regional geography was superseded in the 1970s by *positivist analysis, that is to say, by *systematic geography, it is difficult to attribute the causal mechanism for this change to the processes which Kuhn may have identified.

parallel drainage *See* *drainage patterns.

parallel sheeting *See* *pressure release jointing.

parallel slope retreat The evolution of a hillslope where the angle remains constant for each part of the slope. Parallel retreat assumes uniform lateral erosion over the whole hillslope. The length of the slope element is also constant except for the *pediment which increases in length over time. The concept was first introduced by W. Penck, who accepted the process would depend on the efficient removal by streams of slope-foot debris, and that parallel retreat would take place only on parts of a slope. It is now generally conceded that no single model of slope evolution can explain all the features of hillslopes.

parameter A numerical, characteristic of a complete data such as a *mean, *median, *standard deviation, or *variance, set, as opposed to a sample. *See* *parametric statistics.

parametric statistics (parametric tests) Also known as **classical** or **standard tests**, these are statistical tests which make certain assumptions about the *parameters of the full population from which the sample is taken; it is assumed, for example, that the data show a *normal distribution, and that, where populations are compared, they show the same *variance. If these assumptions do not apply, *non-parametric tests must be used. Parametric tests normally involve data expressed in absolute numbers or values rather than ranks; an example is the *Student's t-test.

parasite An animal or plant living in or on another living organism and drawing sustenance directly from it.

parasitic cone A secondary cone on the side of a volcano, fed by conduits branching from the main feeder, or directly from the magma chamber.

parent material The rock or deposit from, and on which, a soil has been formed. The nature of the parent rock will largely determine the nature of the *regolith, and hence the soil texture; thus, basalt tends to produce clay soils while granite generally gives rise to sandy soils.

Pareto optimality A state of affairs where it is not possible to improve the economic lot of some people without making others worse off; a *mercantilist view. The implications of this view in welfare economics are that, once an economy has ceased to grow, it is impossible to increase the wealth of the poor without opposing the **Pareto criterion**; in other words, without making the rich worse off. This then becomes an argument for retaining the *status quo*, even if the distribution of income in society is very uneven. A **Pareto improvement**, however, occurs if resources can be better utilized so that one group's prosperity increases, but not at a cost to another's.

parish Originally, in Britain, an ecclesiastical unit comprising a village and a church with a clergyman in charge. It is now a unit of local government—a **civil parish**—which does not necessarily share the **ecclesiastical parish** boundaries.

park Originally an enclosed area used for hunting, in the eighteenth century the term applied to the grounds of a country house. It now refers to open land used for recreation in a town or city.

pastoralism The breeding and rearing of animals. **Sedentary pastoralism** ranges in scale from the keeping of a small herd to ranching hundreds of stock over a very large area, and may be either *extensive, as in cattle farming in northern Australia, or *intensive, as in producing veal in crates. **Pastoral nomads,** such as the Masai of East Africa, move with their flocks which supply them with food, shelter, and goods for sale.

paternoster lakes A series of elongated lakes in a *glacial trough, dammed by *riegels or by *moraines. The lakes are 'strung' together by rivers, giving the effect of a rosary; hence the name. In Snowdonia, Llyn Gwynant and Llyn Dinas are paternoster lakes in the Nantgwynant valley, but more extensive examples occur in Scandinavia.

path In behavioural urban geography, a channel along which individuals move within a city.

patriarchy A society where women are dominated by men. Sylvia Walby (1990) has distinguished six locations of patriarchal relations: sexuality, the household, male-on-female violence, paid employment, cultural institutions, and the state. These are not mutually exclusive sites of exploitation: historically, the state has tolerated male-on-female violence within the household, for example. Each of these has its own geography; *see* *feminist geography.

Patriarchy is a key concept in Marxist and socialist feminism, although explanations and interpretations, again not necessarily mutually exclusive, vary from the biological (women are weaker) to the economic (women provide domestic support for the working male, and/or a cheap army of reserve labour) to the cultural (masculinity and traditional masculine skills are valued above femininity and traditionally female skills). *See also* *gender.

patterned ground The arrangement of stones into polygons, isolated circles, concentrations of circles known as nets, steps, and stripes. Polygons and circles are more common on level surfaces, stripes generally form on slopes, but there is no delimiting declivity where one ends and the other begins. The patterns are made of coarser stones, separated by much smaller stones known as fines.

The formation of patterned ground has been ascribed to the formation of convection currents within the *active layer, (although some of the rock fragments may be boulders and extremely heavy), and to the sinking of the surface layer on drying, and thus becoming more dense, in summer. Others attribute patterned ground to the upward injection of slow-freezing, waterlogged silts at particular points. See also *involution. Once there has been some sorting between coarse and fine sediments, the coarser sediments would freeze first, doming up the finer areas. Any larger material on the domes would then roll downslope to the coarser areas; thus the sorting would be accentuated. Patterned ground may also be due to *frost heaving, and the drying and shrinking of surface layers. Patterned ground is most common in *periglacial areas, such as Spitzbergen, but polygons can develop during severe winter freezing.

pays In France, a small distinctive *region characterized by a common natural endowment and its own culture, such as Brie in the Paris Basin. Attempts have been made to distinguish pays in other parts of the world to establish particular regions.

peak land-value intersection, PLVI The point in a *CBD, often, but not always, at a road intersection, where land values are at a maximum. *See* *urban land-value surface.

peasant A farmer whose activities are dominated by the family group. The family provides all the labour and the produce is for the family as a whole. Landholdings are small, sometimes owned by the family, but often leased from a landlord. Most of the produce is consumed by the family, but occasional surpluses are sold in the open market. Although peasants have been characterized as backward and resistant to change, peasant strategies can be highly rational in a society where there is little margin for error.

peat A mass of dark brown or black plant material produced when the vegetation of a wet area is partly decomposed. Peat forms where the land is waterlogged and where temperatures are low enough to slow down the decomposition of plant residues. It is, therefore, characteristically found in cold climates, and as a relict feature in temperate zones. It may be dried and used as fuel, both domestically and in power stations, because of its high carbon content, and is widely sold for garden use; so much so that peat bogs are now under threat.

ped In a soil, an aggregate of silt, sand, and clay of characteristic shape. Peds vary in size even within the same soil *horizon. They result from the forces of attraction between soil particles, and involve the formation of hydrogen and ionic bonds. Peds may be further developed by plant roots, by *polysaccharide gums secreted by soil fauna, and by alternate wetting and drying, or freezing and thawing. *See* *soil structure.

pedalfer Any soil high in aluminium (Al) and iron (Fe), and from which the bases such as calcium and magnesium carbonates have been *leached. Pedalfers generally occur in regions with an annual rainfall of more than 600 mm.

pediment A low-angled plain, with a gradient of less than 7°, found at the foot of a mountain, especially in semi-arid and arid regions. Pediments appear as more or less wide terraces on the borders of the Great Basin, for example in southern California and New Mexico.
   Pediment slopes appear to be at their straightest and most gentle in areas of extreme aridity. Many are concave in longitudinal profile. Usually there is a clear *break of slope between the gently sloping pediment and the steeper regions of the slope above it. Pediments vary in area from tens of square metres to hundreds of square kilometres.
   Some geomorphologists see pediments as erosional features left behind by the recession of the scarp above; others stress the work of sheetwash and migrating stream channels; others widen the term to include depositional features. It seems likely that any one pediment may have resulted from a combination of past and present processes; these landforms are the subject of considerable controversy. The coalescence of neighbouring pediments is thought to be the cause of *pediplain formation.

pediplain An extensive *erosion surface, interrupted by the occasional *kopje or *inselberg, and found in semi-arid and savanna landscapes, especially those of Africa. Pediplain formation, or **pediplanation**, is thought to result from the coalescence of neighbouring *pediments. It is regarded as the last stage in an arid *cycle of erosion or **pediplanation cycle**. In the early stages of the cycle, streams are *rejuvenated and start eroding headwards from the coast. Scarps then retreat away from the drainage lines. Compare with *peneplanation. Ultimately, the pediplain comprises an intricate assembly of pediments which slope down to the local drainage systems, and of basins.

pedocal Any soil high in calcium carbonate and magnesium carbonate because *leaching is slight. Pedocals occur in areas with an annual rainfall of under 600 mm.

pedogenesis The formation of soils. **Pedogenic processes** are soil forming processes. The chief pedogenic factors are time, relief, *hydrology, *parent rock, climate, fauna, and flora. These last three have a profound influence on soils. Climate affects the vertical movements of water and minerals which lead to the formation of soil *horizons. Macro-animals, notably earthworms, are the main agents in the mixing of soil materials. Plant roots attract soil water by *osmosis and their vegetation will determine the nature of the plant litter and hence the nature of the humus.

pedology The science of soils: their characteristics, development, and distribution.

pedon A small sample of a soil sufficiently large to show all the characteristics of all its horizons.

pelagic Of marine life, belonging to the upper layers of the sea.

peneplain Literally almost-a-plain, a plain scarcely rising above sea level at the coast, but rising gradually inland towards the occasional residual hills known as *monadnocks. A peneplain is a low-lying *erosion surface, but most existing peneplains are very old, and have been uplifted and *dissected. See *cycle of erosion. **Peneplanation** is the wearing away of the entire landscape, so that the *planation surface evolves over all sections at all times, whereas in *pediplanation the scarps are subject to progressive retreat.

The central part of the Libyan desert has been described as a 'desert peneplain'; the oldest and most enduring part of the desert, where arid processes are complete.

peninsula A piece of land jutting into, and almost surrounded by, the sea.

perception See *environmental perception.

perched water-table A partly saturated, isolated, confined *aquifer underlain by an impermeable rock with the main *water-table below the two.

percolation The filtering of water downwards through soil and through the *bedding planes, *joints, and pores of a *permeable rock. A **percolation gauge** measures the quantity of water moving in this way.

percolines An underground network of water seepage zones. Old root channels, soil cracks and animal burrows are enlarged by *interflow so that a *dendritic pattern of drainage is formed below the ground surface. Percolines are important in the development of *throughflow.

perennial stream A watercourse which flows throughout the year.

perfect competition A hypothetical state of affairs under which a good is sold. Under conditions of perfect competition, there are many suppliers, each of whom is responsible for only a small number of total sales; there is a perfectly *elastic supply of the *factors of production; there is no collusion between suppliers, and buyers and sellers are fully aware of the prices being charged throughout the market. This is an unlikely state of affairs; *imperfect competition is much more common.

perforation kame *See* *kame.

periglacial Originally referring to the processes and landforms of areas bordering on *ice caps, this term has been extended to refer to any area with a *tundra climate, such as mountainous areas in mid-latitudes, or where frost processes are active and *permafrost occurs in some form. In consequence, as much as 20% of the earth's surface falls into this category.

    Periglacial climates are arid, with temperatures below 0 °C for at least 6 months, and summers warm enough to allow surface melting to a depth of approximately 1 m.

    Periglacial deposits include

    1. those formed by fragmentation due to *freeze–thaw: blockfields, *scree, *stone streams;

    2. those formed through solifluction: *solifluction gravels, lobes, and terraces;

    3. those formed through *aeolian deposition. *See* loess.

    Periglacial processes include *abrasion, *freeze–thaw, *nivation and *solifluction, and are responsible for three processes: the formation of new deposits, the modification of the structures of existing unconsolidated deposits, and the modification of existing landforms by *mass movement. It is suggested that mass movement accelerates under *periglacial conditions, and solifluction can give rise to turf-banked terraces and turf-banked lobes. In each case, the turf limits the extent of solifluction. The crucial factor in periglacial weathering is snow fall, which may protect ground against weathering processes.

period 1. A unit of geological time; the subdivision of an era. Thus the Cretaceous period is part of the Mesozoic era.

    2. The time taken for successive wave crests to pass a given point.

periodic market A trading market held on one or more days of each week and on the same days of the week. The markets served by the traders can be seen as forming, in total, the necessary *threshold populations for goods where one settlement alone would not provide the necessary custom for the trader, although there is enormous variation in the way traders work within **periodic market systems**.

Another way of explaining such a system is to note that a repeated movement of traders nearer to their customers may be more efficient than the infrequent movement of individual customers to a permanent market; or to recognize that the periodic market may bring extra shopping opportunities to the inhabitants of rural areas.

periphery The edge, or margin. *See* *core–periphery.

permafrost Areas of rock and soil where temperatures have been below freezing point for at least two years. Permafrost need contain no ice; a sub-zero temperature is the sole qualification. Any water present need not be frozen since the presence of dissolved minerals lowers the freezing point of water. In **epigenetic permafrost** ground ice develops mainly in upper parts, vertical freezing dominates, and cryogenic textures develop as water migrates under pressure to the freezing front, so that pressure is exerted on the ground leading to deformation. *See* *involution. In **syngenic permafrost** the ground ice is regularly distributed throughout the whole thickness of the permafrost; cryogenic textures develop as permafrost grows upwards, and sediments above are not contorted. The growth of permafrost is **permafrost aggradation**, which decreases the thickness of the active layer and may be caused by the freezing of *taliks. It is responsible for the formation of *pingos. The decline of permafrost is **permafrost degradation**, which plays a key role in the development of *thermokarst. Permafrost is a very sensitive system; small mistakes in constructing buildings in this environment can have catastrophic effects because of *thermo-erosion and thermo-abrasion. Permafrost features are well preserved in the chalklands of southern England.

The permafrost zones of *periglacial areas are of two types: **continuous permafrost** is present in all localities apart from small, localized thawed zones, while **discontinuous permafrost** exists as small, scattered areas of permanently frozen ground. A *frost table marks the upper limit of permafrost, which is overlain by the *active layer. *See also* *talik.

permanent snow line Above this line, the winter fall of snow exceeds the snow which is melted in summer; below it, melting is greater. This line varies with latitude.

permeable In geomorphology, allowing water to pass through along *bedding planes, cracks, fissures, and joints, and through rock *pores. Permeable rocks also have the capacity to be saturated by water.

Permian The latest *period of *Paleozoic time stretching approximately from 280 to 225 million years BP.

personal construct theory This suggests that humans are continually constructing and testing their own, individual images of reality, and that investigations should be based on the personal constructs of the people involved, and not on those of the researcher, i.e. that a **personal construct technique** should be used. For example, in a study investigating shopping centres the respondent is given three shopping centres and asked to identify the two which are felt to be most similar and to justify this choice. The criteria on which this judgement is based are then ascertained and built into the investigation so that the qualities of the centres are constructed from people's experience rather than imposed by the preconceptions of the investigator.

personal space The zone around individuals which they reserve for themselves. Personal space, as opposed to intimate space, is usually that reserved for a normal conversational voice and for friendly interaction. The extent of a personal space around an individual is reckoned to be 1–1.5 m for an Anglo-Saxon but closer contact may be acceptable in other cultures.

pervious A rock which may be non-*porous, but still allows water through via cracks and fissures, although not through pores within the rock. Compare with *permeable.

ph scale A scale, running from 1 to 14, for expressing how acid or alkaline a solution is. A strong acid, with a high concentration of hydrogen ions has a pH of 1–3, a neutral solution has a pH of 7, and a strongly alkaline solution has a pH of 10–14.

phacolith An elongated dome of intrusive igneous rock usually located beneath the crest of an anticline or the trough of a syncline.

phenomenal environment The natural and cultural environment which lies beyond us; the 'real' world, as opposed to the 'perceived' world. William Kirk (*Geography* 48, 1963) suggested that the facts of this phenomenal environment, this 'real world', are filtered through our cultural values, so that they become part of our *behavioural environment. Some writers reject this dualistic view.

Phillips curve A negative exponential curve set out by the New Zealand economist A. W. H. Phillips (1914–75), demonstrating the relationship, based on empirical evidence, between the percentage change in wages and the level of unemployment. The belief was that high wages cause high inflation, and the lower the rate of unemployment, the higher the rate of inflation. Conversely, as unemployment increases, the increase in wages declines, and the higher the level of unemployment, the lower the rate of inflation. Much British government policy has been based on this assumption—most famously in the statement that the use of high unemployment as a strategy to reduce inflation was 'a price worth paying'.

photic zone Those upper levels of a body of water which are penetrated by light.

photochemical smog Nitrogen dioxide (NO_2) is emitted from petrol engines. Ultraviolet light splits this into nitric oxide (NO) and monatomic oxygen (O). This oxygen reacts with free atmospheric oxygen (O_2) to form *ozone (O_3) which is irritating to the lungs. The ozone also reacts with the nitric oxide to make further nitrogen dioxide in a dangerous feedback loop. Furthermore, the hydro-carbons emitted from the burning of *fossil fuels react with some of the monatomic oxygen to form photochemical smog. Photochemical smog is most common in areas like Mexico City or Los Angeles where the sunshine is strong and long-lasting and where car use is high. While it is less visible than ordinary *smog (although sunsets may be tinged with purple and green), photochemical smog can irritate eyes and lungs and damage plants.

photosynthesis The chemical process by which green plants make organic compounds from atmospheric carbon dioxide and water, in the presence of sunlight. Since virtually all other forms of life are directly or indirectly dependent on green plants for food, photosynthesis is the basis for all life on earth.

phreatic Referring to *groundwater situated below the *water-table. The **phreatic zone** is permanently saturated.
    A **phreatic eruption** of a volcano is one in which *meteoric water is mixed with the lava. This water may be given off as a *geyser or as steam.

phreatophyte A class of desert plant with very long tap roots which develop to reach the *phreatic zone.

physical geography The branch of geography which deals with the natural features of the earth's surface. There is some difference of opinion on the scope of physical geography; while *geomorphology, *meteorology, *climatology, *biogeography, and *hydrology are included, soils and oceanography are often omitted from its study.

physiography The study of landforms and processes in physical geography.

physiological density Also known as nutritional density, this is population density in inhabited and cultivated areas.

physiological drought A condition of soil water being sufficient, but temporarily unavailable, as when the water is frozen, or when the rate of *evapotranspiration exceeds the rate of uptake of water by a plant. In the latter case, the plant will wilt in the daytime, but recover overnight, when evapotranspiration ceases.

pictogram A map of distributions where pictorial symbols such as motor cars or soldiers are placed on the location of the phenomenon mapped. The symbols may be drawn to some scale to indicate the sizes of the distribution, but this can be very misleading if they are scaled to the height of the symbol, as the accompanying increase in breadth misrepresents the actual dimensions.

pie chart A circle divided into sectors. The circle represents the total value, and the sectors are proportional to each value within the total.

piedmont Located or developed at the foot of mountains. A **piedmont flat** is a slightly undulating, residual landscape formed around a mountainous upland; W. Penck cited the Harz, surrounding the highlands of the Brocken, as a typical example.

piggy-back principle An integrated system of road and rail transport over distance which combines the benefits of both. Goods are packed in *containers which are taken by container lorries to the railway where they are easily transferred to the railcars for the major part of their journey. They are then taken by road to their destination. In this way, the accessibility of road travel is combined with the lower costs of long-distance railway travel.

pileus A shallow, auxiliary cloud in the shape of a cap above, or attached to, the top of a *cumuliform cloud.

pingo 1. open system pingo A large *ice mound formed under *periglacial conditions, so called because it is formed from an unfrozen pocket confined by approaching *permafrost. Mackay postulated that such pingoes developed on former lakes. For a time, the lake prevents the formation of *permafrost below it, but as it fills, and shallows, permafrost forms, starting at the edges and working inwards, exerting pressure on the saturated material beneath the lake. This causes bulging at the weakest point, where the permafrost is still thin, and it is here that an ice lens develops, formed from the water in the unfrozen pocket.
   2. closed system pingo An ice mound, as above, but formed as water, under *artesian pressure within or below permafrost, causes it to buckle upwards.
   In both cases, pingos collapse when the inner ice lens melts, leaving a depression surrounded by ramparts.
   Pingos are common in subpolar areas, such as the Mackenzie delta of Canada.

pioneer advance The movement of new human settlement beyond the present line of occupancy. Pioneer advance is slow, difficult, and, in modern times, expensive, so that it is often underwritten by governments.

FIGURE 44: *Pingo*

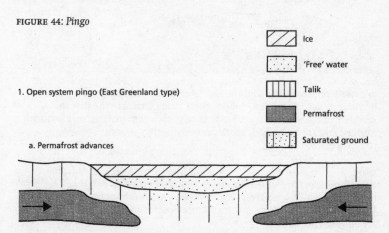

Ice

'Free' water

Talik

Permafrost

Saturated ground

1. Open system pingo (East Greenland type)

a. Permafrost advances

FIGURE 44: *Pingo (continued)*

b. 'Free' water freezes and expands, lake floor mounds up, lake water drains away

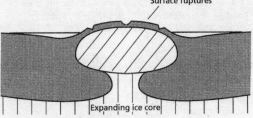

2. Closed system pingo (Mackenzie type)

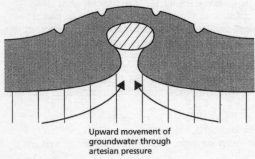

Upward movement of
groundwater through
artesian pressure

3. Collapsed pingo (open or closed system)

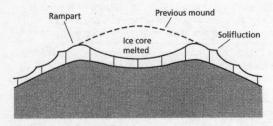

pioneer community The earliest *sere developing on a raw site. In a
*psammosere, salt-tolerant species such as sea lyme are the first to
establish themselves. Over time, dead material from pioneer species forms
humus, and the soil which results is colonized by other species, such as
marram grass.

pipe 1. A volcanic channel or conduit filled with solidified *magma.
Sometimes the hard pipe rock is exposed after erosion.

   2. In *hydrology, a subsurface channel, often near-horizontal, through
which water passes. Pipes can transfer water underground as a rapid route
for subsurface storm flow.

pipe eruption *See* *central vent eruption.

pipeline A steel or plastic tube used to transport gases, liquids, and slurries—mixtures of solids and water. Pipelines have a limited range of uses as they operate from one fixed point to another and can carry a restricted range of goods. Construction costs are high but running costs are relatively low.

pipkrakes In *periglacial environments, needle-like crystals of ice which develop below individual particles, or soil aggregates, which are better conductors. They grow when frosty nights alternate with morning thaws, especially over several days of this cycle. Pipkrakes grow at right angles to the face of soil or weathered rock and prise off the material above them. They also assist in the downhill creep of material.

pitot tube anemometer *See* *anemometer.

pixel In *remote sensing, an element of a picture; the basic unit from which an image may be built up. Pixels can be taken from an area of 5 m^2 to 10 km^2, or more. Pixel information for band or brightness varies according to the sensor system used. Decreasing **pixelation**—the use of smaller pixels—produces a sharper image. A LANDSAT scene of 185×185 km contains nine million pixels, each having a brightness between 0 and 225.

place A particular point on the earth's surface; an identifiable location for a situation imbued with human values. In *humanist(ic) geography, place is a centre endowed with meaning by human beings.

place names The study of the early forms of present place names may indicate the culture which gave the name together with the characteristics of the site. For example, **ey** meaning a *dry point and **ley** meaning a forest, wood, glade, or clearing appear in many place names such as Chelsea and Henley-in-Arden. Place names are used as evidence for the dating of a settlement from which a chronology of settlements may be devised. There are pitfalls; **ham** can mean either village or water-meadow, for example.

place utility The desirability and usefulness of a place to the individual or to a group such as the family. Factors such as housing, finance, *amenity, and the characteristics of the neighbourhood are perceived by the individual or group as being satisfactory or unsatisfactory. In the latter case, stresses may be set up resulting in the desire to move away from the place. Dissatisfaction with one place may lead to *search behaviour for a more satisfactory location.

This concept is difficult to apply since assigning quantitative values to utility is problematic. Even when places are ranked in terms of utility, problems arise from the choice of variables and the weightings given to each; a 1995 ranking of British towns which put Henley-on-Thames first, and ranked the London borough of Lambeth above Durham, was heavily criticized.

Any individual's utility rankings may alter to reflect changes in alternatives, changes in the individual's *action space, and changes in the individual's preferences over time.

placelessness E. Relph claimed that, with mass communication, and increasingly ubiquitous high technology, places become more and more similar, so that locations lose a distinctive 'sense of place'. With increased personal mobility, people are said to identify less with one place; the pull of the home town is slackening. This view is contested; it might be that some meanings are lost as places become increasingly homogenized (similar new architecture, the same chain stores, and so on), but that new meanings are gained. For example, the association of Manchester with a distinctive style of music and club culture is relatively recent.

placer deposit A mineral occurring as an alluvial deposit in the sand and gravel of *alluvial fans and valley floors. Such minerals are generally resistant to corrosion by water. The most important placer deposits are diamonds, gold, and tin.

planation surface The term has been variously used. Some reserve it for a flattish plain resulting from erosion; others use the term as a synonym for any *erosion surface, whether it be a flat or inclined *etchplain, *pediplain, or *peneplain. Few planation surfaces survive because they have been *dissected, but some geomorphologists claim that they can be extrapolated from *accordant summits. *See also* *thermo-planation.

planetary boundary layer The lowest 500 m of the troposphere, which is the layer most strongly influenced by the land or sea beneath it.

planetary winds The major winds of the earth such as the westerlies, *trades, etc., as opposed to local winds.

planeze One of a series of triangular facets facing outward from a conical volcanic peak. The planezes are separated by radiating streams which run down the flanks of the cone.

plankton Minute organisms which drift with the currents in seas and lakes. Plankton includes many microscopic animals and plants including *algae, various animal larvae, and some worms. The animals are zooplankton and the plants are phytoplankton.

planning As practised by local or national government, the direction of development. Proposed changes are scrutinized, and planning permission is only given if the development does not conflict with agreed aims. Planning presupposes an ability to foresee future events and a capability for analysing situations and solving problems.

A developer, refused planning permission, may make a planning appeal to the Secretary of State for the Environment who will consider both sides of the proposal, and may propose an altered plan.

planning blight The adverse effect of a proposed development such as a motorway. The value of housing may drop if such a new development is planned. If the landowner cannot dispose of the property, or cannot make as much use of it as was previously possible, he or she may serve a purchasing notice on the planning department of the local authority. *See* *externality.

planning-system firms Firms which can choose certain courses of action. Rather than responding to the market, they can manipulate and create demand by the use of advertising. Such firms usually have many plants producing diversified goods and are multi-regional if not multinational.

plant 1. In a system, the buildings, machinery, and land into which inputs are made and from which output issues.
   2. In industrial geography, an individual factory producing power or manufactured goods.

plant community An assembly of different species of plants growing together in a particular habitat; the floral component of an *ecosystem. The concept of community can be applied to a range of scales from a small pond to the Amazon rain forest.

plant succession The gradual evolution of a series of plants within a given area. This series of communities occurs in a roughly predictable order while the habitat progressively changes. **Primary succession** is the first occupation of a habitat. **Secondary succession** is the replacement of a community following a disturbance. In **autogenic succession** the plants themselves are the genesis of change; succession is directed from within the ecosystem. In **allogenic succession** the changes are driven by forces outside the ecosystem.

plantation An agricultural system, generally a monoculture, for the production of tropical and subtropical crops, especially bananas, coffee, cocoa, cotton, palm oil, rubber, sisal, spices, sugar, and tea. The corporately owned holdings are large and employ labour on a large scale. Early stage processing often takes place on site. Old-style plantations, generally in Latin America, were developed to support the lavish lifestyle of their owners, but new-style plantations were often developed by colonial powers, and thus may be seen as a spatial expression of *imperialism, and *capitalism. With independence, many Third World countries have nationalized their plantations or redistributed the land.

plastic flow Movement of material, especially rocks and ice, under intense pressure. The material flows like a very *viscous substance and does not revert to its original shape when the pressure is removed. As it moves, *shearing occurs. In ice, plastic flow is due to pressure at depth. Melting and refreezing cause crystals to grow and be drawn out. A thickness of at least 22 m is required for plastic flow to occur in *temperate glaciers.

plastic moulding Ice which moves plastically will flow around and over an obstacle. This may cause deposition in the lee of the obstruction.

plate 330

plate A rigid segment of the earth's crust which can 'float' across the heavier, semi-molten rock below. The plates making up the continents—**continental plates**—are less dense but, at up to 35 km deep, are thicker than those making up the oceans—the **oceanic plates**—which are up to 5 km deep. Thus a plate is a part of the *lithosphere which moves over the plastic *asthenosphere. The boundary of a plate may be a *constructive, *destructive, *conservative, or, more rarely, a *collision margin.

The theory of **plate tectonics** submits that the earth's crust is made up of six large plates: the African, American, Antarctic, Eurasian, Indian, and Pacific plates, and a number of small plates, the chief of which are the Arabian, Caribbean, Cocos, Nasca, Philippine and Scotia plates. The movement of plates causes global changes, such as *continental drift and a remodelling of ocean basins and the creation of major landforms: *oceanic ridges, *fold mountains, *island arcs, and *rift valleys, together with *earthquakes and *volcanoes, which occur at a destructive plate boundary where one plate plunges below another. The causes of plate movement are still the subject of controversy, but it is known that while plates may move away from constructive margins at speeds of up to 6 cm per year, they may be consumed at destructive margins at up to 15 cm per year.

plateau An extensive and relatively flat upland. Some are formed structurally, from resistant and horizontal rocks, or from the outpouring of **plateau lavas** as in the Deccan of India; others are *erosion surfaces.

playa A flat plain in an arid area found at the centre of an inland drainage basin, such as the Salinas Grandes, Jujuy Province, Argentina. Within such an area, lakes frequently form. Evaporation from the playa is high and alluvial flats of saline mud form. The term is also used to describe a lake within such a basin.

Playfair's law *See* *accordant.

Pleistocene An *epoch of the Quaternary period, stretching from the end of the Pliocene, some 2 million years ago to the beginning of the *Holocene. During the Pleistocene, temperatures in the Northern Hemisphere varied between the very cold *glacials and the warmer *interglacials.

plinian eruption A highly explosive volcanic eruption with dense clouds of gas and *tephra being propelled upwards for many kilometres. *Fissure eruptions are common on the flanks of the volcano as are the destruction of the crater walls and the fragmentation of solidified lavas.

plinthite A hard capping or crust at the surface of an unconsolidated soil. The term is used by some to denote a surface layer of *laterite.

plucking The direct removal of loose bedrock by the impact of water or by incorporation into glacier ice. Since the tensile strength of ice is low, plucking generally occurs only when the rock is jointed and weathered. Some writers prefer to use the term *quarrying rather than plucking.

plug flow *See* *Blockschollen flow.

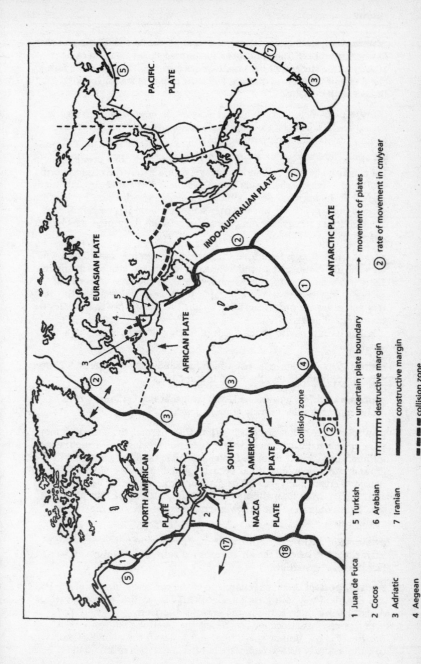

PACIFIC PLATE

ANTARCTIC PLATE

EURASIAN PLATE

INDO-AUSTRALIAN PLATE

AFRICAN PLATE

NORTH AMERICAN PLATE

SOUTH AMERICAN PLATE

NAZCA PLATE

Collision zone

→ movement of plates

② rate of movement in cm/year

‒ ‒ ‒ uncertain plate boundary

ⲧⲧⲧⲧⲧⲧⲧ destructive margin

━━━ constructive margin

▪▪▪▪▪ collision zone

1 Juan de Fuca

2 Cocos

3 Adriatic

4 Aegean

5 Turkish

6 Arabian

7 Iranian

plume An upwelling of molten rock through the *asthenosphere to the lower *lithosphere. The *hot spot thus formed shows up in volcanic activity at the surface. As continents move over the hot spot, there forms a chain of volcanoes at the surface, since the plume is stationary with respect to the *mantle.

plunge pool A pool at the base of a waterfall, often undercutting the sheer rock face. Plunge pools form as a result of *eddying, *hydraulic action, *cavitation, and *pothole erosion.

plural society A society made up of a number of distinct groupings. These may be by race, religion, language, or life styles. The original term envisaged a society where each group lived separately, meeting only for trade, but the term has been extended to societies with a number of ethnic strands, such as Britain. In this latter sense, the term has been criticized, for it indicates an equality of opportunity between the groups which does not exist.

pluralism 1. Robert Dahl used this term to denote any situation in which no particular political, cultural, ethnic, or ideological group is dominant. There is often competition between rival groups and the state or local authority may be seen as the arbitrator. It has been asserted that this is the way that cities are run, rather than by an élite. The theory thus relates to the nature of power.
  2. The term may also be used to signify the cultural diversity of a *plural society.

pluton A mass of igneous rock which has solidified underground. Plutons vary in size from *batholiths to *sills and *dikes.

plutonic rock An igneous rock, such as gabbro or *granite, which has cooled and crystallized slowly at great depth.

pluvial A time of heavier precipitation than normal. Evidence that more than one pluvial has brought wetter conditions to deserts during the last million years include plant and human remains and the presence of groundwater in aquifers. It has been suggested that these pluvials have helped to create many current desert landforms, including wadis. Pluvials and *interpluvials have alternated many times. It may be that pluvials in the *Pleistocene were due to lower temperatures affecting the *hydrological cycle.

pneumatic action Of, or acting by means of, wind or trapped air An example of the latter is the air compressed between a sea wave and a cliff face. *See also* *cavitation.

podzol, podsol A soil characteristic of the coniferous forests of the USSR and Canada. These soils have an ash-coloured layer just below the surface. A hard layer is often found in the lower, B *horizon.
  In podzols, *translocation has meant the *leaching out from the A horizon of clays, *humic acids, iron, and *alluvial compounds. These constituents may then accumulate to form a hardpan or iron band.

podzolization The formation of a podzol. This occurs when severe *leaching leaves the upper *horizon virtually depleted of all soil constituents except quartz grains. Clay minerals in the A horizon decompose by reaction with *humic acids and form soluble salts. The leached material from the A horizon is deposited in the B horizon as a humus-rich horizon band or as a hard layer of *sesquioxides.

poikilotherm An organism which has its body heat regulated by the temperature of its surroundings; **poikilothermy** is the state of being cold-blooded.

point In *Geographic Information Systems, a zero-dimensional *spatial object, specified by a set of *co-ordinates. Point data have each data element associated with a single location.

point-bar deposit The accumulation of fluvial sediment at the *slip-off slope on the inside of a *meander.

polar Applying to those parts of the earth close to the poles.
   Air masses originating over source regions in the mid-latitudes (40°–60°) are termed **polar air masses, P**, and are characteristically cold. They are not to be confused with Arctic or Antarctic air masses.

polar front The discontinuous, variable *front which forms over the north Atlantic and north Pacific, where polar maritime air meets tropical maritime air. The formation of *mid-latitude depressions at the polar front is connected with the development of troughs in the polar front *jet stream, a band of high-velocity winds in the wider *Rossby waves. The polar front jet stream, and hence the polar front, moves southwards in winter and northwards in summer.

polar high A mass of cold, heavy air, bringing high pressure at high latitudes.

polar low *See* *cold low.

polar orbital satellite A satellite used for *remote sensing, which orbits the earth along the meridians of *longitude.

polarization effect *See* *backwash effect.

polder The Dutch term for land which has been reclaimed from the sea, lakes, or river deltas. The land is bounded by a dike, is drained, and is maintained by pumping.

pole The North and South Poles are at either end of the earth's axis, around which the earth rotates. The **magnetic poles**, which are indicated by the needle of a compass, vary in their location over time. At present, magnetic north lies near Prince of Wales Island, Canada and magnetic south is in South Victoria Land, Antarctica. *See also* *geomagnetism.

political geography One of the three divisions of human geography—the others are economic and social geography—this is the geographical analysis of political studies. It is concerned, among other things, with the spatial expression of political ideals, the consequences of decision-making by a political entity, and with those geographical factors which influence political activities and problems.

polje In Slovenia and Croatia, any enclosed or nearly enclosed valley of any origin. More specifically, in *karst terminology, a large enclosed basin, up to 65 km in length and 10 km in width, with a flat floor and steep sides. Streams can be ephemeral or permanent; usually the water drains into *streamsinks. Most poljes are aligned with underlying structures such as folds, faults, and troughs, and most poljes have a long, complex history. The Cuges polje, in Provence, is 5 km long and 2 km wide, formed in soluble Jurassic and Cretaceous limestones.

poll tax *See* *community charge.

pollen analysis The detection of past climates from the different types of pollen grain preserved in lakes, peats, and muds. *See* *palynology.

pollution A substance which causes an undesirable change in the physical, chemical, or biological characteristics of the natural environment. Although there are some natural pollutants such as volcanoes, pollution generally occurs because of human activity. Biodegradable pollutants, like sewage, cause no permanent damage if they are adequately dispersed, but non-biodegradable pollutants, such as lead, may be concentrated as they move up the food chain. Within Western Europe, air pollution, associated with basic industries such as oil refining, chemicals, and iron and steel, as well as with the internal combustion engine, is probably the principal offender, followed by water and land pollution. Other forms of environmental pollution include noise and the emission of heat into waterways which may damage aquatic life. Present-day problems of pollution include *acid rain and the burning of *fossil fuels to produce excessive carbon dioxide.

pollution dome A mass of polluted air in and above a city or industrial complex which is prevented from rising by the presence of an *inversion above it. Winds may elongate the dome into a pollution plume.

polygon In *Geographic Information Systems, a *line enclosing an area.

polygons Through the process of *frost cracking, *periglacial surfaces may exhibit polygonal areas of ground separated by cracks. If these cracks fill with debris, the ground rises around the wedges when the permafrost expands. These are **sand-wedge polygons**. In more humid environments, if the cracks are filled with ice, the feature is an **ice-wedge polygon**.

polysaccharide gum The sticky by-product of the decomposition of roots by micro-organisms, which can bind soil minerals into aggregates, such as *peds.

pool and riffle The alternating sequence of deep pools and shallow riffles along the relatively straight course of a river. The distance between the pools is 5–7 times the channel width. It has been suggested that pool and riffle development is the precursor of meanders but supporting evidence is not conclusive.

population 1. In ecology, a group of individuals of the same species within a *community.
   2. In statistics, the entire and complete collection of individuals under consideration, from which a *sample may be taken. These individuals need not necessarily be living organisms.

population density The ratio of a population to a given unit of area. Crude density is simply the number of people living per unit area and can be very misleading. Britain and Sri Lanka have similar crude densities— around $220/km^2$—but very different living standards. Accordingly, densities may be plotted using different criteria.
   Nutritional density is based on the ratio between total population and inhabited areas. This is thought by some to be an indication of living standards. **Occupational density** is the density of a particular occupation, for example farmers, over the total area of the country, and **room density** is the average number of people per room in a given area.

population dynamics The study of the numbers of populations and the variations of these numbers in time and space. A demographer will study numbers of people; an ecologist will study the numbers of organisms of different species and their numerical relationship to each other.

population equation The future size of a population depends on a range of functions. Thus:

$$P_{t+1} = P_t + (B - D) + (I - E)$$

where P_t and P_{t+1} are the sizes of population in an area at two different points in time, t and $t+1$ are those points, B is the birth rate, D is the death rate, I is the immigration, and E is the emigration.

population geography The study of human populations; their composition, growth, distribution, and migratory movements with an emphasis on the last two. It is concerned with the study of demographic processes which affect the environment, but differs from demography in that it is concerned with the spatial expression of such processes. *See* *demography.

population potential The accessibility of people from a given point; that is, a measurement of how near people are to a point. The population potential at one place is the sum of the ratios of population at all other points to the distances from the place in question to those points. Thus:

$$\text{population potential}, V_1 = \sum_{j=1}^{n} \frac{P_j}{d_{ij}} = \frac{P_1}{d_{i1}} + \frac{P_2}{d_{i2}} + \frac{P_3}{d_{i3}} \ldots + \frac{P_n}{d_{in}}$$

where the population potential (V_1) at point i is the summation ($\sum$) of n populations (j) accessible to the point i divided by their distance (d_{ij}) to that point. Transport costs may be used instead of distance. Note the similarity with the equation for *market potential.

population problems Between about 9000 years before the present until about AD 1800 it is estimated that world population grew from 5–10 million to 800 million, with an average growth rate of 0.1% per annum. Between 1800 and 1980, world population grew to 4.4 billion. World population is set to double every 35–45 years. If this rate of growth is sustained, there would be standing room only in the next century. The problem is more acute in the Third World because a large proportion of the population is very young. The eventual reproductive capacity of such regions could result in an expansion of population which would end in *Malthusian disasters.

population projection The prediction of future populations based on the present age–sex structure, and with the present rates of fertility, mortality, and migration. The simplest projections are based on extrapolations of current and past trends, but a set of very differing projections can be calculated, based on a series of differing assumptions— for example, that current rates of increase will be maintained/will increase/ will decrease.

population pyramid *See* *age–sex pyramid.

population studies The primary analysis and then attempted explanation of demographic patterns and processes. Attempts are made to link spatial variations in the distribution and composition of migration and population change with variations in the nature of places.

pore In geomorphology, a minute opening in a rock or soil, through which fluids may pass. Porous rocks allow water to pass through or be stored within them. **Pore water pressure** is the pressure applied by water in the pores to soil and rock particles. When the rock or soil is saturated, pore water pressure can be so great that slope failure results.

pork-barrel effect Government expenditure aimed at gaining votes. Government contracts can be of vital importance to employment within a region and may increase the popularity of an administration, as elected members of a government divert spending to the areas they represent. Investigations of the geography of federal investment, for example in the allocation of clothing contracts to the US army in the immediate post-war period, have revealed the significance of the pork-barrel effect for certain local economies.

porosity The ratio of the volume of pores to the volume of matter within a rock or soil, expressed as a percentage.

port Defined as a place where ships may anchor to load and unload cargo, a port may be classified by its function. **Terminal ports** are the final destination of cargo-carrying ships. Some specialized ports handle predominantly one particular type of traffic and include *container ports, ferry ports, fishing ports, and naval ports. **Ports of call** lie between terminal ports and may handle part of the cargo of a vessel.

portage The overland transport of a boat and/or its goods from one navigable waterway to another.

positive discrimination A policy designed to favour some deprived region or minority and to redress, at least in part, *uneven development. Policies of positive discrimination have been criticized for treating the effects of inequality rather than tackling its causes. Other criticisms made are that not all of the minority or region needs help, and that many deprived people are outside the catchment area. None the less, the *EU, for example, still uses schemes of positive discrimination, through the European Regional Development Fund.

positivism The belief that an understanding of phenomena is solely grounded on sense data; what cannot be tested empirically cannot be regarded as proven. Positivism has no value judgements, only statements which can be tested scientifically.
    The tests for the validity of a statement are

- statements must be grounded on observation
- observations (e.g. from experiments) must be repeatable
- experiments should all use the scientific method agreed on by the entire scientific community.

To this basis have been added the concepts of **logical positivism**, that a tautology is a form of verifiable statement—an analytic statement—as opposed to a synthetic statement which can be scientifically tested. Positivism was accepted by the 'new' geographers of the 1950s onwards as it was argued that human behaviour followed certain 'laws' which could be used to predict events. Thus, the *gravity model is widely used in transport planning.
    In recent years, geographers have moved away from this vision of themselves as social 'scientists', perhaps because the status of science as 'value-free' has been challenged, as have the claims that the 'laws' of social science (and indeed, the natural sciences) are universally applicable, and because logical positivist geography excluded values, meanings, and interpretations.

possibilism A view of the environment as a range of opportunities from which the individual may choose. This choice is based on the individual's needs and norms. It grants that the range of choices may be limited by the environment, but allows choices to be made, rather than thinking on *deterministic lines.

post-Fordism A system of production characterized by flexibility both of labour and machinery; by the vertical break-up of large corporations, by better use of links between firms so that subcontracting is increasingly used, and by *just-in-time production. Just as *Fordism is associated by geographers with a distinctive spatial pattern of economic activity, post-

Fordism is associated with agglomeration, which will simplify interaction between linked forms of economic activity.

post-industrial city A city exhibiting the characteristics of a *post-industrial society. Service industries dominate with a strongly developed *quaternary sector and *footloose industries abound, often on pleasant open space at the edge of the city. Post-industrial cities are also characterized by large areas of office blocks and buildings for local government administration. These cities often exhibit marked inequality of income distribution because of the contrasts between those who are appropriately skilled—professionals, managers, administrators, and those in high technology service industries—and the poorly paid service workers who look after their needs, together with the unemployed. The former can afford high house prices, and, in fact, contribute to them; the latter cannot.

post-industrial society A post-industrial society has five primary characteristics: the domination of service, rather than manufacturing, industry, the pre-eminence of the professional and technical classes, the central place of theoretical knowledge as a source of innovations, the dominating influence of technology, and levels of urbanization higher than anywhere else in the world. At present, the development of a post-industrial society is linked only with very advanced economies, if it exists at all. *See* *de-industrialization.

postmodernism 1. An architectural style which is a composite of past styles, characterized by a variety of colours, stylistic details from many periods, and what is claimed to be a return to a vernacular style.
2. A philosophical stance which claims that it is impossible to make grand statements—meta-narratives—about the structures of society or about historic causation because everything we perceive, express, and interpret is influenced by our gender, class, and culture. No one interpretation is superior to another. It has brought to geographers a recognition that space, place, and scale are social constructs, not external givens. *See* *structuralism. Of particular interest to geographers is the way that time and space have been 'compressed' by modern transport systems, especially by jumbo jets: nowhere is very far away any more. As a result, cultures are transformed.

Some geographers claim that postmodernism challenges the dominance of time and history in social theories and instead stresses the significance of geography and spatiality.

Postmodernism has also been linked with late capitalism, and with *post-Fordism, but these links are, of course, meta-narratives; the very interpretations which postmodernists reject.

The postmodern tradition also stresses and, indeed, champions difference, and this is a strand which has been welcomed by *feminist geographers, who would claim that geography has been speaking in an authoritarian, masculinist voice for too long.

potential evapotranspiration, PET The maximum continual loss of water by evaporation and transpiration, at a given temperature, given a sufficient supply of water. PET often outstrips actual evapotranspiration.

pothole Loosely, a vertical cave system. More precisely, a more or less circular hole in the bedrock of a river. The hole enlarges because pebbles inside it collide with the bedrock as the water swirls.

power Defined as the ability to do or act, this is also seen as the influence of an individual or group upon another. Power within a society is worked out in its economic, social, and political life; *capitalism is controlled by a minority who dominate the *factors of production over a majority which does not. In a *centrally planned economy it is the state which dominates, supposedly reflecting the will of the people. The power structure of a society is reflected in its social organization and in its economy. These in turn have their own spatial expression.

pragmatism An interpretation of the meaning and the justification of beliefs in terms of their practical effects or contents. The method of reasoning is by *induction.

prairie A large area, found outside the tropics, with grassland and occasional trees as natural vegetation. The term was originally applied to the prairies of North America where rainfall is low and summer temperatures are high. Similar conditions are found in the South American *pampas, the Russian *steppes, and the South African *veld.

prairie soil A soil of the wetter prairies, resembling *chernozem in its high humus content and its development under grassland. However, increased rainfall means that prairie soils are *leached of calcium. This soil therefore has no calcium nodules and is slightly acid.

Precambrian The oldest *era in earth's history dating from about 4600 million years BP.

precipitation In *meteorology, the deposition of moisture from the atmosphere onto the earth's surface. This may be in the form of *rain, *hail, *frost, *fog, *sleet, or *snow. Precipitation develops in two stages. Initially, cloud droplets grow around nuclei through condensation and diffusion. In clouds warmer than −10 °C, the larger droplets then grow by collision and coalescence with the smaller ones. In colder clouds the *Bergeron–Findeisen mechanism is thought to operate, probably in conjunction with the growth of ice crystals through *accretion, as supercooled water droplets freeze on impact with the ice, and aggregation, as smaller ice crystals stick to larger ones. Much precipitation begins in the form of ice crystals, develops into snow flakes, but melts as it falls, to become rain.

predator An organism which takes other live organisms as its food and thereby removes the prey individuals from a population.

predator–prey relationships In theory, there should be an equilibrium between predators and prey. Thus, when predators are scarce, the numbers of prey should rise. Predators would respond by reproducing more and, possibly, by changing their hunting habits. As the population of predators rises, more prey are killed and their numbers fall. Many of the predators then die; thus numbers of predators and prey oscillate between two extremes.

The oscillation predicted above is rarely reproduced in laboratory experiments and is not easy to find in the wild. This is because predator numbers are not solely dependent upon the number of prey available. Furthermore, there must be an opportunity for some prey to avoid attack, otherwise extinction of both species may result. Predation will have no effect on numbers of prey if the individuals caught are beyond reproductive age. Lastly, prey are often sought by more than one predator.

There is a suggestion that predation allows more species to survive. It is argued that predation frees some part of every *niche giving more room for more species. This suggestion has been confirmed in a number of field observations.

pre-industrial city A model of the pre-capitalist city, as advanced by Gideon Sjöberg (1960). The city centre is occupied by a small élite, who, because of their association with political and religious power, control the political, religious, administrative, and social functions of the city, and who are catered for by their domestics. The lower classes, including merchants, occupy the concentric zone surrounding the centre, and the outcasts are consigned to the outer edges of the city. The concept of the exclusive social core has been questioned by Jr. Vance, who notes the many craft quarters at the heart of the pre-industrial city, each characterized by a vertical structuring of space. Many geographers reject attempts to find cross-cultural similarities in *urban social geography.

preservation Preservation has been defined as the protection of human features in the landscape, as opposed to *conservation which is concerned with the protection of the natural landscape. This distinction is not always made.

pressure gradient Also known as barometric gradient, this is the rate of change in *atmospheric pressure between two areas, providing a force which moves air from *high to *low in an effort to even up the unequal mass distribution of the air. On a global scale, the most powerful pressure gradients are in a *meridional direction, caused by meridional disparities in *insolation. It is routine in meteorology to show this horizontal distribution of pressure in terms of the height of *isobaric surfaces, such as the 500 mb level, above ground.

The **pressure gradient wind** is the movement of air in response to pressure differences, blowing from high to low. It is modified, however, by the action of the *Coriolis force. Where the isobars are close together, there is a steep pressure gradient, and winds are strong. Widely spaced isobars indicate a gentle pressure gradient, and winds are generally light.

pressure melting point In glaciology, the temperature, often well below 0 °C, at which ice under pressure will melt. *Warm glaciers have bases at or above the pressure melting point of their ice.

pressure release Also known as *dilatation*, this is the expansion of a rock formed under pressure when that pressure is released. Thus, a glacier may remove the *overburden, and the revealed rock 'bursts' open. Some writers attribute the splitting of granite *tors to pressure release. *See* *inselbergs.

prevailing wind The most frequent winds within a specified period. In *mid-latitudes, for example, most winds are westerly, with an azimuthal bearing of between 181 and 359°.

price The money for which a commodity or service is bought or sold. The **price mechanism** is the way in which supply and demand can regulate economic activities. The mechanism can be spontaneous—'in the market'— or can reflect deliberate governmental adjustments. **Pricing policies** are the arrangements whereby prices of commodities to the consumer are determined. In the past, prices could either be *f.o.b. or *c.i.f., but commodities today are increasingly sold at the *uniform delivered price. Under capitalism, producers may collaborate to maintain artificially high prices. Under socialism, prices are set centrally by the state.

primary industry, primary activity Economic activity, such as fishing, forestry, and mining and quarrying, concerned with the extraction of natural resources. All such activities fall into the **primary sector**.

primary urbanization Urbanization which results from forces arising within a country as a spatial expression of a region's culture, as in the wave of urbanization in UK which resulted from the *industrial revolution. Compare with *secondary urbanization.

primate city The largest city within a nation which dominates the country not solely in size—being more than twice as large as the second city, as in London and Birmingham, UK—but also in terms of influence. The development of **primacy** is not fully understood but some researchers have suggested that the importance of the primate city tends to decline as the economy grows and that, therefore, primacy tends to occur in less developed nations. However, the rise of the primate city may be encouraged by colonialism, as it occurs often at the major port. Capital cities of past empires tend to be over-large. *See* *rank-size rule.

primeur crop A crop of fruit, vegetables, or flowers grown and sold out of season or early in the season.

primogeniture Inheritance by the oldest son.

principal components analysis, p.c.a. A statistical technique of changing the many variables in a data matrix so that the new components are correlated with the original components but not with each other; that is, so that they are now independent of each other. It is a technique used to change a set of original variables into a number of basic dimensions, explained in non-technical language by P. R. Gould (*Trans. IBG*, 1967).

Principal components analysis differs from *factor analysis in that there is no reduction in the number of variables after the transformation.

Geographers use principal components analysis to reorganize or simplify a data set, and to identify groups of intercorrelated variables. Many p.c.a. software packages are available.

principle of least effort The thinking behind *movement-minimization procedures.

private sector That part of a national economy which is not owned by the state. Compare with *public sector.

probabilism *Possibilism sees individuals or groups making choices within the scope of the environment. Probabilism suggests that some choices are a good deal more likely than others.

probability The likelihood of an event occurring. In statistics, probability, p, is expressed as a number ranging from 0—absolute impossibility—to 1—absolute certainty. It may also be expressed as a percentage.

$p = 0.05$ is the 95% level
$p = 0.01$ is the 99% level
$p = 0.001$ is the 99.9% level

proclimax An arrested point of a succession which does not develop into a *climax community because of repeated disturbances, for example, from fire. Overgrazing may lead to a proclimax as animals leave unpalatable plants which are then overrepresented.

producer An organism which can fix energy from the sun and transform it by *photosynthesis into food.

producer goods Also known as **capital goods**, these are the goods, such as machinery and equipment, needed in the production of *consumer goods.

production In ecology, the increase of body mass as food is converted into new living material.

productivity The output of an economic activity seen in terms of the economic inputs such as capital, labour, and raw materials. Some writers argue that this economic view is too narrow and that social and environmental 'costs' must also be considered. Furthermore, the cost of raw materials can also be seen as the depletion of finite resources.

productivity rating An estimate of an area's ability to support plant growth. Productivity, as suggested by Paterson, increases with the length of the growing season, the average temperature of the warmest month, the average precipitation, and the amount of solar radiation. Paterson postulated six grades of productivity and produced a world map indicating the occurrence of these grades. This map reflects the plant life that a climatic zone could support whereas the actual vegetation may differ from this.

product-moment correlation coefficient A statistical test for assessing the degree of correlation between two data sets, x and y. The formula for the test is given as:

$$r = \frac{\frac{1}{n}\sum\left(x - \bar{x}\right)\left(y - \bar{y}\right)}{\sigma x \cdot \sigma y}$$

where σx and σy are the standard deviations of the respective data sets.

Coefficients run from +1—perfect positive correlation—through 0—no correlation—to −1—perfect negative correlation. A further test is necessary to determine the statistical significance of a particular correlation. This is done by expressing the correlation coefficient r in terms of the *Student's t-statistic:

$$t = \frac{r\sqrt{n-2}}{\sqrt{1-r^2}}$$

and then reading off the value of t at the correct degrees of freedom from a graph.

profit surface Variations in profit shown as a three-dimensional surface, derived from subtracting the relevant *cost surface from the corresponding *revenue surface, a calculation of the utmost difficulty.

proglacial Situated in front of a glacier. A **proglacial lake** is formed between the terminus of the ice and the higher ground which is often in the form of a terminal *moraine.

progradation The accumulation of beach material which leads to an extension of the beach seawards. When there is an excess in the supply of sediment, a beach will prograde. Progradation is a feature of, for example, *delta and *mangrove coasts.

programming region A region which is designed to serve a particular purpose, such as a multi-purpose river project or a depressed region which requires a regional policy as an attempt to solve its problems.

projection A technique for transforming the three-dimensional sphere of the earth into the two dimensions of a map. There are four aspects of the map to be considered: area, distance, direction, and shape, and it is impossible to recreate them all in the same map. *See* *map projection.

propulsive industry A vigorous, fast-growing industry characterized by a high level of technology, expert management, and an extensive market, for example, the manufacture of semiconductors. Propulsive industries are instrumental in promoting growth in the other industries to which they are linked.

protection Procedures adopted by a government to favour domestic goods by imposing *quotas or *tariffs on foreign imports. Governments adopt protectionism in order to help the country become self-sufficient, to protect new industries, or as a bargaining tool.

protoindustrialization The phase in a peasant society as rural industries develop in advance of industrialization.

psychic income The enjoyment, which cannot be measured in financial terms, that people derive from location in a particular place. The example usually cited is an entrepreneur who chooses his factory site near a favourite golf-course in order to receive enjoyment as well as gaining profit from his industry. Although no manufacturer will readily operate at a loss, many industrialists will locate away from the optimum point in order to benefit from psychic income.

psychologism The explanation of social phenomena wholly in terms of the mental characteristics of the individuals concerned. Psychologism thus overlooks the economic, social, political, and environmental influences which act on the individual.

public goods Goods freely available, either naturally, like air, or from the state, like education in most developed countries. **Pure public goods**, like defence, are provided for all. However, pure public goods may be distributed impurely, as when an area with a high crime rate has a higher level of policing. **Impure public goods**, like libraries, are provided at particular locations, so they are more accessible to some than to others.

public sector That part of a national economy owned and controlled by the state. It includes nationalized industries, national and local government services, and public corporations. Compare with *private sector.

pull factor A positive factor exerted by the locality towards which people move. Pull factors have included: the granting of new land for farmers (the Prairies and Great Plains), assisted passages and other government inducements (Australia), freedom of speech or religion (America in the eighteenth century), and material inducements (Hong Kong). People moving for material gain are currently termed *economic migrants. See* *push factor.

pumice A very light, fine-grained, and cellular rock produced when the froth on the surface of lava solidifies.

pumped storage scheme Electricity cannot be stored, so when demand is low, at night, some can be used to pump water from a lower to a higher reservoir. At peak demand, the water is allowed to fall back to the lower level, passing through turbines which turn generators. One such scheme operates at Blaenau Ffestiniog, North Wales. The scheme incurs a net loss of energy, but uses electricity which would not otherwise be utilized.

push factor In *migration, any adverse factor which causes movement away from the place of residence. Examples of pushes include: famine, changes in land tenure (the Highland Clearances, 1790–1850), political persecution (Tamil separatists, Sri Lanka, 1989), mechanization which made agricultural workers redundant (but *see* *rural depopulation) and which made factory products cheaper than those of cottage industry. Relatively few migrations are spurred by push factors alone.

push moraine *See* *moraine.

puy The French term for a volcanic neck, revealed by differential erosion. The type location is the Puy de Dôme. The Hohentweil, Hegau, Germany is a further example.

pyramid of numbers, ecological pyramid A diagram of a *food chain which shows each *trophic level as a horizontal bar. The bars are centred about a vertical axis and the levels are drawn in proportion to the *biomass at each level. There is a big step between *producers and primary *consumers, but thereafter the steps are smaller. Generally speaking, the animals on the higher levels are larger and rarer than animals lower down the pyramid.

pyramidal peak Synonym for *horn or aiguille.

pyroclast A fragment of solidified lava, ejected during explosive volcanic eruptions. Classification of pyroclasts is by size. Fragments less than 4 mm across are ash; compacted ash is *tuff*. Material between 4 and 32 mm is *lapili* and fragments larger than 32 mm are *blocks*. Collectively, these fragments are *tephra. Pyroclasts formed from lava produce *volcanic bombs and volcanic *breccia. **Pyroclastic flows** are also known as *nuées ardentes. They result from the bursting of gas bubbles within the magma, which fragments the lava. Eventually, a dense cloud of fragments is thrown out to form a mixture of hot gases, volcanic fragments, crystals, ash, pumice, and shards of glass.

Q

quadrat A small, usually square, frame used in sampling, notably in
*biogeography and *ecology. Quadrats are placed systematically or at
random over the area to be studied and the vegetation occurring within
the quadrat is recorded.

quadrat analysis A statistical technique for analysing distributions. The
area to be analysed is divided into cells of equal size and the number of
points occurring within each cell is determined. This distribution is then
compared with a hypothetical, or expected distribution based on the
theory being investigated. It should be noted that the size and shape of the
cells may influence the observed distribution.

qualitative Concerned with meaning, rather than with measurement.
The emphasis is on subjective understanding, communication, and
empathy, rather than on prediction and control, and it is a tenet that
there is no separate, unique, 'real' world. Qualitative methods vary, and
are generally based on *empirical research, but there is some discussion
over the extent to which the researcher should intervene, and much
awareness of the way in which any research process will affect the subjects
of the investigation.

quality of life The degree of well-being felt by an individual about his or
her life-style. Preferences vary, but most assessments of the quality of life
consider *amenity, together with social benefits such as health, welfare,
and education, and economic aspects such as income and taxation.

quantification The numerical measurement of processes or features. The
data derived from this type of exercise may be used in what is argued to be
a more objective analysis than is produced by *qualitative methods. For
the relevance of quantitative methods in geography, *see* *quantitative
revolution.

quantitative revolution Until the early 1950s, *geography had used
descriptive, *qualitative methods. Fred Schaefer (AAAG, 1953), is widely
credited with initiating a move to seek 'laws' which would explain
geographical phenomena, particularly within the field of human
geography, thereby introducing the shift to *logical positivism and the
concept of geography as a science of *spatial distribution. This movement
to *locational analysis was seen as a revolution in geography, and was
accompanied by extensive *quantification. From that time, *quantitative
methods were introduced; at first slowly, with an emphasis on hypothesis
testing, using statistical techniques like the *chi square test, but later
employing the ideas of *social physics to employ mathematical models and
more sophisticated statistical analyses; the fashion was for geography to
become a spatial science, and the work of thinkers like Walter Christaller
(1933), August *Lösch (1954), Johann *von Thünen (1826), and Alfred

Weber (1909) was the inspiration. Some commentators think that this 'number crunching' was popular because it could present itself as being politically untainted, others that it gave geographers increased status.

The heyday of *logical positivism and the quantitative revolution was short-lived. Quantification was attacked for being unrealistic and bloodless, turning humans into automata, for being too deterministic, and for ignoring the importance of subjective experience. *See* *paradigm.

quarrying The removal of rock which has broken up by *jointing to form blocks. The agents of erosion may be ice, rivers, or the sea. Quarrying by ice occurs when ice is frozen to rock. Since the tensile strength of ice is low, quarrying is not possible unless the rock is shattered.

quartiles The percentiles which divide any ordered distribution into four. The 25th is the **lower quartile** and the 75th percentile is the **upper quartile**.

The **quartile deviation** is the value of the *interquartile range, divided by 2.

Quaternary The most recent *period of the *Cenozoic era. During the Pleistocene *epoch of this time, from about 1.8 million years BP, to some 10 000 years ago, much of Britain's glacial and periglacial scenery evolved.

quaternary industry, quaternary activity, quaternary sector
Economic activity concerned with information; its acquisition, manipulation, and transmission. Into this category fall law, finance, education, research, and the media.

quota A fixed level indicating the maximum amount of imported goods or persons which the state will allow in—**import quota**—or out—**export quota**—in a given period of time. It may also refer to limits on production in an economy.

R

R and D, research and development Industrial innovation requires research which seeks to apply new discoveries to industrial processes. These discoveries are then developed as factory systems. Most large industries have R and D facilities, and often have access to independent R and D from educational establishments and government programmes. The concentration of *high-tech enterprises along the British M4 motorway is attributed in part to the research establishments along that axis. *See* *science park.

radial drainage *See* *drainage patterns.

radial plan *See* *finger plan.

radial–concentric plan The street pattern of a settlement where a number of roads radiate from the centre and are cut through by a series of circular roads which form rings round the centre.

radiation Energy travelling in the form of electromagnetic waves. These may be X-rays, ultraviolet, visible, infra-red, microwaves, or radio waves.

radiation fog *See* *fog.

radiative forcing The increase in the trapping of outgoing *terrestrial radiation by *greenhouse gases.

radical geography A description of the geographical writing which began to appear in the 1970s, and which was based on a *Marxist geographical analysis. Major topics of concern are health, hunger, poverty, and crime, and the aim is not simply to analyse what is happening, but also to advocate change.

  With the changes in Eastern Europe and the USSR in the late 1980s and early 1990s, the value of a Marxian analysis became less attractive to some, but others continue to study geography from the perspective of political economy. *See* *Marxist geography.

radiometer A passive *remote sensor, which is sensitive to *terrestrial radiation of one or more wavelengths of the visible and infra-red. The radiometer scans each area line by line, and sends the information to the ground station in digital form.

radiosonde A free-flying balloon carrying meteorological instruments. The balloon climbs to a height of 20–30 km above mean sea level, sending information from these sensors to ground stations, whereupon it bursts, returning the equipment to the ground. The progress of the radiosonde is tracked by radar.

radius of curvature In a *meander, the mean distance from the centre of the curve to points at the edge of the meander.

rain A form of precipitation consisting of water droplets ranging from 1 to 5 mm in diameter. The type of rain produced reflects the circumstances in which it formed. A mass of warm air rising at a warm front will develop layered clouds and produce steady rain. Air forced to rise quickly at cold fronts will bring heavier rain. These are both examples of frontal rain. Convection rain occurs when warm, unstable air rises rapidly. Air forced to rise over mountains may form orographic (relief) rain. *See also* *Bergeron–Findeisen theory and *coalescence theory.

rain forest An area of luxuriant forest which has developed where rainfall exceeds 1000 mm per annum. Rain forest does develop in temperate latitudes, for example, in the state of Washington, USA, but most of the world's rain forest is found within the tropics. *See* *tropical rain forest.

rain gauge **Standard meteorological gauges** have a funnelled aperture of 150–170 cm, and are set 30 cm above ground level. **Tilting syphon gauges** empty when the bucket is full, usually for each 0.2 mm of rain, and a trace can register each 'tilt', thus giving a fairly continuous record of precipitation. Rain gauges must be sited in as representative a location as possible, but the choice of location is difficult, since many precipitation events are highly localized.

rain shadow An area of relatively low rainfall to the *lee side of uplands. The incoming air has been forced to rise over the highland, causing precipitation on the windward side, and thus decreasing the water content of the air which descends on the lee side. If there is a deep layer of cloud on the windward side, it is deepened by the enforced rise, and its rate of precipitation enhanced, which may increase the rain shadow effect.
   The descending air is subject to *adiabatic warming, and this increases its capacity to hold much of the remaining water vapour thus further reducing rain on the lee side.

rainbow An arch of the visible parts of the spectrum caused by the reflection and refraction of sunlight within raindrops.

raindrop erosion The dislodging of soil particles by large drops of rain. The particles are pushed into the soil spaces, helping to secure the soil surface against *infiltration and thereby increasing *run-off. Raindrop erosion is most active in tropical, subtropical, and semi-arid environments, particularly where rainfall is intense and the ground is free of vegetation.

rainfall intensity The rate at which rain falls, usually measured in millimetres per hour. Intense rainfall is associated with *convectional rain, notably in thunderstorms and tropical regions, where intensity may be over 100 mm per hour. (British rainfall intensity is normally of the order of 2 mm per hour.) The intensity of rainfall is normally inversely proportional to its duration.
   The reaction of a river to a storm event is linked to rainfall intensity; intense rain has a greater impact on the ground (see raindrop erosion), so that run-off is usually rapid.

rainfall run-off The overland and downslope flow of rainwater into channelled flow when the rock or soil is saturated.

rainsplash The impact of raindrops on the soil may break down soil *peds, loosen soil particles, and cause *turbulence in the *sheet wash of water flowing downslope.

raised beach A former *beach, recognizable by beach deposits and marine shells, which now stands above sea level some metres inland. Where land is rising because of *isostasy, several raised beaches may be seen at different levels; classic British examples are the raised beaches of Scotland which range from 6 to 14 m above present sea level. Raised beaches are often marked by rock platforms backed by dead *cliffs.

ramparts The high sides of *ice wedge polygons; as the ice wedges grow, they push up the soil at the edges.

ranching Large-scale and *extensive cattle rearing, best developed on temperate grasslands, like the *pampas of Argentina and Uruguay.

random Haphazard, without a regular pattern, with an equal chance of any event or location occurring. A **random number table** consists of a series of numbers taken entirely at random, generated by chance.
   Random sampling uses such numbers to select individual units. The numbers are used as co-ordinates on a grid system of the area under consideration.

Randstad Holland The highly urbanized area of the Netherlands dominated by Rotterdam, the Hague, IJmuiden, Amsterdam, and Utrecht.

range In statistics, the difference between the two extreme values in a data set.

range of a good or service The maximum distance an individual will travel to obtain a given good or service. An illustration is given by the distance people will travel to buy a pint of milk as opposed to the journey for an Old Master drawing. *See* *high order goods and services, *convenience good, *shopping good. There is confusion, however, in the fact that many trips are motivated by the need to purchase more than one commodity.

range, township, and section The division of land west of the Appalachians adopted by the US Government in 1785. **Townships** are squares of 36 square miles, and a series of townships constitutes a **range**. Each township may be divided into **sections** of one square mile. All divisions are related to a base *meridian.

ranker An *intrazonal soil, not yet fully developed. This soil is shallow, with an A *horizon directly on top of non-calcareous rock.

ranking of towns Many attempts have been made to devise a *hierarchy of towns based on the belief that one could be found. One approach is to rate cities by the number of retail outlets, but this is difficult since a 'shop' might be a large jewellers or a corner store. One refinement of this is to give different weightings to different functions and then calculate a total for each settlement. *Multivariate analysis has also been used, but the data are not always available and are difficult to collect. Other methods emphasize the interaction between a settlement and its field (*see* *centrality) and rank settlements by the size of their *spheres of influence; yet this method, too, has its difficulties. *Graph theory has also been used to rank towns.

rank-size rule Settlements in a country may be *ranked in order of their size. The 'rule' states that, if the population of a town is multiplied by its rank, the sum will equal the population of the highest ranked city. In other words, the population of a town ranked n will be $1/n$th of the size of the largest city—the fifth town, by rank, will have a population one-fifth of the first.

On normal graph paper, the plot of cities and their size and rank will appear as a concave curve. When plotted on logarithmic scales, the graph may emerge as a straight line. This is the **rank-size pattern**.

It is usually possible to relate the ranks and sizes of the central places in country by using a *regression analysis:

$$\log P_k = \log P_1 - b \log k$$

where P_1 is the population of the largest city or town, P_k is the population of the kth town by rank, and b is a coefficient which must be established empirically for each investigation. The greater the value of b, the steeper the slope, and the greater the *primacy of the largest city or town. Many developing countries show a sharp fall from the largest, *primate city to the other cities, and this is known as the **primate rule**. Why the relationship should occur, and why the value of b varies has not yet been explained.

Other variations occur: the *binary pattern* shows a flat upper section and a steeper lower section, giving a concave curve. This pattern is shown in federal countries such as Australia. A further modification is the *stepped order* where a number of settlements may be found at each level with each place resembling others in size and function.

Raoult's law This states that the presence of a solute will lower *saturation vapour pressure. *See* *condensation nuclei.

rapids Areas of greatly disturbed water across a river. Unlike waterfalls, rapids are the result of a continuous and relatively gentle slope, rather than a sudden vertical drop.

rare gases Also known as noble or inert gases, these are uncommon gases such as neon or krypton which are chemically unreactive.

raster In *Geographic Information Systems, a grid square. Raster data are *spatial data expressed as a matrix of cells, with spatial order indicated in the ordering of the cells. A raster map is a map stored as a regular array of cells; a raster scanner records an image by breaking it into *pixels. **Raster**

to vector conversion consists of changing an image made up of cells (rasters) into one made up of *lines and *polygons.

rates A form of local taxation raised in Britain on all property, domestic or industrial, based on the **ratable value** of the property. The ratable value was assessed by the local authority—rates varied nationally—and a tax had to be paid in relation to that value. The system of domestic rates was superseded by the *community charge or poll tax, which was, in turn, replaced by the council tax.

ratio scale A scale which shows data which have been converted into another form. For example, an occupational group, such as teachers, may be shown as a percentage of the total workforce.

Ravenstein's 'laws' of migration These were formulated by E. G. Ravenstein (1885) and state that:
1. Most migration is over a short distance.
2. Migration occurs in steps.
3. Long-range migrants usually move to urban areas.
4. Each migration produces a movement in the opposite direction (although not necessarily of the same volume).
5. Rural dwellers are more migratory than urban dwellers.
6. Within their own country females are more migratory than males, but males are more migratory over long distances.
7. Most migrants are adults.
8. Large towns grow more by migration than by natural increase.
9. Migration increases with economic development.
10. Migration is mostly due to economic causes.

Ravenstein's findings stimulated an enormous volume of work, and, although the 'laws' have been adjusted by succeeding researchers, they have not been totally rejected. Observations of each 'law' as applied to Britain in the 1980s, for example, show that with respect to point 1, over half the moves made annually in England and Wales were in the same local authority area; and to point 3, that the largest urban centres received the highest number of immigrants. For an elaboration of point 6, *see* *chain migration, and for point 9, *see* *mobility. *See also* *gravity model.

reaction time The time between any kind of change and the response it elicits in a system.

recession The decline in river flow after a storm event has passed.

recession limb That part of a *hydrograph which records the fall in *discharge after the river has reached a peak of flow due to a storm event. The line of the recession limb is controlled by the amount of water stored in the basin and the way it is held in the *catchment area.

recessional moraine *See* *moraine.

reclamation The process of creating usable land from waste, flooded, or derelict land. *See* *drainage, *polder, *urban development corporation.

recovery rate The time taken for a *diffusion wave to re-form having been blocked by a *diffusion barrier. The recovery rate of a wave front is directly related to both the type and size of the barrier it encounters.

recreation An activity undertaken for the pleasure or satisfaction which it gives to the individual. Some would restrict the term to activity away from the home. Geographical studies of recreation include demands for, and movement to, recreational facilities, and assessments of the impact of recreation on the landscape. *See* *tourism. **Recreation carrying capacity** is the amount of recreation which a site can take without any deterioration of its qualities.

rectangular drainage *See* *drainage patterns.

recycling The reuse of *renewable resources in an effort to maximize their value, reduce waste, and reduce environmental disturbance. In Britain, the 1980s saw an increase in recycling, notably with regard to glass and paper. The recycling of scrap metal is a major source of metal in refineries from Ghana to the Netherlands.

red rain The washing out of fine dust particles over mid-latitudes. For Europe, the major sources of dust are the Sahara and the fringes of the Sahel. This dust is picked up by air streams which then rise by convection, or over relief barriers, into the upper *troposphere. The dust is carried north and west until it is washed out by precipitation. Red rain is generally alkaline because of the high calcium content of the Saharan dust.

redevelopment The demolition of old buildings and the creation of new buildings on the same site. Redevelopment can solve existing problems of congestion and poor design but, for residential areas in particular, it is seen to be wasteful of resources, destroying communities, and creating urban deserts until building takes place. Some redevelopment schemes have been built on too large a scale, and individuals are 'lost' in the concrete.

Post-war Britain has seen much city-centre redevelopment. The old city centres evolved without motor vehicles, and the population of the city was smaller. Many old city centres are therefore inadequate. Redevelopment is based on the provision of CBD functions and often segregates vehicular traffic from pedestrians.

redlining The practice by banks and building societies of identifying those poorer districts of a city, often the *inner city, where buildings are seriously in decline, and refusing to grant mortgages for the purchase of property within those districts. The result is that people within those areas may find it impossible to borrow to improve their homes, and the process of improvement, as newcomers move into an area, is blocked. Lending institutions have denied that redlining takes place, but there is clear evidence that it has.

reduction The loss of oxygen from a compound. For example, the *sesquioxide ferric oxide can be reduced to the monoxide ferrous oxide by bacteria. *See* *gleying. A more precise definition of reduction is that it represents a gain of electrons in a compound.

refugee A refugee is defined by the UN Convention relating to the Status of Refugees, 1951, and the UN Protocol, 1976, as a person who, owing to a well-founded fear of persecution for reasons of race, religion, nationality, membership of a particular social group or political opinion, is outside his or her country of nationality and who is unable or unwilling to return. *See* *asylum migration.

reg A stony *desert pavement. *See* *hammada.

regelation When ice is under pressure, its melting point is lower than 0 °C. When the pressure lessens, the melting point rises so that refreezing occurs. This is regelation.

regime A recurring pattern, as in the seasonal pattern of climates or the yearly fluctuations in the volume of a river or a glacier.

region Any tract of the earth's surface with either natural or man-made characteristics which mark it off as being different from the areas around it.
   Many attempts have been made to distinguish regional boundaries. The French *pays* were taken as a model for the demarcation of regions but few researchers can agree to the boundaries of almost any region, such as the Great Plains, because different criteria are used to determine the extent of **multiple-feature regions**. **Single-feature regions**, like areas in the USA of covered bridges, are easier to demarcate. Many regions have clearly distinguishable cores but the regional characteristics diminish with increasing distance from the core. These are **nodal regions**. *See also* *areal differentiation, or *chorography.

regional geography The study in geography of *regions and of their distinctive qualities. A precondition of this study is the recognition of a region, its naming, and the delimitation of its boundaries. One approach has been to identify 'natural' regions while another was to establish economic regions based on agriculture and/or industry. Often there was an intimation of a link between the two types of region. Once the keystone of geography, the status of regional geography has been in decline since the 1950s, but *areal differentiation, which may be seen as a branch of regional geography, has grown in importance in recent years.

regional inequality A disparity between the standards of living applying within a nation. It is difficult to quantify the prosperity or poverty of a region, but there are two basic indicators. The first is unemployment, which has been used in Britain as a symptom since the 1920s. Most UK regional policy has concerned the alleviation of unemployment. The second indicator is per capita income, which in Britain generally falls to the north and west. Other factors indicating disparity include the type of industry and its growth or decline, numbers of young people in further education, housing standards, and the quality of the environment. Some would assert that economic development brings about regional inequality. *See* *uneven development.

regional multiplier The stimulation of economic growth by growth itself. As secondary industries develop they create a demand for raw materials and goods. Thus, machinery is made from steel and this stimulates steel manufacturing while the development of the steel industry requires more machinery. As manufacturing industry prospers, more jobs arise in service industries.

regional policy A policy, adopted by government, aimed at redressing *uneven development within a country. The incentives for a government to tackle regional imbalance include: a desire to alleviate regional unrest, the wish to unite party representatives from poorer, as well as richer regions, a yearning for social justice, the need to check out-migration from disadvantaged regions, and the ambition to use fully the human resources and plant of a declining area.

  Measures include: improving the *infrastructure; building new towns to move people away from poor housing stock and to stimulate the construction industry; and providing inducements to new industry to locate in the area in the form of tax incentives, grants and subsidies, and the provision of purpose-built factories.

  Recent thinking, however, has argued that disadvantaged regions will be regions of cheap labour which will ultimately attract investment without government intervention or expenditure, and there has been a shift in Britain from assistance at a regional level to assistance to smaller, well-defined units. *See* *enterprise zone, *urban development corporation.

regional science An interdisciplinary study which concentrates on the integrated analysis of economic and social phenomena in a regional setting. It seeks to understand regional change, to anticipate change, and to plan future regional development. This study is particularly associated with the work of Isard and draws heavily on mathematical models.

regionalism A move to foster or protect an indigenous culture in a particular region. This may be a formal move, made by the state as it creates administrative or planning regions, or an informal move for some degree of independence arising from a gut feeling, based on territory, of a minority group.

regionalization The demarcation of regions such that there is little variation within each region while each region is sharply distinct from the others. The bases for regionalization vary so much that different regions may be delimited according to the criteria used.

register of population A record of the major events in the lives of a population: births, deaths, marriages, divorces, and adoptions. In most European countries, registers were kept before censuses were held and are probably more reliable for historic investigations.

Registrar General's classification of occupations, UK A number of classifications exists, but the broadest one groups occupations into five

socio-economic classes with the implication that occupation is a
meaningful indicator of social welfare.

    I. Professional occupations—e.g. doctors and lawyers.
   II. Managerial and lower professional occupations—e.g. managers and
       teachers.
  IIIN. Non-manual skilled occupations—e.g. office workers.
  IIIM. Manual skilled occupations—e.g. bricklayers, coalminers.
   IV. Semi-skilled occupations—e.g. postal workers.
    V. Unskilled occupations—e.g. porters, dustmen.

regolith A general term for the unconsolidated, *weathered, broken rock
debris, mineral grains, and superficial deposits which overlie the unaltered
bedrock. The depth of the regolith varies with the intensity and duration
of the weathering process; within the tropics it may be hundreds of metres
deep. Soil is simply regolith with added organic material.

regression line In an investigation of two variables, where one variable
is dependent on the other, a regression line is a 'best fit' line through a
series of points on a graph, showing the form of the relationship between
two sets of data. This line can be drawn by eye, but individuals'
perceptions of the location of the best-fit line can vary widely, so the best-
fit line is often calculated using the **least squares method**. The aim of this
method is to ensure that the sum of the squares of the deviations of all the
points from the line is at a minimum.

By convention, y is the dependent variable; that is, it is the variable
whose values are being predicted from the independent variable x. The
appropriate regression line involves the regression of y on x. This is
particularly important, because a regression of x on y would give a
different line. The regression line is described by:

$$y_c = a + bx$$

where y_c is the computed value of the independent variable, a is the y
intercept (the value of y_c when $x = 0$), calculated from

$$a = \bar{y} - b\bar{x},$$

and equation b is the slope of the regression line, calculated from

$$b = \frac{n\left(\sum xy\right) - \left(\sum x\right)\left(\sum y\right)}{n\left(\sum x^2\right) - \left(\sum x\right)^2}$$

where n is the number of paired variables.

The regression line may be used to calculate the values of one data set,
given the values for the other.

rehabilitation The installation of modern amenities and the repairing of
old houses which are structurally sound. Rehabilitation is a method of
improving the housing stock of a city without destroying existing
neighbourhoods, and many local authorities give grants for the

rehabilitation of individual houses. It may be that rehabilitation only postpones *redevelopment as the refurbished houses will decay over time.

Reilly's law The principle that the flow of trade to one of two neighbouring cities is in direct proportion to their populations and in inverse proportion to the square of the distances to those cities. For a settlement between two cities, *a* and *b*, trade to those cities may be expressed as:

$$\frac{R_a}{R_b} = \frac{P_a}{P_b}\left(\frac{d_b}{d_a}\right)$$

where R_a and R_b are the volumes of trade to town *a* and town *b*, P_a and P_b are their respective populations, and d_a and d_b are the distances from the place under consideration to towns *a* and *b*.

rejuvenation The renewed vigour of a once active process. The term is generally applied to streams and rivers which regain energy due to the uplift of land through *isostasy or by a fall in the *base level. Rejuvenation may also apply to the resumption of movement along an old fault line.

relative humidity, *U* The ratio of the actual vapour density (which indicates the amount of water vapour present in the air) to the theoretical maximum (saturation) vapour density at the same temperature, expressed as a percentage. This may be expressed as:

$$U = 100 \, e/e'w$$

where *e* is actual vapour pressure and *e'w* *saturation vapour pressure with respect to water at the same temperature.

   Saturated air has a relative humidity of 100%. Air with a relative humidity in excess of 100% is said to be **supersaturated**. Relative humidity is measured with a *hygrometer. It varies both diurnally, with a dawn maximum and an afternoon minimum, and, less conspicuously, *annually, both variations being in opposition to the pattern of temperatures. *See also* *absolute humidity, *dew point.

relative variability The mean deviation as a percentage of the *arithmetic mean. It is useful in comparing two apparently similar data sets.

relaxation time The time taken for any system to re-establish *equilibrium after a change in the factors which control or influence that system, for example, for a slope to regain an equilibrium state after rapid undercutting at the base through human agency.

relevance Within geography, the degree to which its subject and methodology can make a practical contribution to the solution of environmental and social problems. In the 1970s certain geographers felt that there was not enough attention paid to the unequal distribution of goods and resources throughout the population and that geographers should address themselves to questions of human welfare and social justice. *See* *welfare geography.

relict landform A geomorphological feature which was formed under past processes and climatic regimes but still exists as an anomaly in the changed, present-day conditions. *Raised beaches and *corries in Scotland are just two examples, and certain geomorphologists claim that, in some areas, such as hot deserts, the majority of landforms are relict.

relief The shape of the earth's surface. High relief generally denotes large local differences in the height of the land; low relief indicates little variation in altitude.

relief rain *See* *orographic precipitation.

religion, geography of The study of the spatial distribution of organized religions and their territorial development over time.

remembrement A French term for the reordering, consolidation, and enlargement of land holdings.

remote sensing The gathering and recording of information about the earth's surface by methods which do not involve actual contact with the surface under consideration. Remote sensing techniques include photography, infra-red imagery, and radar from aircraft, satellites, and spacecraft.

rendzina A soil rich in humus and calcium carbonate, developed on limestone. The A *horizon is dark and calcareous, but usually thin. The B horizon is absent and the C horizon is chalk or limestone. A rendzina is an example of an *azonal soil, dominated by rock type.

renewable resource A recurrent *resource which is not diminished when used but which will be restored. Examples include tidal and wind energy. Renewable resources may be consumed without endangering future consumption as long as use does not outstrip production of new resources, as in fishing.

rent gap The gap between the actual rent paid for a piece of land and the rent that could be collected if the land had a 'higher' use. The idea is central to, but does not entirely explain, the phenomenon of *gentrification, which is likely to take place in urban areas where the rent gap is wide.

rent gradient The decline in rents with distance from the city centre. It reflects the cost of transport from the outlying districts to the centre. It is suggested, however, that as city centres decline and as the importance of motorways as locational factors increases, the traditional gradient of rent from city centre to outer suburbs might be reversed.

replacement rate The degree to which a population is replacing itself, based on the ratio of the number of female babies to the number of women of childbearing age. This is the gross reproduction rate. The net reproduction rate is defined as the average number of daughters born to mothers during their reproductive years while allowing for mortality. If the number is over 1, the population will grow, if below 1, the population will diminish.

repose slope A slope with a steepness regulated by the *angle of repose of the superficial debris on it.

reserves The proportion of resources, notably mineral resources, which can be extracted using the prevailing technology.

residential differentiation, residential segregation The evolution of distinct neighbourhoods, recognizable by their characteristic socio-economic and/or ethnic identity. Studies in Chicago in the 1950s showed segregation by occupational class, and demonstrated that the most highly segregated groups were at the bottom and top of the socio-economic scale. An important reason for residential clustering is the desire of the members to preserve their own group identity or life-style, and to give the social group a cohesive political voice.

For some groups, residential segregation may be maintained by high land values, by *redlining, or other discriminatory practices. Ethnic minorities with extended families may seek out areas of cheap, large housing or may group together for protection. *See also* *bid-rent theory, *sector theory, *neighbourhood, *social area.

residual In statistics, the difference between an actual, observed value and a value predicted by a *regression. A **positive residual** is where the observed value exceeds the computed value; a **negative residual** is the opposite.

resolution In *remote sensing, the sharpness of the image transmitted from a satellite. It depends on the number of pixels.

resource Some component which fulfils people's needs. Resources may be man-made—labour, skills, finance, capital, and technology—or natural—ores, water, soil, natural vegetation, or even climate. The perception of a resource may vary through time; coal was of little significance to Neolithic man, while flint was of great importance. Such resources depend on relevant technology. Other resources, like landscapes and ecosystems, may be permanently valued whatever the technology.

Resources can be *renewable—**flow resources**—or *non-renewable—**stock resources**.

resource allocation The assessment of the value of a resource or of the effects of exploiting a resource. The quantity of a resource may be determined in absolute terms, such as area, or in terms of the ability of man to utilize it. Values of resources may also be seen in social terms, although these are difficult to cost.

resource management A form of decision-making concerned with the allocation and conservation of *natural resources. The main emphases are on: an understanding of the processes involved in the exploitation of resources, the analysis of the allocation of resources, and the development and evaluation of management strategies in resource allocation. It is therefore a cross-disciplinary study, concerned with the complex relationships which govern resource exploitation and allocation. Sustainable development and environmental protection are major goals. *See* *Environmental Impact Assessment.

resource-frontier region A newly colonized region at the periphery of a country which is brought into production for the first time. *See* *core–periphery model.

restructuring A change in the economic make-up of a country. It may involve: reordering production to achieve *economies of scale; a switch of investment from one sector to another (*see* *de-industrialization); a change in the spatial distribution of industry; or a change in the economic system, as in Mikhail Gorbachev's *perestroika*.

Restructuring may be necessary in a declining economy in order to promote growth. It may also be imposed by an authority, like the World Bank, to improve the ailing economy of a nation which is in debt to that authority. Restructuring in this case usually involves major cuts in the public sector. *See* *structural adjustment.

retrogressive approach A way of attempting to understand the past by studying the present. It is argued that the analysis of past landscapes requires that the present landscape be studied. Compare with *retrospective approach.

retrospective approach The study of present-day landscapes in the light of the past landscapes. This approach would put *historical geography at the heart of all contemporary geography, and some geographers see the landscape as a reflection of all the geographies that have gone before it, although only traces of some might persist. Compare with *retrogressive approach.

return cargo If a vehicle must return empty after making a delivery, the freight rate must be high enough to cover both journeys. Any cargo utilizing the return journey can negotiate favourable freight rates.

return flow Water which has seeped through the soil as *interflow but which backs up the hillslope when it has reached a saturated layer.

return period The length of time between events of a given magnitude. This can be calculated for many natural phenomena including droughts, floods, and earthquakes.

Calculation of the return period is only possible if continuous records have been kept over a number of years.

revenue surface A three-dimensional 'contour model' representing the variation in revenue over an area. There are two 'horizontal' axes: the first from left to right, the second stretching at 60° from the first to represent the land surface stretching away from the observer; these illustrate distance, while the vertical axis shows the spatial variations in revenue. *See* *cost surface.

Reynolds number (R$_e$) Four factors combine to determine whether the flow of water within a channel is *turbulent or *laminar: the density, velocity, and viscosity of the water, and the hydraulic radius of the channel. Since the density of water is 1, the Reynolds number expresses this combination as:

$$R_e = \frac{VR}{\mu}$$

where V = velocity of the liquid, R = hydraulic radius, and μ = viscosity of the liquid. The Reynolds number is a dimensionless quality; in streams, the maximum number for laminar flow is between 500 and 600, depending on temperature, and at high Reynold's numbers, above 2000 to 25000, flow is turbulent.

rhizosphere The immediate environment of plant root surfaces.

ria The seaward end of a river valley which has been flooded as a result of a rise in sea level. Cork harbour in Ireland is a ria formed by the drowning of the River Lee. The name is from the type location in Galicia, Spain. Compare with *fiord.

ribbon development A built-up area along a main road running outwards from the city centre. Such a location combines the attraction of cheaper land away from the city centre with high accessibility and the chance of attracting trade from passing traffic. The line of buildings on each side of the road may be only one plot deep. Ribbon development is characteristic of many Mid-Western settlements in the USA, and was also a trait of much inter-war development in Britain.

Richter scale A scale of the *magnitude of earthquakes, ranging from 0 to (in theory) 10. On this scale a value of 2 can just be felt as a tremor. Damage to buildings occurs for values of over 6, and the largest shock ever recorded had a magnitude of 8.9.
   The scale is logarithmic and is related to the amplitude of the ground wave and its duration. *See also* *Mercalli scale for measurements of earthquake *intensity.

ridge and furrow A set of parallel ridges and depressions formed during the period of strip cultivation in the Middle Ages. As the land was ploughed always to the same pattern, the plough then threw up earth to make ridges which often survive in the present landscape. Ridge and furrow landscapes are still clearly visible in the landscape of midland England, and parts of lowland Germany, where they are called *Raine*, or *Anwande*.

riegel An outcrop of rock forming a bar across a *glacial trough. Riegels usually develop when outcrops of resistant rock cross the trough, and are often separated by flattish areas, so that they often act as a dam to impound the waters of a lake. *See* *paternoster lake. As a result of *abrasion, the upstream sides are *striated and rounded, while plucking makes the downstream side rough and jagged. *See* *roche moutonnée.

rift valley Also known as a *graben*, this is a long strip of country let down between normal faults, or between a parallel series of step faults. The Valle de Cibão Graben, Hispaniola has a length of 250 km and a maximum width of 40 km—roughly the same as the Rhine rift valley.

Rift valleys only form in brittle, resistant rocks, since more plastic rocks will thin and deform under pressure. They are among the largest, structurally controlled landforms, and the biggest terrestrial rift valley system is the East African system, at 3000 km long.

*Plate tectonic theory suggests that rift valleys are the result of large-scale doming above a mantle *plume, followed by fracturing along the crest of the dome as plates diverge. Many rifts, like the Rhine rift valley, have the Y-shaped pattern characteristic of a triple junction indicating that they arise from plate separation, but some have been attributed to the presence of *hot spots.

rills Small *channels, between 5 and 2000 mm in width, and very closely spaced. They develop well in areas with heavy rainfall, especially upon weaker rocks, such as volcanic ash. **Shoestring rills** cut into the soil in a system of long, parallel lines. Rills are often seasonal features and few contain enough *load to be regarded as miniature rivers. Rills are more likely to be formed by solution than by *abrasion. They may widen and deepen to form *gullies.

rime *See* *frost.

rime ice Ice which forms when supercooled water droplets in the air freeze directly onto a glacier.

ring In *Geographic Information Systems, a sequence of closed *links, *strings, or chains, which do not intersect.

ring city A city created along the outer side of a circular routeway while there is open space at the centre of the ring. Randstad Holland is an example.

ring dyke *See* *dike.

rip current A strong current moving seawards in the near-shore zone. Various causal mechanisms have been suggested:

- the sudden entry of a tidal stream into shallow water
- the meeting of two tidal streams
- the accumulation by strong winds and waves of a large body of water at the top of the beach, the return flow of which creates the rip current.

riparian Relating to a river bank. Owners of land crossed or bounded by a river have **riparian rights** to use the river for domestic purposes, for the watering of livestock, for generating power, and for recreational purposes.

rising limb That section of a river *hydrograph which covers the beginning of the increased discharge until the maximum flow.

risk The likelihood of possible outcomes as a result of a particular action or reaction. Technically speaking, the likely outcomes of risks can be assessed as a series of different odds, while there is no calculation of probabilities in *uncertainty. *See* *hazard.

river capture *See* *capture.

river terrace A bench-like feature running along a valley side, roughly parallel with the valley walls. Most terraces form when a river's erosional capacity increases so that it cuts down through its *flood plain. Many river valleys have been subject to alternating phases of *aggradation and *dissection such that a series of terraces has developed. These are **cut and fill terraces**, formed as erosion alternates with deposition. Two similar terraces on each side of a river are **paired terraces**. These occur at times of elevation of the land surface or when downcutting is greater than *lateral erosion. **Unpaired terraces** usually form when lateral erosion dominates.

riviera A term taken from the Riviera coast from Marseille in southern France to Genoa in Italy, now used to describe any coastline of outstanding natural beauty.

robber-economy The exploitation of resources which takes no account of provision for the future. Compare with *sustainable development.

roche moutonnée A rock shaped by two major glacial processes. The *stoss and the central sections of the rock are streamlined, and *abrasion is the dominant process. The down-ice, lee side is rugged and steep, and is thought to have undergone *quarrying. Roches moutonnées may be over 100 metres in height and up to 1 km in length, but they are usually much smaller.

rock Material made of mineral particles bonded together. Rock is a hard, elastic substance which does not significantly soften on immersion in water.

rock fall *See* *fall.

rock flour Silt- and clay-sized particles of debris formed from grinding due to *abrasion within and at the base of a glacier. Streams arising from areas of glacial abrasion have a grey-green colour as rock flour is carried within them, in suspension.

rock glacier A very slowly moving river of equiangular rock debris, whose interstices are probably ice-filled.

rollers In hydrology, *see* *eddy.

room density Also known as the occupancy rate, this is the number of

persons in a house per unit habitable room. (Kitchens and bathrooms are
excluded.) It is a widely used index since it is an easily calculated and
sensitive indication of housing provision, where any density of over one
person per room indicates overcrowding.

Ro-Ro system A drive on–drive off ferry system (*r*oll on, *r*oll off), used on
routes across the English Channel.

Rossby waves Long ridges and troughs in the westerly movements of
the upper air, with a wavelength of around 2000 km, discovered by C. J.
Rossby in 1939. Four to six waves girdle the Northern Hemisphere at any
one time. Some are a response to relief barriers, like those east of the
Rockies, and east of the Himalayas. Rossby waves are also thought to be a
reaction to the unequal heating of the earth's surface. *See* *dishpan
experiment.

   Rossby waves are intimately connected with the formation of
*depressions and *anticyclones in the middle latitudes of the Northern
Hemisphere. As air travels west–east into a trough, it slows down, and piles
up, causing *convergence just ahead of the ridge which follows.
Convergence in the upper air causes a downflow to the ground, creating
high pressure systems at ground level, below and just ahead of troughs in
the Rossby waves. As air leaves the trough, and its passage straightens out,
air speeds pick up, and the air moves very fast as it swings round the outer
arc of the ridge. Air then diverges just ahead of the next trough.
*Divergence in the upper air causes low pressure systems at ground level
below and just ahead of ridges.

   At times, the waves are few, and shallow; a pattern known as a high
zonal index. In other cases, the flow becomes markedly *meridional; a
pattern known as a low zonal index. Upon occasion, the waves break down
into a series of cells. *See* *blocking.

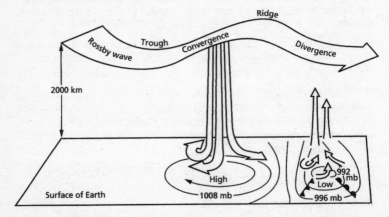

FIGURE 46: *Rossby wave*

Rostow's model of growth W. W. Rostow (1978) saw economic growth as occurring in five stages. Initially, technology is primitive and social structures are rigid and hierarchical. Production per capita is low and change is rare. This is the 'traditional society'.

The second stage is a transition stage—the 'preconditions for take-off'. Here, possibly because of outside stimuli, investment rises, the *infrastructure begins to be developed, and there is growth in the agricultural and industrial sectors. It is upon these bases that the next stage, 'take-off' occurs. This is a short period of time during which the economy and society are transformed. Investments and savings rise, and new industries grow in the *primary and the manufacturing sectors.

Growth gives rise to the 'drive to maturity'. Industrial development now diversifies, imports fall, and investment is still high. The final stage—the 'age of high mass consumption'—is reached as consumer goods are of increasing importance, real income rises and the welfare state is established. The Rostow model assumes that capitalism is the underlying structure of the society, and it does not seek to explain the sequence of changes. In addition, the terminology is somewhat vague.

rotational slip The semi-circular motion of a mass of rock and/or soil as it moves downslope along a concave face. Evidence for rotational slip in *cirques comes from the dirt bands observed in cirque ice, which become progressively steeper from the back wall, but then flatten towards the cirque mouth. The basic mechanism for rotational slip would seem to be the imbalance between *accumulation at the head of the glacier and *ablation at the snout, which steepens the gradient of the glacier.

r-selection, k-selection Two major strategies may be adopted for the survival of a species. r-selected plants (where r stands for maximum increase) respond swiftly to favourable conditions with most of their energies devoted to rapid maturity and reproduction. *See* *opportunist species. k-selected plants survive by putting their energies into persistence. *See* *equilibrium species. If r- and k- are envisaged as being two ends of a continuum, most species have some of the characteristics of both to a greater or lesser extent along the continuum. It is suggested that over time an r-strategist species may develop k-strategist tendencies. As an r-strategist plant reproduces, the space available for seeding becomes less, so that rapid development is no longer important and persistence is required.

rubification The change of soil colour to yellow or red. This occurs in warm climates where intense weathering liberates iron. This iron attaches to clay minerals and, combined with some *lessivage, rubifies the soil.

rudaceous Coarse-grained sedimentary rock, either consolidated as in *conglomerate or unconsolidated as in *till.

run-off The movement over ground of rain water. Run-off occurs when the rainfall is very heavy and when the rocks and soil can absorb no more.

rural In, of, or suggesting, the country. In practice, it is difficult to distinguish truly rural areas because of the blurring of the *rural–urban fringe and the increase of commuting whereby rural inhabitants work in cities. Perhaps the clearest indication of **rurality** in society is the distance to large *urban centres.

rural community A group of people living in the same rural place who have common ties with each other and with their location. It is the smallest social group which caters for the daily social life of the inhabitants.

rural depopulation The decrease in population of rural areas, whether by migration or falling birth rates as young people move away. It has been argued that the mechanization of agriculture leads to rural unemployment, and hence depopulation. Some writers argue that the move to the cities took place before mechanization and that machines were, therefore, needed to supplement the dwindling workforce.

rural geography The study of the rural landscapes of the developed world. It includes the origin, development, and distribution of rural settlement, *rural depopulation, the causes and consequences of agricultural change, patterns of recreational use of the countryside, tourism, planning, and the growing influence on rural areas of urban dwellers.

rural planning The management of rural areas with regard to some objective such as the maintenance and improvement of rural living standards, or easy access to jobs and social, economic, and welfare services. The most usual aspects of rural planning are maintenance of rural landscape, developing the recreational use of the countryside, and the planning of populations, settlements, and amenities. Also studied are rural problems such as lack of access to amenities and services, substandard housing, and rural unemployment.

rural–urban continuum The belief that between the truly rural and the truly urban are many 'shades of grey'; if we actually look along a scale from the single isolated farm all the way to the *megalopolis, we do not find any clear boundaries between hamlets, villages, towns, and cities. The change is seen as a continuum, and it applies to the way people live as much as to the nature of the settlements they live in.

    The concept of the continuum has been attacked as being simplistic, and overgeneralized, not least because many geographers have detected village-type communities within large cities.

rural–urban fringe The transition zone between the city and its suburbs, and the countryside. Certain types of land use are characteristic of this zone: garden centres, country parks, riding-stables, golf-courses, sewage works and airports are common, and these are neither truly urban nor truly rural uses. They do, however, give an urban air to the countryside, an air which can be cited as an argument for further development—since the zone is not really 'countryside', it need not be preserved. *See* *rural–urban continuum.

rustbelt *See* snowbelt.

S

saeter In Scandinavia, an upland pasture, usually used in summer only.

Sahel With a name implying the edge of the desert, the **Sahelian zone** borders the southern Sahara. The vegetation is more varied and continuous than in true desert, with scattered grasses, shrubs, and trees. The vegetation density generally increases towards the southern margins, and after rains there is an extensive grass cover. With annual rainfall between 200 and 400 mm, *pastoralism, often *nomadic, is the predominant agricultural system, but rainfall is unreliable; wetter periods, such as the 1950s and early 1960s, encourage an increase in livestock numbers to the point of overstocking so that severe droughts, as in the early 1970s and 1980s, bring huge losses of livestock, crop failures, and famine.

sakia A simple apparatus used for lifting water to irrigation canals by means of a circular chain of buckets set vertically. Usually a beast of burden trudges round in a circle to turn a wheel which is geared to the vertical chain.

saline Salty. **Saline soils**, such as the *solonchak, acquire their sodium, potassium, and magnesium salts from the evaporation of soil water which leads to natural *salinization. **Salinity** is the amount of salt present in a solution, usually expressed in parts per thousand by weight.

salinization The build-up of salts at or near the surface of a soil. In hot, dry climates, *evapotranspiration exceeds *precipitation, so that surface water evaporates rapidly. This causes the soil moisture, together with its dissolved salts, to come to the surface by capillary action. This water then evaporates, leaving behind a crust of salts on the surface. This process occurs naturally in desert soils, but the incorrect use of irrigation in arid lands can cause salinization, which is a problem, for example, from the San Joaquin valley of California to the Punjab.

In coastal districts, salinization can occur when sea water percolates into the soil as a result of the over-pumping of groundwater.

salt marsh Siltation in estuaries or sheltered bays may create mud flats, many of which become vegetated. It is this vegetation that traps silt particles and, to some extent, consolidates them. As the marsh develops, *halophytes, such as marsh samphire and sea aster in Britain, pave the way for less hardy specimens. The marsh becomes part of the coast land.

The salt marshes of tidal estuaries have a very high biological *productivity, but in economic terms they are valued as grazing land, potentially reclaimable land for industry, or for waste disposal. In the tropics, it is *mangrove swamps which are created by a similar mechanism.

salt weathering A form of weathering, especially important in hot deserts, which combines *physical weathering (*crystal growth) with some *chemical weathering (*hydration). When saline solutions in rock pores and joints begin to crystallize, stresses are set up within the rock, and surface flaking or *granular disintegration result.

The salts concerned may come from earlier chemical *weathering or may have been carried inland from the sea by spray, snow, or rain. Salt weathering arising from saline *groundwater also attacks built structures in arid lands; concrete, in particular, is liable to disintegration, so that foundations which penetrate to zones near the groundwater need special sheathing, and cement must be salt-free. **Salt blisters**, which form under thin, heat-attracting, bitumen surfaces like roads, car parks, and runways, can heave the surface into small domes, or cause cracks to develop.

saltation The bouncing of material from and along a river bed or a land surface. The impact of a falling sand grain may splash other grains upwards so that a chain of saltating particles may be set up. Saltation upwards is the result of *lift forces; the downward movement occurs when lift is no longer effective and the particle is subject to drag and gravity.

sample A portion of the full *population taken to be a worthwhile and meaningful representation of that population. A **systematic sample**, often used along a transect, selects survey points that are equally spaced over the area under investigation. A **random sample**, commonly used in vegetation studies, selects points at random intervals, the co-ordinates being taken from a table of random numbers. In **stratified sampling** the area under study is divided into different segments by the student. For example, a survey area may be divided into different geological regions or a residential area may be divided into detached, semi-detached, and terraced housing. Within each zone, the sample points are generated from a random number table and the number of points sampled in each zone correspond with the proportion of the total area that each zone represents.

The size of the sample must also be considered. If the data are widely spread, more samples are needed than if the values are clustered. A running mean can be calculated from the data, and when the addition of more measurements does not change this mean very greatly, enough data have been measured.

sand Particles of rock with diameters ranging from 0.06 mm to 2.00 mm in diameter. Most sands are formed of the mineral, quartz. Sandy soils are loose, non-plastic and permeable, and have little capacity to hold water.

sand dune A hill or ridge of sand accumulated and sorted by wind action. Once a dune is formed, sand will settle on it rather than on bare surfaces. This is because the friction of the sandy surface is enough to slow down the wind, which then sheds some of its *load.

Dunes formed in the *lee of some obstacle are **topographic dunes**: *lunettes* are formed in the lee of a *deflation hollow, *nebkhas* form in the lee of bushes, **wind shadow dunes** form in the lee of hills and plateaux, although some dunes form windward of such topographic obstacles.

Parabolic dunes are hairpin-shaped with the bend pointing downwind and originate around patches of vegetation.

Wind direction is significant. Where the direction is very changeable, **star dunes** form. The linear **seif dunes** form when two prevailing winds alternate, either daily or seasonally. When the supply of sand is limited, barchans form with the horns pointing downwind. **Barchans** may be reshaped into dome dunes by strong winds. **Barchanoid** dunes are undulating, continuous cross-wind dunes which may grade into long transverse dunes. *See also* *coastal dunes.

sand dune stabilization Advancing sand dunes may threaten farmland, settlements, and airports; sand encroachment in Africa, for example, has caused problems at the margins of oases, on irrigated land in arid areas, on agricultural land on the banks of the Nile and Niger, and on deltaic lands in clay regions such as the Gash and Tokar deltas of Sudan. Techniques for stabilizing dunes include:

- the use of vegetation to bind the grains together
- spreading gravel over the surface
- spreading chemical adhesive over the dune
- spraying the surface with oil
- erecting slatted fences, or walls to divert the wind and protect structures.

sandstone A sedimentary rock composed of compacted and cemented sand. The sand grains are chiefly quartz and feldspar.

sandur, sandar (pl.) A sheet, or gently sloping fan, of *outwash sands and gravels. Sandar in fan form are commonly found in glacial troughs, such as the Hooker Valley, South Island, New Zealand.

sand-wedge polygons *See* *frost cracking.

Santa Ana *See* *local winds.

sapping The breaking down and undermining of part of a hillslope such that small slips occur. Mechanisms include: undercutting by wave action, undercutting at the foot of a *bluff along a river, and *freeze–thaw at the base of a *bergschrund.

saprolite Chemically rotted rock *in situ*. The word is often used for the lower segment of a *weathering profile.

saprophyte An organism which feeds on dead plant or animal material. Most saprophytes are fungi or bacteria. They are important in nutrient cycles as they bring about decay and liberate nutrients for plant growth.

sastrugi Ridges of ice particles, transported by wind, and lying across an ice sheet. They are orientated at right angles to the prevailing wind.

satellite town A town designed to house the overspill population of a major city, but located well beyond the limits of that city, and operating as a discrete, self-contained entity. Most of the early *new towns were satellites of London.

satisficer A decision-maker whose aim is to make a choice which is acceptable rather than optimal, possibly because it is impossible to locate on the maximum-profit site (because of prior occupancy, planning restrictions), logically impossible to pinpoint that site, or because the decision-maker lacks the necessary knowledge and/or ability. See *behavioural matrix, for a model of the way knowledge and decision-making ability intersect.

It may be that maximization of profit is not the only goal. The decision-maker may be inclined to maximize *psychic income, that is the satisfaction which may come to managers from causes other than financial ones, such as *amenity. Other satisficers are loath to undertake the somewhat hazardous process of seeking the optimum because they are averse to uncertainty. Others may be guided by moral concerns.

saturated adiabatic lapse rate, SALR *See* *adiabatic lapse rate.

saturated mixing ratio lines Lines of constant *humidity mixing ratio plotted against height and pressure on a *tephigram.

saturated zone overland flow, saturation overland flow *See* *overland flow.

saturation, saturated air There is a constant exchange of water molecules between liquid water, or ice, and the air, as *evaporation and *condensation take place. Saturation vapour pressure, (*es*) depicts a balance in the air between condensation and evaporation. *es* is greater in water droplets than in sheet water, and lower in impure water than in pure. *See* *Raoult's Law. Where *es* in an air parcel is greater than *e*, the *ambient *vapour pressure, there will be net evaporation; where *e* is greater than *es*, there will be net condensation.

savanna Broad belts of tropical grassland flanking each side of the equatorial forest of Africa and South America. These belts are associated with the sinking of high-level equatorial air on its return to the *inter-tropical convergence zone. Such descent leads to an adiabatic temperature increase so that rainfall is slight. Trees are modified to minimize water loss; they have small flat leaves and are often thorny.

The boundaries of the savanna are far from clear; there is a gradual change from tall grasses, 1–3 m high with scattered trees, to grassy woodland, and finally to the rain forest. However, where savanna vegetation has been repeatedly burnt, there can be a sharp division between this and the equatorial forest which is less easily fired.

The African savannas have been major areas for the extension of agriculture since the 1960s, but wide variations in annual rainfall have been a major problem and, although schemes such as the Kariba and Volta dams have been developed to remedy water shortages, they have created new habitats for diseases and pests.

scale A level of representation. Traditionally, this has applied to *cartography, where scale is the ratio between map distance and distance on the ground. Thus, a *representative fraction of 1 : 25 000 indicates that one centimetre (or inch) on the map represents 25 000 centimetres (or inches) on the ground.

Geographers may also refer to the scale of an investigation or study, such as local, regional, or national, and may additionally be concerned with the connections between events on a local, regional, and national scale; in a classic argument, P. J. Taylor (1985) asserted that the process of modern capitalist *accumulation is experienced locally, justified nationally, and organized globally. Other scale-related terms include micro-scale, *meso-scale and *macro-scale.

scarp retreat A synonym for *parallel retreat.

scarp slope A steep slope; the steeper ridge of a *cuesta. *See* *fault scarp.

scatter diagram A pictorial way of representing data to see if there is a relationship between two sets of measurements, such as how river speed and meander depth may be associated with each other. Two scales, one for each type of variable, are drawn at right angles to each other and each set of data is plotted. If the dots appear to fall onto a diagonal line, an association is indicated. The statistical significance of that association may then be evaluated by using statistical tests. *See* *regression.

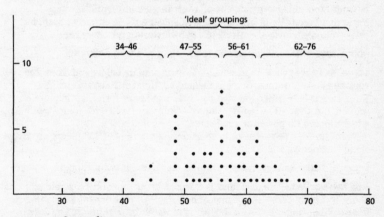

FIGURE 47: *Scatter diagram*

scattering In *meteorology, the dispersal of incoming sunlight by air molecules, which scatter shorter (blue) wavelengths better than longer (red) ones. The cloudless sky we see is thus enhanced in the blue wavelengths by scattering from other paths.

schist A *metamorphic rock, finer-grained than *gneiss, and characteristically with broad, wavy bands, which are not *bedding planes, but sorted zones of minerals such as mica. Schists are foliated and are easily broken into flakes. They form from the *metamorphosis of slates and shales.

science park An area of industrial development set up in collaboration with, and in close proximity to, a centre of higher education, in order for industry to capitalize on academic research, for jobs to be created locally, and for the educational institute to generate income. It is hoped technology will not be the only transfer, and that academics will become aware of business methods. Oxford, UK, is among the many university towns to have a science park.

scientific method An approach to problem solving. The first stage is identifying the nature of the problem and the second stage is the formation of a hypothesis as the potential answer. Information must then be collected and classified according to the limits defined by the question. In the fourth stage the hypothesis is tested against the real world and attempts are made to establish laws. These are easier to establish in the physical sciences than in the social sciences. A combination of laws produces a theory which defines and explains the problem. *See* *positivism.

Scirocco *See* *local winds.

scoria A volcanic rock made of sharp rock fragments and full of air pockets once occupied by gases. **Scoria cones**, also known as ash cones, may be built around the central vent of a volcano; Mount Quill, in the Netherlands Antilles is an example of a volcano made entirely of ash, but such volcanoes, formed entirely of *pyroclasts, are relatively rare.

scour and fill The erosion and later filling of a water channel.

scree Shattered rock fragments which accumulate below and from free rock faces and summits. Scree is formed by *freeze–thaw, and its formation is very much affected by *lithology; scree-forming rocks are jointed and thus allow penetration of water, but rocks with joints too widely spaced do not form good scree.

The term may be extended to the slope, commonly of 35°, made up of these fragments.

scud In *meteorology, fractocumulus. *See* *cloud classification.

sea breeze When coastal land is warmed by the morning sun, the air above it rapidly becomes warmer and more *buoyant than the air above the neighbouring sea. This rising air is replaced by cool, moist air, drawn in from above the sea. A circulatory system develops, with warm air rising over the land and descending over the sea to flow back onshore as a gentle sea breeze. The boundary between sea and land air—the **sea breeze front—** can mark sharp contrasts in temperature and humidity, and may move tens of kilometres inland. At night, the effect tends to reverse; *see* *land breeze. Sea breezes greatly modify coastal climates and can bring cooler conditions in the afternoon to areas with hot, humid climates.

sea-floor spreading The creation of new crust as *magma rises up at a *constructive *plate margin. The magma pushes the plates apart creating new oceanic crust and pushing away the far end of the plate. *See* *oceanic ridge, *magnetic stripe.

seamount A mountain on the ocean floor which does not break the water surface. A flat-topped seamount is a *guyot.

search behaviour The way in which an individual or entity reacts to information by selecting one of a set of alternatives to solve the problem of location, especially in the context of *migration. The decision-maker has an awareness of a set of places from which to choose. The choice depends on the degree to which information is available and on the decision-maker's ability to evaluate the information. (See *behavioural matrix.) Some decision-makers comb through the options thoroughly; others are not thorough and display a certain irrationality in their choice. The choice made may be incorrect if based on inexact, partial data and superficial analysis.

secession The transfer of part of the territory and population of one state to another, whether pre-existing, or newly created. Some **secessionist** movements are strongly contested and ultimately is resisted, as in the Biafran war of the 1960s; some are resisted but eventually granted, as in the creation of the Irish Free State in 1922; some take place peacefully, as in the separation of the Czech Republic and Slovakia in 1995.

second home A property occasionally used by a household whose normal place of residence is elsewhere. Second homes are usually found in rural areas where they are used for recreation and leisure. To some extent, second homes can bring increased custom to a rural area, but in some cases the purchase of second homes by outsiders can drive up house prices beyond the pockets of local residents, and this may cause resentment. For this reason, the Welsh Sons of Gwynedd initiated a campaign of arson against English-owned second homes in North Wales in the late 1970s and early 1980s.

secondary air mass An *air mass which has been modified by the passage of time or by its movement to an area differing from the source region. Some schemes of classification differentiate between a k air mass which is colder than the surface over which it is moving and a w air mass which is warmer. In general, a k air mass is inclined to *instability, with gusty, turbulent winds, while w air masses have *stable or *inversion conditions, with *stratus clouds.

secondary industry, secondary sector The creation of finished products from raw materials; that is, manufacturing. This activity often involves several stages. While secondary industry accounted for 33.2% of all EC employment in 1987, employment in manufacturing is generally falling in the *advanced economies; only 12.8% of the Japanese labour force was engaged in manufacturing and construction in 1988. See *post-industrial.

secondary urbanization Urbanization which results from forces which are external to a country, such as foreign, colonial settlement. For example, pre-colonial Australia had no urban settlement.

sector principle The principle on which claims to territory in the Arctic and Antarctic are made. The territory is shared out in the form of arbitrary sectors, each one having an apex at the poles and including an outer area bounded by the coast. The principle has worked well in establishing control of the Arctic ice, but there are disputed claims over the land mass of the Antarctic which may possibly yield mineral resources.

sector theory The view that housing areas in a city develop in sectors along the lines of communication, from the CBD outwards. High quality areas run along roads and also reflect the incidence of higher ground. Industrial sectors develop along canals and railways, away from high quality housing. Thus a high status residential area will spread out along the lines of the sector by the addition of new belts of housing beyond the outer arc of the city. Once contrasts in land use have developed in a sector near to the city, these contrasts will be perpetuated as the city grows. This theory was advanced by Homer Hoyt (1939) as an alternative to Burgess' *concentric model, and was based on residential rent patterns in the USA. *See also* *Mann's model.

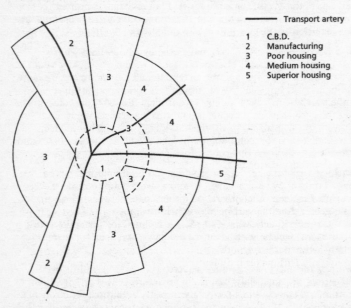

FIGURE 48: *Sector theory*

sedentary Fixed, not moving, as in **sedentary agriculture** where the farmer and the fields are permanently settled. Compare with *shifting cultivation.

sediment Material which has separated and settled out from the medium—wind, water, or ice—which originally carried it. For pluvial sediments the ability of a river to carry sediment depends on particle size as well as the river discharge. *See also* *load. **Sediment yield** is the total mass of sediment in suspension or as *bedload which reaches the exit of a drainage basin. High sediment yields may reflect the discharge of the river basin but the nature of the catchment area, be it of weak or resistant rock, farmed or urban land, is also important.

sedimentary rock A rock composed of sediments, usually with a layered appearance; see *bedding plane. The sediments come mostly from pre-existing rocks which have been broken up and then transported by water, wind, or glacier ice.

Rocks formed from such sediments are *clastic sedimentary rocks and may be subdivided by size into three groups: *argillaceous, *arenaceous, and *rudaceous.

sedimentation The process of *deposition of sediments in a variety of environments; often used to describe the blocking of an aquatic system by the deposition of sediment. Sedimentation may choke reservoirs and raise the river bed by the deposition of silt.

seed-bed location A location with the necessary requirements for the encouragement of new growth. These may be: good transport links, skilled labour, cheap labour, or cheap property, and seed-bed locations were at first characteristically at the edge of the CBD, although recent urban development policy has made sites like these less available. *Enterprise zones and *science parks are examples of publicly-sponsored seed-bed locations.

seepage In hydrology, an oozing out of water.

segmented economy An economy characterized by a variety of firms ranging from multinationals to small workshops, as opposed to an **atomistic economy** which is made up of a host of small firms.

segmented labour A labour force made up of two or more types of worker. Thus, a *dual labour market may be composed of skilled and unskilled segments, with workers from the latter unable to break in to the skilled section. Movement of labour from one type to another is not easy, especially between one type of skilled labour and another, distinctly different, skill. This makes it difficult to *restructure the economy.

segregation The separation of the subgroups of a large population, particularly into distinct residential areas. This segregation may be based on grounds of income, race, religion, or language. *See* *index of segregation.

Early urban geographers suggested that incoming groups to the city went through a series of stages, from initial contact to complete *assimilation, and they used *indices of segregation to monitor the process. In this context, segregation was assumed to be undesirable. More recently, geographers have been looking at the way in which segregation might contribute to *community feeling, or class formation, or the way in which *residential segregation contributes to racism.

seif *See* *sand dune.

seine fishing A type of fishing where a long net is used to surround shoals of fish. When the two ends of the net meet the net is hauled onto the boat.

seismic Of an earthquake. The **seismic focus** or **seismic origin** is the point of origin of the earthquake within the crust. The resulting shocks are **seismic waves**, which may be recorded and measured by a **seismograph**. **Seismology** is the study of earthquakes, and of other earth movements, such as those caused by humans.

seismic tomography The interpretation of *earthquake waves in order to discover the nature of the internal structure of the earth, including the patterns of flow in the *mantle. For example, seismic waves travel more rapidly through the cold regions of the earth's interior than the warm; inferences can therefore be made about interior temperatures by timing the velocity of waves.

selective logging The felling, at intervals, of the mature trees in a forest of mixed age. This type of forest management mimics natural processes in that the canopy is maintained while timber is produced.

selective migration This may be spontaneous as when a particular age–sex group may constitute most of the migrants in a country. Most countries now practise selective migration by demanding qualifications, such as skills, youth, and health, for immigrants.

semiotics The ways in which signs and meanings are created, decoded, and transformed. For geographers, these signs may be in the landscape; landscapes may be 'read' in different ways, and may become part of the political process. *See* *iconography.

sense of place Either the intrinsic character of a place, or the meaning people give to it, but, more often, a mixture of both. Some places are distinctive through their physical appearance, like the Old Man of Hoy; others are distinctive, but have value attached to them, like the white cliffs of Dover.

Less striking places have meaning and value attached to them because they are 'home', and it is argued that attachment to a place increases with the distinctiveness of that place. Planners use this argument by consciously creating or preserving memorable and singular structures to make a space distinctively different. The Cardiff Bay Development scheme has done this, first by preserving the best of the old buildings, and even relocating one— the Norwegian church. All this is done to encourage in the residents an attachment to that place.

A final element is our own experience of that place; if you had been desperately unhappy in central London, it might be that the sight of Trafalgar square would reawaken a sense of misery in you.

sensible temperature Not the temperature recorded on a thermometer, but the temperature as felt by the individual. Humidity is an important factor; most people feel the heat more in 'muggy' weather, and feel the cold more when it is 'raw'. Wind speed is also significant, *see* *wind chill.

separatism The ambition of a minority to form its own sovereign state. *See* *nationalism, *secession.

sequent occupance The succeeding stages of human inhabitation over time on one site. Each stage is seen as being established by its predecessor, although the sequence will almost certainly be interrupted by outside forces. This concept, developed by D. Whittlesey (*AAAG*, 1929), owes much to *human ecology, but is more complex, because it envisages interruptions and transformations, so there is no suggestion of an 'ideal' sequence, or *succession, as in *ecology; instead, the idea is that each stage contains within it the seeds of the next.

serac A mass of glacier ice, formed between crevasses and most often found at a sudden increase in the slope of the glacier.

sere A particular and easily recognized stage of an ecological succession; thus a primary stage is a **prisere**, a **hydrosere** develops in water, a **psammosere** on a sand dune, and a **xerosere** in an arid location. These seres will be moderated by succession. If a plant community is interfered with, perhaps by felling or burning, the resulting secondary stage is known as a **sub-sere**.

service industry Also known as *tertiary industry, this is any of those economic activities, including wholesaling, transport, and retailing, concerned with the distribution and consumption of goods and services. To these may be added administration and, possibly, the provision of information, although some define the latter as *quaternary industry.

sesquioxide A compound formed by the oxidation of a trivalent element, with a ratio of $2:3$. For example, iron sesquioxide, (Fe_2O_3).

set-aside grant Within the EU, a gift of money from the European Parliament to persuade a farmer to take land out of agricultural production. The purpose is to cut down on the creation of surpluses. It is not unusual for farmers simply to increase yields and, therefore, maintain production on the land they continue to farm, especially since it is common practice to leave the less productive strips around the edge of each field unfarmed, rather than abandoning a whole field.

The introduction of set-aside has also led to a proliferation of golf courses within the EU.

settlement In human geography, any form of human habitation from a single house to the largest city.

settlement hierarchy A division of settlements into ranks, usually according to the size of the population. In *central place theory, large, high-order settlements provide *high-order goods and services. In theory, the greater the rank of a settlement, the more goods and services it provides. *See* *rank-size rule.

settlement pattern The nature of the distribution of settlements. Some settlement patterns may be seen as a reflection of cultural traditions. For example, the isolated farmstead is typical of North Wales (*see* *dispersed settlement), while the *nucleated village is typical of lowland England. The technique of *nearest neighbour analysis may be used to test for any regularities in settlement patterns.

severe local storm A storm with torrential rain, large *hail, high winds, and nearly continuous *thunder and lightning. Its violent nature stems from the accumulation and subsequent discharge of great *convective instability. This may be achieved when very warm, moist air is held for some time beneath an inversion, until the inversion breaks down. Severe local storms may be *multicell.

shadow price A price used in *cost–benefit analysis to value intangible items like clean air.

shaduf A simple apparatus used for lifting water for irrigation by means of a bucket, and a lever to raise and lower it. When above the bank, the shaduf may be swung round ready to feed water onto the land or into a trough.

shale A fine-grained sedimentary rock formed when layers of clay are compressed by the weight of overlying rocks. Shales have a layered structure and are easily split along the *bedding planes.

shanty town *See* *squatter settlement.

shape index A statistic used to quantify the shape of any unit of area. R. J. Chorley and P. Haggett (1969) expressed this statistic as:

$$\text{shape index} = \frac{1.27A}{L^2}$$

where A = area of shape in km^2
 and L = the length of the longest axis in km.
A value of 1.0 expresses maximum compaction, where the shape is circular. As the shape is elongated, the less compact is the slope, and the lower the value of the index.

share-cropping A type of farming whereby the tenant pays his rent to the landowner in produce rather than in cash. The landlord often provides seeds, stock, and equipment in return for a fixed proportion of the output. Share-cropping usually shows low yields in comparison with owner-occupied farms or cash tenancies since the incentives are less.
  In the United States, share-cropping replaced the plantation system after the Civil War. Farm labourers, usually black, were allocated land in return for a share of the cash crop. Economic control was maintained by creating a class of landless tenants, by keeping the 'share' retained at subsistence level, and by encouraging indebtedness through company stores; social control took the form of segregation, violence, and paternalism.

shear *See* *wind shear.

shear box An apparatus which can determine the resistance of a rock or soil to shearing. The soil is placed into a layered box. Whilst a normal force is applied to the top layer, the bottom layer is pulled out sideways. The *shear strength of the soil is the force which needs to be applied to deform the sample.

shear plane The face along which *shearing occurs.

shear strength The ability of a rock or soil to withstand *shearing.

shearing The deformation of a material so that its layers move laterally over each other. In geology, shearing bends, twists, and draws out rocks along a fault or *thrust plane. Such shearing is sometimes accompanied by shattering or crushing of the rock near the fault. A shearing force acts parallel to a plane rather than perpendicularly. Shear stress is the force or forces applied tangentially to the surface of a body and causing bending, twisting, or drawing out of that body.

sheet erosion A very slow-acting form of erosion whereby a thin film of water—**sheet wash**—transports soil particles by rolling them along the ground. Compare with *rill action and *gullying. Some geomorphologists believe sheet erosion to be common on the upper and lower parts of a slope, while rill action dominates on the steeper mid-slope.

sheeting The splitting of the outer layers of rock, probably as a consequence of *pressure release resulting from the erosion of overlying material. Other factors include *exfoliation and the swelling of the outer layers of rock as they are moistened by rain or dew. Sheeting is generally parallel to the land surface.

shield The very old, rigid core of relatively stable rocks within a continent, such as the Fennoscandian Shield of northern Europe, or the Laurentian Shield of Canada. This is usually a *Precambrian nuclear mass around which, and to some extent upon which, younger sedimentary rocks have been deposited.

shield volcano A volcano formed of successive eruptions of free-flowing *lava which creates a gently sloping, broad dome tens of kilometres across and more than one kilometre high. Shield volcanoes are characteristic of the Hawaiian chain; Manua Kea is an example, with a basal diameter of around 200 km, 4000 m beneath the sea.

shift share analysis A method of estimating the relative importance of different elements in any growth or decline of regional industrial employment. This change could be due to the national rate of change in manufacturing, or the industrial structure of the region itself and its locational advantages or disadvantages.

The **differential share** compares the differences in a region between the actual employment in a particular industry and the employment it would have had if it had changed at the national rate for that industry; that is, the difference is between the observed change and the expected change, if employment had followed the national pattern.

The **regional share** indicates what would have happened if the region had maintained its share of total national manufacturing employment. The **structural shift** estimates the change expected in a region if each industry in the region changes at its own national rate. Shift share analysis has been strongly criticized, partly because it does not explain why some sectors grow and/or decline, or why industries move into, and out of, an area. It is, however, a useful way of starting to look at industrial change in an area, and the data are easy to acquire.

shifting cultivation In this agricultural system, a patch of land is cleared, crops are grown, and the patch is then deserted until the soil regains its fertility. Bush fallowing is a practice similar to that of shifting cultivation but involves no change of residence, grows crops for longer, and has a shorter fallow.

There are many varieties of shifting cultivation, but as a rule it is characterized by a large diversity in crops; in Africa, the crops grown by shifting cultivators include bananas, plantains, cassava, beans, peppers, rice, maize, and millet. Since the nutrient supply is constantly decreasing under cultivation, yields decrease with time, and eventually the farmer must clear and cultivate new land to meet her/his needs. In the African rain forest, the usual period of cultivation is from two to four years, and, depending on the properties of the soil, a fallow period of 8 to 19 years is required to regenerate soil fertility.

Shimbel index The number of *edges connecting any *node by the shortest possible routes to all other nodes on a network. It is, therefore, a measure of the *accessibility of that node.

shingle Pebbles on a beach, rounded by *abrasion, and reduced in size by *attrition.

shire An administrative district formed in England by the Anglo-Saxons, generally for the purposes of taxation. It was superseded by the Norman counties, some of which retain the term 'shire'.

shopping centre B. J. Berry (1967) proposed a hierarchy of shopping centres: CBD, regional shopping centre, community shopping centre, neighbourhood shopping centre, convenience shopping centre. Shopping centres may thus be regarded as *central places, and their spatial distribution is of considerable interest to urban geographers.

shopping goods *High order goods which are relatively rare purchases. More time is usually taken over their selection, and the customer may travel long distances for a particular purchase. See *central place theory.

shopping mall See *suburbanization.

shore The land adjoining a large body of water or next to the sea. The **backshore** is the part normally above the high water mark but still influenced by the sea. The **foreshore** covers the area between the high and low tide marks and is exposed at low tide. The **nearshore** is seaward of the foreshore and ends at the breaking point of the waves. The **offshore**, in coastal geomorphology, is the zone seaward of the breakers but in which material is moved by the waves.

shore platform A very gently sloping platform extending seaward from the base of a cliff. Platforms widen as the cliffs retreat. They are subject to *salt weathering, alternate wetting and drying, *water-level weathering, and processes of erosion such as *quarrying, *hydraulic action, *pneumatic action, and *abrasion. The best-known (and least well understood) example is the *Strandflat* of Norway.

It is argued that a platform of over 800 m in width cannot have been formed by these forces alone, hence the term 'shore platform' rather than 'wave-cut platform'. Thus, changes in erosional processes and in sea level must have taken place.

sial The continental crust, dominated by minerals rich in silica and aluminium.

sidewalk farming In the USA, the cultivation of a holding some distance away from the urban area where the farmer lives. It is generally restricted to cereal crops.

significance test A statistical test aimed at demonstrating the probability that observed patterns cannot be explained by chance. The significance level is the level at which it is decided to reject the *null hypothesis. Most statistics, like the *correlation coefficient and *chi-square, have their own statistical table. The result of the calculation is compared with the value on the table for the appropriate *degrees of freedom.

A significant result has a 1 in 20 (5%) probability of the observation occurring by chance, a highly significant result has a 1 in 100 (1%) probability, and a very highly significant result has a 1 in 1000 (0.1%) probability.

silage Green crops, such as grass and clover, which are compressed, fermented, and stored for use as animal fodder.

silcrete A *duricrust cemented with silica.

sill An *intrusion of igneous rock which spreads along bedding planes in a nearly horizontal sheet. This level sheet may be up to 300 m in thickness. The best-known British example is the Great Whin Sill, which runs across the Pennines.

silt Fine grains of soil minerals ranging between *clay and *sand in size. Silts are often laid down by rivers when the flood water is quiet. The diameter of silt particles ranges from 0.002 mm to 0.06 mm.

Silurian A *period of *Paleozoic time stretching approximately from 430 to 395 million years BP.

sima The lower part of the continental crust and the oceanic crust, dominated by silica and magnesium.

sink hole In limestone topography, a roughly circular depression into which drain one or more streams. It is known in Britain as a **swallow hole** and sometimes used as a synonym for a *doline.

sinter A deposit of minerals, notably of silica and sulphates, precipitated in layered deposits from the gases released in an area of volcanic activity. A stepped series of sinter bowls, known as **sinter terraces** can result from this deposition. Sinter encrustations have entirely covered the buildings of the ancient city of Hierapolis, near Denizli, in Turkey.

sinuosity The amount that a river *meanders within its valley, calculated by dividing total stream length by valley length.

site The position of a structure or object in physical, local terms such as a river terrace.

Site of Special Scientific Interest, SSSI A site in the UK which is of particular importance because of its geology, topography, or ecology. SSSIs are graded in terms of importance from 1 to 4. Although planning permission for the development of an SSSI is granted only after consultation with the Nature Conservancy Council, it is not unusual for SSSIs to be built on; a section of the M3 was completed across the SSSI of Twyford Down, Hampshire.

situation The location of a phenomenon, such as a town, in relation to other phenomena, such as other towns. Compare with *site.

skid row In the USA, that section of a city whose population has a large number of drop-outs, derelicts, and petty criminals. Such areas are said to have developed from cheaper accommodation for young, male in-migrants.

slaking In geomorphology, the disintegration of fine-grained rocks. It has been suggested that slaking is the result of alternate wetting and drying of the rock, but the mechanism is not fully understood.

slash and burn The clearing of land, usually tropical, where the trees are cut down, the land is cleared of most of the trunks, and the rest of the vegetation is fired. It is common practice in *shifting cultivation, and the ash formed acts as a fertilizer.

slate A weak sedimentary rock, easily split along thin layers of bedding, formed by the compression of shales by the overlying rocks.

sleet Very wet snow, or a mixture of rain and snow, usually developing when temperatures are just above freezing point.

slickenside A rock surface which has been scratched or polished by the effects of friction during structural changes.

slide A form of *mass movement in which material slides in a relatively straight plane. Slides are the most common form of mass wasting, and usually have a length much greater than the depth of the moving material. The sliding mass generally breaks into many blocks as it moves. A modern example of a slide was that which moved down the slopes of the Vaiont Valley, in the Italian Alps in 1971. The debris fell into a reservoir, displacing 250 000 000 m^2 of water, which swept down the valley in a flood wave, killing 2600 people. Slides are triggered by high water pressure, but the competence of the rock is also significant. Compare with *slump.

slip-off slope The relatively gentle slope at the inner edge of a meander. This is the site of *point-bar deposits.

slope In geomorphology, any slanting surface of the earth's crust, above or below sea level. Slope studies refer mostly to *hillslopes.

slope convection In *meteorology, large-scale, slightly tilted air flow, weaker than *cumuliform convection, and notably linked with the growth of extra-tropical *cyclones.

slope elements The differing parts of a slope. These may be convex, straight, or concave; the shape of a slope is an expression of the predominant processes acting on it.

Convex slope elements are generally at the top of a slope. These usually gentle slopes are formed by soil *creep and *rainsplash. Downslope from the convex segment, there may be a **straight slope element** of bare rock. This is the *free face (fall face). Other straight slopes develop where mass movement is the dominant geomorphological process. Slopes with shallow

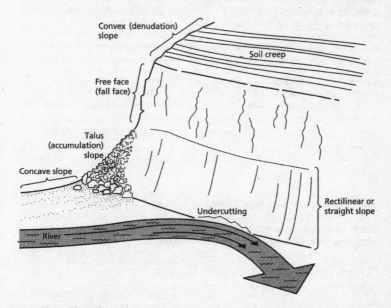

FIGURE 49: *Slope element*

debris and steep slopes undercut by rivers or waves both tend to have straight segments. *Talus slopes are often straight.

Concave slope elements are either due to increased water erosion downslope or to larger pieces of debris rolling further than the rest. **Compound slope profiles** may exhibit some or all of these segments and may show a repetition of certain segments. *See also* *downwearing, *parallel retreat.

slope wash The downslope movement of sediment by an almost continuous film of water.

slope winds *Anabatic and *katabatic winds.

slum An area of poor housing, often characterized by multi-occupance and overcrowding. Schools are poor, items sold in local shops are relatively expensive, and sanitation inadequate. Slum populations often exhibit high concentrations of drug abusers, alcoholics, criminals, and vandals.

slum clearance The demolition of substandard housing, usually accompanied by *rehabilitation and *redevelopment. Some schemes involve rebuilding on the same site, as in London's Barbican, while other clearances have relocated the population at the edge of the city, as in the Roehampton estate in south London.

slump A form of *mass movement where rock and soil move downwards along a concave face. The rock or soil rotates backwards as it moves in a *rotational slip. Slumps are most common in thick *regoliths and large mudstone rock units, but can also occur in hard rock which has been shattered. They differ from *slides because they always have *shear planes which are concave, while the latter have relatively straight shear planes.

smog A combination of smoke and fog. The fog occurs naturally; the 'smoke' is introduced into the atmosphere by the activities of man. After the five-day long period of smog in London in 1952, smoke abatement measures were introduced in Britain. British cities are still estimated to lose between 20 and 55% of incoming solar radiation from November to March through smog.

Furthermore, chain reactions occur in association with exhaust gases, notably in areas of intense car use such as Los Angeles. Toxic gases are formed. *See also* *photochemical smog.

SMSA, Standard Metropolitan Statistical Area An urban area of the USA. This can be a town of 50 000 people or two towns, each with more than 15 000 people, and together totalling more than 50 000, or a county with more than 75% of its population working in industry. In addition to these three categories are areas which seem, by employment, commuting, or population density, to be urban rather than rural.

snow Frozen vapour from the atmosphere. A snow crystal is an ice crystal up to 5 mm across, variously shaped as a prism, plate, star, or needle. Snow crystals fall from stratiform clouds when the low-level air is several

385   **social Darwinism**

degrees below freezing point and the air above is colder. When the
low-level air is near 0 °C, snow crystals aggregate to form snowflakes.

snowbelt Also known as the rustbelt, or the frostbelt, this term applies
to the states of the north-east USA, such as Michigan, Illinois, Indiana,
Ohio, New Jersey, New York, and Pennsylvania, which are experiencing
major out-migration to the *sunbelt of the southern and western states.
   During the 1970s the concept of a population shift from the snowbelt to
the sunbelt received a good deal of attention, but later research shows that
job losses in the USA have not been confined to the snowbelt, and that
economic and demographic indicators vary widely within both belts; the
western parts of the snowbelt actually experienced a gain in employment
in the 1960s and 1970s.

snowline The level at which, with altitude, snow becomes a seasonal or
permanent feature. This varies with *aspect and latitude.

social anthropology The study of people in a social context with a
strong historical bias. It tends to be concerned with the cultures and
societies of the, as yet, non-industrial world.

social area analysis The analysis of a city to define **social areas**—urban
areas which contain people of similar living standards, ethnic background,
and life-style. Three constructs have been used to differentiate urban areas.
First is social rank. As it changes, the distribution of skills changes from
manual to semi-skilled and skilled white collar jobs. The second factor is
urbanization, which weakens the importance of the family unit as it
increases. The third is segregation which sees a redistribution of
population as it proceeds. Variables are chosen for the three constructs: for
example, occupation, education, and rent for social rank; fertility and
number of working women for urbanization; and isolation of racial groups
for segregation. These variables are then combined to form categories for
residential areas, such as low social rank, high urbanization, and high
segregation.
   Social area analysis was developed by E. Shevky and W. Bell (1955), and
has now largely been replaced by *factorial ecology.

social capital Assets, like roads, schools, and hospitals, which belong to
society rather than to individuals.

social cleavage The spatial division of society into distinct groups.

social costs In economics, the total costs of any action. These costs are
made up of private costs, which are met by the individuals concerned, and
indirect costs, which are borne by third parties.

social Darwinism The application of the concept of evolution to the
development of human societies over time. It is an idea which
emphasizes the struggle for existence of each society, and the survival
of the fittest of them. Such ideas have been used to justify naked
capitalism, and have been extended to defend power politics,
imperialism, and war.

social distance The perceived distance between social strata, as in different socio-economic, racial, or ethnic groups. This is usually measured by the amount of contact between groups, such as through friendship and marriage. This distance may have arisen spontaneously, as certain groups prefer to 'keep themselves apart' but is often imposed on one group by a dominant group; the charter group, for example, may keep a distance between it and a minority group, through discriminatory practices.

social ecology An alternative term for *human ecology.

social formation The prevailing pattern of class structure which goes hand in hand with a particular *mode of production. That is to say, there will be one type of class structure associated with capitalism and another, quite different type, associated with communism.

social geography Originally this was defined as the study of the spatial patterns of social, as distinct from political and economic, factors. The subject may now be subdivided into three categories. The first lies in the spatial expression of capitalism; the city has a social structure as an expression of class structures which are reflected in its morphology. Another aspect stresses the 'alternative' view of human geography which studies the response of the economically disadvantaged rather than the successful. A third category emphasizes *welfare geography.

social justice The distribution of the benefits and the hardships in society, together with the way they are allocated. Geographers are particularly concerned with the spatial expression of social justice; where do the advantaged and disadvantaged groups live, why do they live there, and what is the connection between their place of residence and their future advantage or disadvantage. Such questions can be asked on local, regional, national, and global scales. Also of interest to geographers is the way in which moral systems vary spatially and whether, with the rise of *postmodernism, we can look for universal principles of social justice.

social network The cluster of relatives, family, and neighbours to which an individual or family is connected. Such groupings often share the same values and goals.

social physics A view of human society which seeks analogies from the world of physics to aggregate human behaviour. Perhaps the best-known example of this is the *gravity model which sees the attraction of a town for the surrounding population as being proportional to its population and inversely proportional to the distance away from the town. This is analogous to the gravitational attraction of a physical body in Newtonian physics.

social polarization The results of *segregation within a society such that the ends of the social spectrum consist of large social groupings which are very different from each other.

social space The combined use and perception of space by distinct social groups, as opposed to *personal space. Social space provides an environmental framework for the behaviour of the group, such as a community, or *natural area.

social statistics Information, judged by a government to be of public interest, about people: their birth, lives, and death. Social statistics include the vital statistics of birth, death, and fertility, together with wealth, income, living standards, occupation, and education.

social well-being A state of affairs where the basic needs of the populace are met. This is a society where income levels are high enough to cover basic wants, where there is no poverty, where unemployment is insignificant, where there is easy access to social, medical, and educational services, and where everyone is treated with dignity and consideration. Many attempts have been made to quantify social well-being. *See* *territorial social indicators.

socialism A social system based on equality and social justice, once linked with common ownership of the *means of production and distribution, but now become more fluid. Some writers consider that socialism is achieved when the major part of the means of production is owned by the state. In communist theory, socialism is the first stage on the road to full communism. It differs from communism in that it is attached to ethical and democratic values and because it allows both common and state ownership.

sociation A unit of plant communities; the smallest area of ground in which the full range of plant types for that community can be found.

socio-economic groups *See* *Registrar General's classification of occupations.

sociology The study of societies; both the description of social phenomena and the evolution of a conceptual scheme for these phenomena. Different strands may be recognized: curiosity about how a society hangs together, theories of social evolution, and the interpretation of these theories.

softwood Easily worked wood obtained largely from fast-growing coniferous trees such as pine, spruce, and fir.

soil The naturally occurring, unconsolidated, upper layer of the ground consisting of weathered rock which supplies mineral particles, together with humus; the most common medium for plant growth. The five major factors affecting the formation of a soil are: climate, relief, parent material, vegetation, and time.

soil association 1. In Britain, a group of *soil series developed on a similar parent material or on a combination of rocks.
   2. In the USA, an area in which different soils occur in a characteristic fashion, or a landscape which has characteristic kinds, proportions, and distributions of component soils.

soil classification An ordering of soil types. The simplest arrangement distinguishes between *pedocals, rich in calcium carbonate, and *pedalfers, low in calcium but high in compounds of aluminium and iron. The Great Soil Groups are *zonal, *azonal, and *intrazonal. Of the *zonal soils, *podzols are found beneath coniferous forest, and *latosols develop in warm, moist conditions. *Chernozems, *prairie soils, and *chestnut soils are formed beneath grassland. Grey and red *desert soils occur in hot, arid areas, and *tundra soils form in *periglacial environments.

*Intrazonal soils include *peats, saline *solonchaks, and alkaline *solonetz. Azonal soils include *alluvium and *sands. *See also* *US soil classification.

soil creep *See* *creep.

soil erosion The removal of the soil by wind and water and by the *mass movement of soil downslope. The wind erosion is by *deflation; water erosion takes place in gullies, *rills, or by *sheet wash; downslope mass movement ranges from soil *creep to *landslides.

Accelerated soil erosion is erosion increased by human activity. The causes of such erosion include the removal of wind-breaks, such as hedges (a common cause of soil loss in East Anglia, for example), deforestation (held to be the cause of soil erosion in Nepal, or the Algerian Atlas, for example) and the exposure of bare earth, either by arable farming (continuous cropping was a major reason for the formation of the *Dust Bowl), or by over-grazing (as in the *Sahel or in Zululand—*see* *donga). Fire, war, urbanization, and strip-mining also accelerate the erosion of the soil. *See also* *gullies.

soil horizon *See* *horizon.

soil moisture Moisture is held in the *capillary soil pores. The **soil moisture budget** is the balance of water in the soil; this is the net result of the combined effects of precipitation (P) and *potential evapotranspiration (PE).

When PE exceeds P, there will be a phase when soil moisture is used up, after which, with the continued excess of PE over P, there will be a **soil moisture deficit**. This deficit may be remedied if, during wetter seasons, P exceeds PE, thus inducing soil moisture recharge. This may then be followed by a **soil moisture surplus**. *See also* *moisture index.

soil pore Any open space within the soil framework. The porosity of a soil is judged by the percentage of pore space. Water will not drain freely through the fine **capillary pores**, with an average pore space of less than 0.03 mm, and which retain water through surface tension, but drains freely through the larger **non-capillary pores**.

soil profile A vertical series of soil *horizons from the ground surface to the parent rock. The profile results from the *translocation of soil constituents, and the horizons vary in their degree of separation. A soil is classified according to the arrangement of its horizons.

soil series A group of soils formed from the same parent rock and having similar *horizons and *soil profiles, but with varying characteristics according to their location.

soil structure The way in which sand, silt, clay, and humus bond together to form *peds. Four major structural forms are recognized: block-like, platey, prism-like, and spheroidal. Platey structures are formed of thin, horizontal layers. Prism-like structures are called columnar where the tops are rounded, and prismatic where the tops are level. Spheroidal structures are called *crumbs if highly water absorbent, and granular if only moderately so.

soil texture The make-up of the soil according to the proportions of sand, silt, and clay present. Twelve different textural classes are recognized, and the structure of the soil can be determined when the percentage of these three soil constituents are plotted on a ternary diagram.

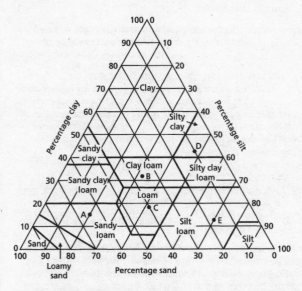

FIGURE 50: *Soil texture*

sol brun lessivée A type of *brown earth from which some clay has been *leached, forming an E *horizon.

solar constant The rate per unit area at which *solar radiation reaches the outer margin of the earth's atmosphere. This fixes the energy supply for the *atmospheric heat engine. Despite its name, the solar constant probably varies slightly over time.

solar energy Any energy source based directly on the sun's radiation. Solar heat is trapped by an absorbent material, usually a black metal panel. The heat is then transferred to pipes which carry warmed air or water. In another method, the sun's rays may be centred on to one spot where the concentrated rays heat up a liquid in order to power a generator. The sun's radiation may be used also in solar cells which convert it into electricity. The chief advantage of solar energy is that, to all intents and purposes, it is inexhaustible. Its disadvantages include the fact that when it is most needed for heating purposes, the days are short, the intensity of the rays is low, and the sun is often obscured by cloud.

solar radiation The electromagnetic waves emitted by the sun, varying in wavelength from long-wave radio waves, through infra-red waves and visible light, to ultraviolet waves, X-rays and gamma radiation. Earth gets only 0.0005% of the sun's radiation. Most solar radiation passes straight through the *atmosphere without warming it, but it is received and absorbed by the earth.

solarimeter *See* *sunshine recorder.

solfatara A vent through which steam and volcanic gases are emitted. Some writers reserve the term for vents through which steam only is emitted.

solifluction Literally meaning soil flowage, this is the slow, downslope movement of water-saturated debris in *periglacial regions and other areas with cold climates. Rates of movement may be 0.9 cm yr^{-1} on gentle slopes, 12–25 cm yr^{-1} on steeper slopes. Solifluction is more rapid where the active layer is deep, or on equator-facing slopes where permafrost thaws more rapidly (see *aspect) and develops due to the oversaturation of soils by water from melting ground ice.

   Solifluction terraces are minute terraces, with heights and widths of about 1 m, and with rough bedding, known as tumultuous bedding, if the material was very fluid. **Solifluction lobes** are deposits of waste which have formed bulges in the slope profile without breaking the surface. Solifluction may develop in areas without *permafrost; winter freezing of the sub-surface layer may be sufficient.

solifluction gravel Also known as *coombe rock*, *head*, *taele gravel*, and *warp*, this is a heterogeneous, unsorted mixture of angular rock fragments in a matrix of clay or silt. Many such gravels form featureless spreads at the foot of higher ground, but they can take fan-like forms.

solonchak An *intrazonal saline soil found in hot, arid climates. Evaporation of soil moisture brings saline groundwater to the surface where it, too, evaporates. The sodium and calcium chlorides and sulphates which have been *translocated remain as a grey surface crust.

solonetz An *intrazonal, formerly saline soil. Periodic rainfall has leached the salts from the surface layer and these accumulate in the B *horizon.

solstice The time (21 June or 22 December) at which the overhead sun is furthest from the equator and appears to stand still before returning towards the equator. The longest day occurs at the **summer solstice**; the shortest day at the **winter solstice**.

solum In soil science, the layers above the parent material. This part of the earth's surface is strongly influenced by climate and vegetation.

solution In geomorphology, the process whereby a fluid, usually water or *carbonic acid, as the solvent, picks up and dissolves particles of a solid (the solute). *See* *carbonation, *aggressivity.

solution mining A mining technique whereby low-grade ores are injected with a solvent. The solvent leaches out the metal in solution which is then pumped to the surface. The metal is then extracted from the solution.

sorting In geomorphology, the deposition of sediments in order of size. With wind- and water-borne sediments, the larger particles are usually dropped first. *Varve clays are especially good examples of sorting.

souming In *crofting, the right of grazing on common pastures.

source pricing *f.o.b. pricing.

source region The region from which an *air mass derives its properties.

Southerly Burster *See* *local winds.

sovereignty The authority of a state which, according to international law, is autonomous and not subject to legal control by other states or to the obligations of international law.

spa A type of resort having mineral springs which are, or have been, thought to have curative properties. The term comes from the town of Spa, in Belgium, and is used in some British place names: Leamington Spa and Droitwich Spa are two examples.

space The extent of an area, usually expressed in terms of the earth's surface. From this meaning, derives the term **spatial**; and spatial relationships are at the heart of geography. It is important to distinguish between **absolute space**, which refers to clearly distinct, real and objective space, and **relative space**, which is space as perceived by a person or society and concerns the relationship between events and between aspects of events.

space–cost curve A graph showing the variation in costs along one dimension across a specified, or generalized area (that is, a section through a *cost surface) which can relate to the costs of a single input or to total costs. If the costs rise steeply from the lowest point, locational choice is restricted, while shallow curves allow more leeway in the choice of location for a particular industry. A combination of a space–cost and space–revenue curve can be used to establish the *spatial margins of production (*see* for an illustration).

space–revenue curve A graph showing the variation in revenue along one dimension across a specified, or generalized area, that is, a section through a *revenue surface indicating, in one dimension, the revenue to be earned from a given volume of sales. A combination of a space–revenue and space–cost curve can be used to establish the *spatial margins of production.

space–time forecasting model A model which attempts to forecast changes in variables over time and space. This type of model is usually a *regression based on earlier values and takes into account the lag caused by *diffusion. Such models are often of the 'black box' type which does not explain changes but is, none the less, used for prediction.

spaghetti model In *Geographic Information Systems, a simple *vector data model which is a line by line translation of a paper map. Such a model is inefficient, since any relationships within it are not inherent, but must be derived through computers.

spatial To do with geographic (not outer) space; with distribution or location across a landscape or surface. In *Geographic Information Systems, a **spatial reference** is a co-ordinate reference in two or three dimensions, or a codified name, which links information to a unique point on the earth's surface; hence spatially referenced.

spatial analysis A type of geographical analysis which seeks to explain patterns of human behaviour and its spatial expression in terms of mathematics and geometry, that is, locational analysis. Many of the models are grounded in micro-economics and predict the spatial patterns which should occur, in, for example, the growth of networks and urban systems, given a number of preconditions such as the *isotropic plain, *movement minimization, and profit maximization. It is based on the tenet that *economic man is responsible for the development of the landscape, and is therefore subject to the usual criticisms of that concept, such as the lack of free will.

spatial autocorrelation A clustering pattern in the spatial distribution of some variable which seems to be due to the very fact that the occurrences are physically close together, that is, that they are in geographical proximity. They are not independent of each other, but somehow linked. In other words, the data are **spatially dependent**. Spatial autocorrelation is widespread: rich people move to areas where other rich people live; people only go to parties because other people go, and so on. If the values in the cluster are more alike than would be due to random processes, there exists a **positive autocorrelation**; if they are less alike than would occur through random processes, there exists a **negative autocorrelation**.

In illustration of this key point in the difficulty of using standard statistical techniques in geography, consider a *random sample of a population, selected by using random number co-ordinates as applied to a map. The sample will not be random unless the individuals are also randomly distributed over the map area, yet a chief concern of, say, urban geographers is the way in which individuals cluster together.

Most standard statistical tests rest on the assumption that the observations made are independent of each other (i.e. are *not* autocorrelated), so techniques in **spatial statistics** have been, and are still being, developed in order to investigate spatial autocorrelations.

spatial co-variation The study of two or more geographic distributions which vary over the same area, such as unemployment and crime. A close 'fit' of two variations shows that the phenomena are associated by area. Statistical methods can be used to determine these associations.

spatial diffusion *See* *diffusion.

spatial-growth model This model is based on the concepts of W. W. Rostow and E. Taaffe and is the spatial expression of *Rostow's stages of economic growth. In stage I, most villages are untouched by change, subsistence agriculture is the rule, and only a few isolated ports have contact overseas. Stage II is analogous with Rostow's 'take off'. Some of the ports expand while others stagnate, and communications develop to the interior. These developments are the infrastructure necessary to economic growth. In stage III, 'the drive to maturity', growth takes place at the larger ports and connections are made between inland centres. The interior centres continue to grow and it is suggested that the *primate city is located inland as it reflects a shift to domestic rather than export markets.

spatial-interaction theory The view that the movement of persons between places can be expressed in terms of the attributes, such as population or employment rates, of each place. This theory is based on the *gravity model and, although it can be made to explain spatial interactions, it has no theoretical underpinning; its validity is, therefore, restricted.

spatial margin The points at a distance from a factory where costs are equal to revenue and no profit is made. Beyond the spatial margin, costs are such that the producer would make a loss if goods were transported there. A producer may locate his factory anywhere within the boundaries of the spatial margin and operate at a profit. In some industries, a spatial margin approach is rather meaningless because the entire country lies within them.

The industrialist may well have other factors—like *amenity—influencing his choice of location. If the benefits gained from locating near amenities but still within the spatial margin compensate for extra profits at the optimum location, an entrepreneur may choose to locate there. Other reasons for locating within the spatial margins but not at the optimum location may be the residence of the founder, the availability of factory space, sociability on the part of the entrepreneur, and the support of local authorities or central government. *See also* *cost surface.

spatial monopoly The monopoly of a good enjoyed by a supplier over a marked area where no competitor exists. If there is only one supplier of a good or service, and transport costs are passed on to the customer, the area over which the supplier has a monopoly will be bounded by a series of

a. costs vary spatially; revenue constant

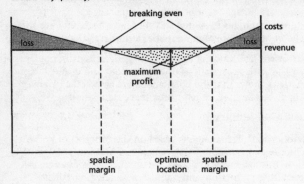

b. revenue varies spatially; costs constant

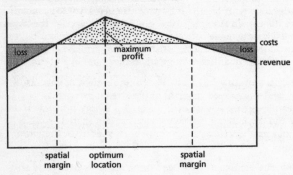

c. costs and revenue vary spatially

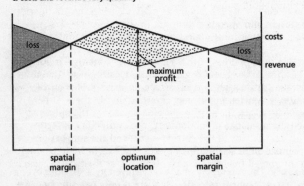

FIGURE 51: *Spatial margin*

points at which the price of the commodity is equal to the price charged by a competitor.

Spatial monopolies can also occur when firms agree to carve up the market spatially, or when the monopoly arises through the privatization of public utilities, as in the case of the British Water Boards.

spatial preference The choice of one spatial alternative, such as a housing area, a holiday resort, or a shopping centre, rather than another. **Repressed preference** occurs where the preference cannot be acted upon—you might prefer to go to Jamaica rather than Brighton, but if you can't afford to indulge your preference, it has to be repressed. In the case of an **absolute preference**, no alternative is considered—it's Jamaica or nothing! With **relative preference**, choices need to be made—Brighton now, or Jamaica in three years when you've saved for it. Relative preferences can be *manifest*, where a choice has been made, or *latent*, where the individual is aware of a possible choice but does not yet need to make that choice.

spatial science A study of *human geography which considers *space to be a fundamental factor in the way society operates and the way individual people behave. Geography was seen as a science of spatial distribution, that is to say, of locational analysis. This interpretation of the subject was closely linked with the *quantitative revolution.

Spearman's rank correlation coefficient Also known as Spearman's rho, the meaning of this coefficient is the same as that of the *product-moment correlation coefficient. The two sets of variables are ranked separately and the differences in rank, d, are calculated for each pair of variables. The equation is:

$$r = 1 - \frac{6\sum d^2}{\left(n^3 - n\right)}$$

where n is the number of paired variables.

specialization index A quantitative measure to indicate degrees of industrial specialization in a given area.

$$I = \sqrt{P_1^2 + P_2^2 + P_3^2 + \ldots P_n^2}$$

where I is the index of specialization and P_1, P_2, and P_3 are the percentage of total employment of each industry in turn.

An index of 70 or over indicates a high degree of specialization in comparison with 55 or below for a diversified area. *See* *diversification.

species A *population or series of populations in which the individual members can interbreed freely with each other, but not with other species.

species–area relationship The relationship between the numbers of different plant and animal species and the area they inhabit. Generally speaking, the number of species present increases with the increase in area of a community, although the rate of increase in species numbers slows down as tracts become successively larger.

specific humidity The actual mass of water vapour present in a kilogram of moist air; in general terms, the mixing ratio.

speleothem A collective noun for depositional features such as *stalactites and *stalagmites. Most speleothems are made of calcareous rock, but columns of other material such as gypsum or silica may be found.

sphere of influence Initially a region influenced by a colonial power but not directly colonized; this is an outmoded concept. The term is increasingly used as a synonym for the *urban field of a city.

spheroidal weathering The weathering of jointed rocks along the joints by water, such that shells of decayed rock surround isolated, unaltered corestones of unweathered rock.

spit A ridge of sand running away from the coast, usually with a curved seaward end. Spits grow in the prevailing direction of *longshore drift. Their ends are curved by the action of waves coming from different directions.

spodosol A soil of the *US soil classification. *See also* *podzol.

spontaneous settlement An unsatisfactory term for a *squatter settlement, since it indicates that squatter settlements are not planned. They often are, although rarely within the bounds of the law.

spread, lateral spread A relatively rare form of *mass movement; a type of *slumping generally only found in the clayey sediment around the edge of *ice sheets.

spread effect An expression used by Gunnar Myrdal (1975) to describe the filtering through of wealth from central, prosperous areas, to *peripheral, less wealthy areas. Thus, increased economic activity at the core may stimulate a demand for more raw materials from the periphery, and technological advance in the core region may be applied to other regions. A belief in the spread effect lies behind the planning of *growth poles; in a sense, the spread effect is the spatial equivalent of trickle-down economics.

spring The point at which water emerges at the land surface. A spring often marks the top of the *water-table, or occurs where a layer of *permeable rock lies above an *impermeable rock layer.

spring-line villages A series of villages at the foot of a scarp through which water percolates until it emerges as a spring at the point where the rock type changes to an impermeable layer. Such springs attracted early settlement.

spring tide *See* *tide.

squall A storm characterized by sudden and violent gusts of wind.

squall line A cluster of storm-bearing *convection cells, each between 2 and 8 km in width, measuring in total some 100 km, located along a non-frontal line, or belt. This storm band may form when cold air overruns warm sector air in an *occlusion, or along an *inversion created by an influx of maritime air over a surface cooled by nocturnal radiation. There may be a link with an upper-air trough. Squall lines occur in West Africa where low-level *monsoon air is overrun by dry, warmer air from the Sahara. In each case, conditions are gusty and rain is heavy.

squatter settlement An area of usually unauthorized, makeshift housing, generally at the edge of a Third World city, and forming up to three-quarters of its area. Local terms include *barrio*, (Spanish—Latin America) *favela* (Portuguese—Brazil), *bustee/bastee* (India), *kampong* (South-East Asia).

Squatter settlements grow because demand for cheap housing outstrips supply. Houses are made from available cheap materials such as packing cases, metal cans, plywood, and cardboard. Sanitation is grossly inadequate, electricity and gas may not be available, and roads are not metalled. In addition, education and medical facilities are severely limited.

Policies to deal with squatter settlements vary. For some time, the city authorities tried to bulldoze the settlements (Nairobi), or expel from the city anyone who did not have an authorizing pass (South Africa). More recently, city authorities have recognized the value of their informal migrants, and have attempted to up-grade squatter settlements by self-help schemes (Lusaka), by granting legal title, or by the provision of sites with limited services around which migrants can construct their houses.

squatting The illegal commandeering of housing or land. Squatting in property, usually in the inner city, is common in the developed world, especially in derelict property; squatting on land at the edge of the city is more common in the developing world. *See* *squatter settlement.

stable ecosystem One which will maintain or return to its original condition following any disturbance.

stable population A population where fertility and mortality are constant. This type of population will show an unvarying age distribution and will grow at a constant rate. Where fertility and mortality are equal, the stable population is stationary.

stability The state of a parcel of air which, if displaced vertically, will return to its original position. Thus, if a parcel of air cools more on rising than the air which surrounds it, it becomes denser than its surroundings and therefore sinks.

The atmosphere is **absolutely stable** when the environmental *lapse rate is less than both the dry and saturated *adiabatic lapse rates. Atmospheric stability is reinforced by *inversions.

stack An isolated islet or pillar of rock standing up from the sea bed close to the shore, such as the Old Harry rocks, Dorset, or the stacks off the Pay de Caux, France. A stack is a residual feature formed when marine erosion attacks a headland. Initially, caves develop; some ultimately meet to form an *arch. With the collapse of the roof of the arch, a stack is left.

stadial A time of glaciation when glaciers advanced and periglacial conditions extended, but not as significantly as in a *glacial.

stage 1. The level of water in a channel. Stage recorders monitor the depth of water at a gauging station. Because there is a relationship between discharge and stage at any point, stage can be used to calculate discharge.
  2. The position reached in a sequence such as *Rostow's stages of growth.
  3. In geology, a *stratigraphic division of rocks formed at the same age, usually having the same fossil assemblage.

stages of growth *See* *Rostow.

stalactite A column of pure limestone hanging from the roof of a cave. It grows as an underground stream deposits its dissolved load of calcium carbonate and it may extend far enough to meet a *stalagmite and thus form a continuous column.

stalagmite A column of pure limestone, formed on the floor of a cave when the dissolved calcium carbonate in the underground water is deposited and the water evaporates as it splashes onto the cave floor. Both *stalactites and stalagmites are common in limestone caves, from the Peak District of Derbyshire, UK, to the Swabian Alb, Germany.

stand An area of vegetation dominated by one species, for example, an oak stand.

standard deviation, σ A measure of the spread of values on each side of the *mean in a data set; a measure of dispersion. It is calculated as the square root of the *variance of a data set. The units of the standard deviation are the same as the units used for the values.

σ may be derived from the equation:

$$\sigma = \frac{\sqrt{\sum (x - \bar{x})^2}}{n}$$

A low standard deviation indicates a close grouping of values about the mean and vice versa.

standard distance Just as the *standard deviation indicates how closely the values in a data set are clustered around the *mean, so standard distance in a spatial distribution indicates how closely the points are clustered around the *mean centre, from the equation:

$$\text{standard distance} = \frac{\sqrt{\sum d^2}}{n}$$

where d is the distance to a given point (co-ordinates x,y) from the mean centre ($\bar{x}, \bar{y}$) and n is the total number of points. It is largely used to compare distributions.

standard error of the mean In statistics, the *standard deviation of a set of means of samples, where all samples are of the same size and selected at random from the same population.

standard industrial classification A grouping of industries classified by a government. The British government recognizes 27 main groups of industry—main order headings—which break down into sub-groups—minimum list headings.

Standard Metropolitan Statistical Areas *See* SMSA.

standardized mortality ratio The ratio of observed to expected deaths.

$$SMR = \frac{\text{observed deaths} \times 100}{\text{expected deaths}}$$

The expected deaths are derived from national figures, while the observed deaths reflect the real conditions. Thus a comparison is made between national and local trends. An SMR of 100 indicates that the age-standardized mortality rate in the group being studied is the same as the overall, or standard population. A ratio less than 100 indicates a higher than average death rate; over 100 is a lower than average one.

standing crop The *biomass present at a given time in a given area.

staple A principal item in an economy. This may be food for domestic consumption as with the potato in early nineteenth-century Ireland, or maize in East Africa. Many Third World exports are based on a staple such as cocoa in Ghana, or sugar in Mauritius. Dependence on an export staple is seen as characteristic of a developing, rather than developed country. The **staple export model** sees the production and export of staples as a trigger to economic growth.

star dune *See* *sand dune.

state A territorial unit with clearly defined and internationally accepted boundaries, having an independent existence and being responsible for its own legal system. The state may be seen as a supplier of public services (education and health, for example), as a regulator of the economy (fixing interest rates, and so on) as a social engineer (education is, after all, a form of social engineering) and acting as a referee between conflicting groups in society (*see* *pluralism). The **theory of the state** looks at the state as a set of institutions: armed forces, government, judicial system, and so on, and asks why societies find it necessary to form the separate instrument we call the state.

State capitalism is an economic system where the government owns and directs large parts of the economy in competition with the private sector. **State socialism** is the ownership, management, and planning of virtually all of the economy by the state.

stationary population *See* *stable population.

steady state A system where input is balanced by output. Thus, a soil might contain a constant amount of water with the 'new' water entering the system being exactly balanced by the 'old' water leaving it. *See* *dynamic equilibrium.

steam fog A shallow, wispy, smoke-like fog formed when cold air passes over warmer water, and is rapidly heated. Convection currents carry moisture upwards, which quickly recondenses to form fog. Steam fog is common in winter over rivers where the air is more than 10° colder than the water. *See* *Arctic sea smoke, *fog.

stem flow *See* *interception.

step An abrupt, but short, break of slope, especially in a *glacial trough. Here, *extrusion flow seems to occur in basins, and *extending flow on steps, so that there is *overdeepening in the basins. Steps tend to be smoothed and striated upstream, as the ice grinds the bedrock, and craggy downstream, because of *quarrying. *See also* *riegel.

steppe The wild grasslands of central Europe and Asia. The natural vegetation has by now been removed or much altered by cultivation and grazing.

stepped leader *See* *lightning.

stepwise migration A type of migration which occurs in a series of movements, for example, from a hamlet to a village, from a village to a town, and from a town to a city, that is, up the hierarchy.

Stevenson screen A box on legs, used to house meteorological instruments. The box is wooden to insulate the apparatus from glare and direct sunlight (the door is orientated polewards), and white-painted to cut down the absorption of *solar radiation, with louvred walls to provide natural ventilation.

stillstand A time of tectonic inactivity between phases of movement. The term is also used to indicate a time when sea level remains constant.

stimulus–response theory The basis of *behaviourism which sees human behaviour as a learned response to stimuli. This approach reduces the environment to a set of stimuli, ignores the way that man creates his reality, and sees the mind as a 'black box' with no need to understand it. Most stimulus–response theory is based on work with animals.

stochastic Governed by the laws of probability. A **stochastic model** describes the sequence of outcomes from a particular, initial event in terms of the probability of each set of developments occurring through time. This means that the same process could produce a variety of results, and, in that aspect, a stochastic model is very different from a *deterministic one.

stock 1. An irregular igneous *intrusion which cuts across the strata of the *country rock. A stock is similar to a *batholith, but is much smaller.

2. The material components of the environment including mass and energy, and biotic or abiotic matter. A stock becomes a *resource when it is of use to man.

stocking rate The number of livestock per unit area.

Stone Age The period, in Britain, from about 25 000 BP to around 20 000 BP. During this time people first made implements and weapons of stone. It is the first major phase of prehistoric culture.

stone pavement In a *periglacial landscape, pavements of large boulders on saturated land in valley floors. It has been suggested that the larger stones have been pushed up to the surface, leaving behind a silty layer. Unlike *felsenmeer, the stones are not thought to be in their original sites.

stone stream A linear arrangement of rocks, seemingly developed through *freeze–thaw.

stone stripes *See* *patterned ground.

stoping The assimilation at depth of *country rock by an igneous *intrusion. The heat of the intrusion melts the country rock which then mingles with the *magma.

store cattle Cattle bought or bred for fattening, thence to be sold to a butcher.

storm beach A beach ridge situated well above the normal limit of high tides, formed as unusually large waves fling ashore shingle, cobbles, and boulders.

storm surge As a *depression passes over the sea, the water is subject to lower *atmospheric pressure than its surroundings, causing the water level to rise and the surrounding water to sink; a fall in pressure of 1 mbar will produce an increase in height of almost 1 cm. Such rises in the sea surface can be compounded by high winds, setting up very long wave motion. In a constricted sea area, such as the North Sea, the resulting storm surge can be devastating, as in February 1953.

Storm surges associated with *hurricanes are common in the relatively confined locations of the Gulf of Mexico and the Bay of Bengal, and are responsible both for many deaths and for the salt contamination of coastal agricultural land.

stoss In geomorphology, facing towards the advancing ice flow. *See* *roche moutonnée.

stoss and lee topography A glaciated landscape where the landforms facing up-glacier show erosion while their lee sides show a degree of protection from glacial erosion. *See also* *crag and tail.

strandflat An extensive *shore platform up to 65 km in width along the coast of Norway. Its formation has been variously attributed to *freeze–thaw, glacial erosion, and periglacial erosion.

Straßendorf A *street village.

stratified Showing distinct layers. Glacial till which shows stratification is often stratified through sorting and redeposition by *meltwater.

stratigraphy The study of the divisions of rocks in time and of the links between similar rocks which occur in different areas. For **stratigraphical column** see *geological column.

strato-, stratus Layered cloud. See *clouds.

stratosphere A layer of the earth's atmosphere, above the *troposphere, 50 km in depth. Within the stratosphere, temperatures remain constant until the 'ceiling' of the stratosphere, the stratopause, is reached.

stratovolcano A cone-shaped volcano with a layered internal structure; for example, one formed of alternating layers of lava and *pyroclasts. Mount Demavend, in the Elburz Mountains of Iran, is an example.

stratum, strata (pl.) In geology, a layer of distinctive deposits with surfaces roughly parallel to those above and below.

stream order The numbering of streams in a network. There are many different methods; the most widely used is that of Strahler. This system classes all unbranched streams as first order streams. When two first order streams meet, the resulting channel is a second order stream. Where two second order streams meet a third order stream results, and so on. Any tributary of a lower order than the main channel is ignored. This system is not ideal. There are other, more appropriate methods but this is the most widely used. See *bifurcation ratio.

streamsink An opening in the earth, usually produced by the solution of limestone, down which surface streams and groundwater disappear.

street village The German *Straßendorf*; a settlement of linear form strung out along a routeway. Most street villages grew up as Dark Age German colonists moved eastwards into the forests of Central Europe, cutting roads as they went.

strength In geomorphology, the resistance of a rock mass to rupture under stress. Intact strength is the strength of a rock with no fissures or *joints. Mass strength is the strength of the rock including joints and fractures.

Strength varies with the following factors, in order of importance: the spacing of joints, the cohesion and *frictional force of the rock, the *dip of any fissures, the state of weathering of the rock, the width of fissures, the movement of water in or out of the rock mass, the continuity of the fissures, and the amount of infilling of soil within the fissures.

stress The force applied to a unit area of a substance measured in newtons per square metre. **Compressive stress** crushes the rock which may collapse as the air pockets within it are compressed. *Tensile stress is a force which tends to pull a rock or soil apart and which may cause fractures and pores to open. A *shear stress deforms a rock or soil by one part sliding over another.

striae See *striation.

striation A long scratch biting into a rock surface. Most **glacial striations** are a result of *abrasion by the fragments incorporated in the ice. These striations are only a few millimetres across.

strike The direction along a sloping stratum which is at right angles to the *dip.

string In *Geographic Information Systems, a sequence of text items or line segments.

string bog In a *periglacial landscape, a marshy area which contains ridges of peat, and, for most of the year, ice within the peat. Peat-forming plants tend to grow in clumps, initiating the ridges which are separated by shallow depressions occupied by ponds and lakes.

strip mining The removal of the *overburden to expose and extract mineral deposits by the use of excavators and drag lines. Permission for strip mining is often granted only if the company replaces the overburden when mining is complete.

Strombolian eruption Volcanic activity, relatively frequent and mild, in which gases escape at intervals, producing small explosions.

structural adjustment Introducing changes to a nation's economy, such as currency devaluation, promotion of exports, and cuts in public services and subsidies. These cuts are usually made in order to qualify for a loan from the *IMF, especially for countries experiencing difficulties in servicing foreign debt.

structuralism An approach to, among other disciplines, human geography, which stresses the structures which underlie human behaviour. Fundamental themes include:

- the underlying elements of the structure remain more or less the same, but the relationships between them alter
- things that appear 'natural' to us, like masculinity and femininity, are actually social constructs
- individuals, too, are the product of relationships.

Thus, what individuals do may be what they are permitted to do by the overall circumstances—structures—in which they operate. These structures are the rules, conventions, and restraints upon which human behaviour is based. For example, within the structure of capitalism, the *optimal location for an industry would be at the point of maximum profits. Within the structure of 'green' politics, the optimal site would be a site where environmental damage is least.

The impact of structuralism on human geography was at its height in the 1970s.

structure The configuration of the rocks of the earth's surface. Structures vary from the small, as in columnar structure, to the large, as in *basin and range structure.

Student's t-test There are two versions of the t-test. The **one-sample Student's t-test** tests a hypothesis by comparing a sample mean ($\bar{x}$) with a hypothesized true mean (μ). The *calculated t* is computed from:

$$\frac{\bar{x} - \mu}{\text{standard error of the mean}}.$$

The difference that would be expected to occur by chance as a result of sampling from a population with the hypothesized true mean is found by reference to tables of Student's t for n^{-1} degrees of freedom, at the appropriate level of confidence; say, 95%, where n is the sample size. This is the *tabulated t*. If *calculated t* is less than *tabulated t*, it is not possible to reject the null hypothesis of no difference between the sample mean and the hypothesized true mean.

The **two-sample Student's t-test** uses a similar technique to compare a two sample means, $\bar{x}_1$, and $\bar{x}_2$ using

$$\frac{\bar{x}_1 - \bar{x}_2}{\text{standard error of the difference}}.$$

The relevant **tabulated t** is found from the tables, using (n_1^{-1}) and (n_2^{-1}) degrees of freedom, where n_1 and n_2 are the two sample sizes.

This *two-tailed test* is useful to geographers, who use it to compare, for example, samples of grasses at two sites with different rainfall regimes in order to establish whether there is a significant height difference between the two samples, that is, to discover whether rainfall influences grass height. It is used for small samples, usually less than thirty, expressed in *interval level measurements.

subaerial Occurring on land, at the earth's surface, as opposed to underwater or underground.

subcontinent *See* continent.

subduction The transformation into *magma of a denser *plate as it dives under another, less dense plate. *See* *ocean trench, *island arc, *Benioff zone.

subduction zone A zone where rocks of an oceanic plate are forced to plunge below much thicker continental crust. As the plate descends it melts and is released into the *magma below the earth's crust. Such a zone is marked by volcanoes and earthquakes. *See* *destructive margin, *plate tectonics.

subglacial At the base of a *glacier, hence **subglacial channel, subglacial stream**, a stream flowing in a tunnel at the bottom of an ice sheet or glacier, usually close to the edge or snout. Subglacial streams are fed by streams on the top of the glacier, which descend through ducts in the ice. These streams can flow at very high pressures, and consequently have considerable erosive power. *See also* *esker.

subgraph In *network analysis, the graphs, or networks, forming an unconnected part of a whole graph or network.

sublimation A direct change of state from a solid to a gas omitting the liquid stage, or of a gas directly to a solid, as in the formation of ice from water vapour. The latter is common when cirrus clouds, formed of ice crystals, grow directly from water vapour by crystallization. Through this process **latent heat of sublimation** is released.

sublittoral zone That zone between the lowest mark of ordinary tides to the end of the *continental shelf.

sub-optimal location A satisfactory, but not *optimal location; that is, occurring inside the *spatial margins. *See* *satisficer.

subsequent streams Rivers running down the *strike of usually weak *strata, or along the line of a fault. Subsequent streams usually run at right angles to *consequent streams of which they are frequently tributaries.

subsere *See* *sere.

subsistence farming A form of agriculture where almost all the produce goes to feed and support the household and is not for sale. Some of the output may be bartered. If there is no market trade in any surplus, the economy is classed as tribal or 'primitive'; if some of the surplus is sold for necessities, such as salt, the economy is classed as 'peasant'. Very few of the former types of subsistence economy remain.

subsoil That part of the soil below the layer normally used in cultivation to the depth to which most plant roots grow. The term is rarely used in soil science, where it would be termed a C *horizon.

subtropical The term is used loosely to refer either to regions which experience some features of *tropical meteorology during part of the year, or to regions of near-tropical climate.
   A more precise definition denotes those areas lying between the Tropic of Cancer and 40° N, and the Tropic of Capricorn and 40° S.

subtropical anticyclones Areas of high pressure brought about when air which has risen in the tropics subsides in *subtropical areas. The air is warmed *adiabatically as it descends; therefore rainfall is unlikely.
   Some authorities think that the subtropical anticyclones are the key to

the world's surface winds, as they affect both the *trade winds and the *westerlies.

suburb, suburbia In theory, one-class communities located at the edge of the city and developed at low rates of housing per hectare, although the homogeneous nature of the suburb has been contested. The provision of open space is a characteristic feature. *See* *suburbanization.

suburbanization The creation of residential areas and, to some extent, industry at the edge of the city. The term suburb usually indicates an area of houses set apart, and in open spaces. Suburbanization in Britain began with the development of mass transport systems: railways, trams and trolley buses, motorbuses, and then mass car ownership, all of which made possible the separation of work and home. It is aided by *decentralizing forces within the city: higher local taxes, pressure on space, natural increase, and congestion and pollution, together with relatively cheap land and higher amenity at the edge of the city, *decentralization of industry, and the freedom of *footloose industries from locational constraints.

   In America, the suburban shopping mall had replaced downtown as the major retail centre for many Americans by the 1980s; a 1981 survey showed that the 20 000 plus malls accounted for over 50% of US retail trade, and suburban growth in the USA reflects this *decentralization. Certain suburbs have high-prestige zip codes; the prestige of an address in, for example, Princeton, Cambridge, Mass., and California's Silicon Valley has attracted many US corporations.

succession, plant succession A series of complexes of plant life at a particular site. In theory, plant succession is viewed as the growth and development of plant life on originally bare earth, with a definite sequence of communities.

sulphur dioxide An atmospheric *trace gas sent into the *atmosphere by natural oxidation of sulphur compounds. Levels have risen rapidly over the last century with the increased burning of sulphur-rich coal and some oils. This biologically destructive gas irritates animal tissues, and is a major cause of lung damage, after smoking. It reacts with rainwater to form weak sulphuric acid; one of the major ingredients of *acid rain.

sun-synchronous satellite A satellite with remote sensors, orbiting 860 km ($\frac{1}{7}$ of the earth's radius) above the earth. The satellite is thereby virtually fixed in relation to the sun. Any location is overflown more or less once in every 12 hours, at predictably varying clock times. *See* *geostationary satellite.

sunbelt In the USA, the southern and western states, such as New Mexico, Arizona, and Florida, which are experiencing major in-migration from the states of the north-east (the snowbelt or frostbelt). Movement to the sunbelt states is based on their resources, their amenity, and, supposedly, cheap non-unionized labour. This movement has been stimulated by federal investment in aerospace and micro-computers.

sunshine recorders The Campbell Stokes recorder is a quartz ball which focuses sunlight onto the paper strip surrounding it on three sides, on its polar side. Sunlight burns a trace in the strip. A solarimeter comprises a quartz or polythene dome, 7 cm in diameter, concentrating heat onto the blackened top of a thermopile, whose lower end registers *ambient temperature.

sunspot A dark patch on the surface of the sun. Sunspots usually occur in clusters and last about two weeks. The number of visible sunspots fluctuates in an eleven-year cycle. It has been suggested that the sun is 1% cooler when it has no spots, and that this variation in solar radiation might affect the climates of the earth.

superadiabatic lapse rate A *lapse rate over 9.8 °C/1000 m (dry *adiabatic lapse rate), rare in the free *atmosphere, but common just above land surfaces emitting strong terrestrial *radiation.

superimposed drainage A pattern of rivers which have been let down onto a very different underlying structure from the one on which they were formed. Thus, the radiating drainage pattern of the English Lake District is thought to be one formed on a dome which has subsequently been removed by erosion, revealing very different geological structures. Also known as epigenetic drainage; examples include the Bayerische Wald section of the Danube valley in southern Germany.

supermarket A self-service shop providing most foodstuffs under one roof and at least 185 m^2 in area. A **superstore** is larger and provides clothing and consumer goods as well as foodstuffs. Most superstores are located at the edge of urban areas where land is cheaper and parking is easy.

superpowers Before 1990, the USA and USSR, both characterized by very great areal extent, large populations, and formidable military power. Since the break-up of the USSR, the USA would claim to be the only superpower.

superstructure According to Marx, the institutions of society; the legal and institutional forms of the social system. These include the state, the law, government and official power, and the body of moral, political, religious, and philosophical beliefs. Marx believed that changes in the economic base would lead to a transformation of the superstructure.

supply curve A representation on a graph of how much of a good will be supplied at a given price. With supply on the horizontal axis and price on the vertical axis, a typical supply curve slopes upwards and to the right because supply tends to increase as the price paid to the supplier increases.

supraglacial On the surface of a glacier.

surf The foam of a wave breaking on a sea shore.

surface boundary layer *See* *boundary layer.

surge phenomena *See* *eddy.

survey analysis The different research methods used to collect and analyse data not available from other sources; generally, data from individuals. These data are usually obtained through sampling and with the use of questionnaires, and are normally checked for consistency before being manipulated and analysed with the help of a computer.

survivorship curve A plot of population figures against time for a group born in the same year, showing how many remain after each year, starting from birth.

suspension The state in which small particles of an insoluble material are evenly distributed within a fluid such as water or air. Particles may be carried upwards when *turbulence outstrips the force of gravity.

sustainable development Development that meets the needs of the present without compromising the ability of future generations to meet their own needs; not simply the use of resources at a rate which could be maintained without diminishing future levels, but development which also takes social implications into account.

sustained-yield resource A resource which is managed such that it may be regarded as renewable. Forestry may be managed as a slow growing but renewable resource of fuel and timber and a sustained-yield recreational resource. Sustaining the recreational appeal of forests as more and more visitors flock in may be more difficult than sustaining the timber flow.

swallow hole A vertical or near-vertical shaft down which a stream disappears in areas of limestone topography.

swash The water moving up a beach from a breaking wave. *See* *backwash.

swidden cultivation *Shifting cultivation. Despite the unstable appearance of the swidden system, since land use changes every two years, it can be said that it is a stable response to the environment as it mimics the exchange of elements occurring naturally.

symbiosis An association of two participants whereby both partners benefit. Thus, flowering plants rely on insects for pollination and the insects feed on their nectar. Lichens are an amalgamation of fungus and algae so close that it is difficult to separate them. Such an interdependence may be termed *mutualism*. Measures of interdependence vary from total to slight.

sympatric Living in the same region. The term is used by ecologists to specify separate species whose territories overlap. Different species can occupy the same geographical location and yet still have individual *niches as they use different parts of the environment.

synchronic analysis The study of the internal linkages of a system at a given point in time. An example from *historical geography is the taking of *cross-sections.

syncline A downfold of rock *strata. *See also* *fold.

synclinorium *See* *fold.

synoptic The term means 'simultaneous' and, in meteorology, covers the weather stations and conditions displayed on a synoptic chart. **Synoptic meteorology** is that branch of *meteorology concerned with a description of large-scale, current weather on a synoptic scale; that is, studying weather phenomena up to 1000 km in size, and lasting about one day, such as a *lee depression.

system Any set of interrelated parts. A system can consist entirely of abstract ideas, but geographers prefer to use the concept in such fields as ecology, hydrology, and geomorphology. An **open system** allows mass and energy to circulate into and out of it; a **closed system** gives and receives energy but not mass.

A system deals with inputs, throughput, and outputs. Systems usually have a negative feedback, i.e. a redress of balance such that a kind of equilibrium is maintained. An example of this is the performance of a hillslope: increased mass movement downslope leads to decreased stream erosion at the base of the slope.

Systems may be studied at all scales and it should be noted that each system is part of a larger system. Thus, an oak-leaf system is part of an oak-tree system which is part of an oak-wood system . . . and so on. It is difficult to establish the boundaries of a system. In this latter example, we must decide where an oak-wood system begins and ends.

In a **cascading system**, a series of small sub-systems are linked from one system to another.

systematic geography The study of a particular element in geography, such as agriculture or settlement, seeking to understand the processes which influence it and the spatial patterns which it causes.

taele gravel An alternative name for a *solifluction gravel.

tafone, tafoni (pl.) A hollow on a sheer face, produced by localized *weathering, mainly through *granular disintegration. Canopy-shaped, overhanging cave roofs are typical. Tafoni are well developed at Les Calanaches in western Corsica.

taiga The predominantly coniferous forest located south of the *tundra in northern continents. *See* *boreal forest.

tail 1. The tapering end of those parts of a frequency distribution away from the arithmetic mean. Statistically, a one-tailed test investigates only one end of a distribution; a two-tailed test investigates both ends.
   2. *See* *crag and tail.

taken-for-granted world A synonym for *life world.

talik Within a *permafrost zone, the layer of unfrozen ground that lies between the permafrost and the seasonally thawed *active layer. Talik most often occurs below rivers and lakes, or where strong springs emerge; open taliks develop under large rivers such as the Yenisey or Lena. Taliks play an important part in *fluvial activity in *periglacial environments since they further *thermo-erosion. *See* *pingo.

talus A *scree slope formed of *frost shattered rock *debris which has fallen from the peaks above. Some writers use the term talus as a synonym for scree; others use the term to indicate the origin and type of slope, which is usually straight, and at an angle of 34–35°. Such slopes often form as a result of the coalescence of a series of **talus cones**, for example, on the south side of the Dachstein Massif, Germany or along the shores of Wastwater, in the English Lake District. The slow, downslope movement of talus is **talus creep**, and is initiated either by the shock of new fragments falling on the scree or by the movement of individual particles resulting from heating and cooling.

tapering 1. Of freight rates, the lowering of transport costs per unit distance with an increasingly long journey. This is because *terminal costs are included in the calculation of rates; as the length of the journey increases, the proportion of the rate accounted for by the terminal costs necessarily falls. In practice, rates are calculated zone by zone, giving a **stepped rate**.
   2. The *distance decay effect where the benefits of a *public good decline with distance from the point of supply.

tariff A list of duties or customs to be paid on imports. **Preferential tariffs** reduce import duties on products of a certain type or origin and **retaliatory tariffs** are levied by a nation whose exports are taxed by a trading partner. Tariffs may be imposed to increase the cost of imported goods in relation to domestic production, thereby reducing the volume of imports and keeping the *balance of payments in credit, or to protect domestic industry from foreign competition through the same mechanism; the term **tariff barrier** is often used. *See* *GATT (General Agreement on Tariffs and Trade), *protectionism.

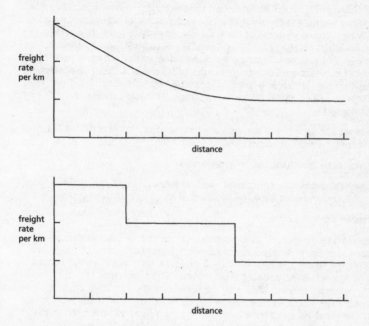

FIGURE 52: *Tapering*

tarn A small mountain lake.

Taylorism A system of production devised by F. W. Taylor (1911), and characterized by the division of factory work into the smallest and simplest jobs while closely co-ordinating the sequence of tasks in order to achieve maximum efficiency, as, for example, on a production line. As a result, skilled managers and technicians oversee semi-skilled or unskilled workers who are engaged in simple, repetitive chores. This system of production has had profound *spatial implications, as large firms often allocate skilled and unskilled jobs to different locations, creating a *division of labour. *See* *Fordism.

tear faults A fault characterized by lateral movement, transverse to the *strike of the rocks.

technoaddiction The reorganization of human societies around technological innovations to such an extent that they become dependent upon them. A recent example is the dependence of *high-energy societies on machines driven by *fossil fuels.

tectonic Of, or concerned with, the processes acting to shape the earth's crust.

teleology This is, in literal terms, the study of purposes, goals, or ends. In other words, the theory that acts, objects, states of affairs, and so on can only be justified in terms of the ends towards which they are directed, or by the functions they fulfil. 'Rightness' is not intrinsic in an action or process, but is dependent on the consequences of the action or process. As a very simple example, the significance of a lioness stalking an antelope only becomes apparent when it is seen as directed to some other purpose, that is, of killing the antelope. In geography, Rostow's theory is teleological.

temperate Describing those locations and climatic types falling between *subtropical and subarctic.

temperate glaciers *See* *warm glacier.

tenant capital The equipment, such as livestock, seed, fertilizers, machinery, and cash supplied by a tenant in an agricultural system.

tensile stress *See* *stress.

tephigram An *aerological diagram, of two axes aligned at 45° to each other. The nearly horizontal lines show *atmospheric pressure, together with the heights at which they are found; the lines at 45° to the isobars show temperatures, running from bottom left to top right.

Superimposed upon these two are three sets of guide-lines: those running from bottom right to top left show the dry *adiabatic lapse rate (DALR); those running from bottom left to top right, at about 60° to the base, show constant humidity (mixing ratio lines); and convex, curved lines, indicate the saturated adiabatic lapse rate (SALR). With the help of these lines, the behaviour of a rising parcel of air may be predicted.

First, the readings from a *radiosonde are plotted to show the environmental lapse rate. Next, from the same source, dew point temperatures are plotted against height. From the temperature and height of the air parcel under investigation, a plot is made of its temperature fall at the DALR, parallel to the DALR guide-lines. Similarly, the dew point change of the air parcel with height is plotted, parallel to the mixing ratio lines. Where these two plots intersect, the parcel will begin to cool at the SALR, and a plot is made of the parcel parallel to the SALR guide-lines. At the point where this plot intersects the environmental lapse rate line (above), the height of the top of any cloud formed by the parcel may be established.

tephra A deposit made of fragments of rock shattered by an explosive volcanic eruption. The material may range in size from so-called 'bombs', which are greater than 32 mm in diameter, to fine dust and *ash. The coarser, heavier particles fall out close to the volcano vent, while,

depending on wind conditions, the finer dust may be carried hundreds of
kilometres.

terminal costs Transport costs incurred by the handling of goods at each
end of a route, or at *break-of-bulk points. If these costs are a major
element in the price of transport, then *line-haul costs are of minor
importance. In such cases, distance is not an important part of transport
costs. See *tapering.

terminal moraine See *moraine.

terms of trade The relationship between the prices of imports and
exports. The trend in this century has been for cheap primary products
and expensive manufactured goods, and—with the exception of oil—most
raw material prices fell very sharply from the mid-1980s. This has
happened because large companies from the rich, industrialized nations
can dominate and structure internal markets in a way that is denied to
small, unorganized Third World commodity producers. This change has
acted adversely on developing countries; for example, African terms of
trade deteriorated by over 30% between 1980 and 1989. It has led to
policies of industrialization, aimed at import substitution, in the Third
World, and to attempts to reduce production in order to increase prices.

ternary diagram A triangular graph used to illustrate the percentages of
three components where the total percentage is 100%. See *soil texture for
an illustration.

terra rossa A red intra-zonal soil developed in Mediterranean regions by
the weathering of limestone. The soil has a clay-loam texture and its red
colouration comes from the dissociation of clay to form iron oxide.
Leaching during the winter rain makes the soil acid.

terrace A bench-like feature. See *river terrace, *terrace cultivation.

terrace cultivation A system of steps, or benches, cut into a hillside.
Terracing of the mountain slopes in Shikoku and Kyushu, Japan, was
introduced in the fifteenth century, as a method of farming on steep
slopes. In Kenya, the *Fanya Juu* method is used; grass strips are planted in
one or two rows along the terrace line, and a channel about 0.5 m wide
and 1 m deep is dug below the grass-line. The material from the channel is
thrown onto the grass strip above, and a ridge is formed. Run-off from the
upper area deposits soil on the ridge, and in time a bench terrace forms, as
the grass strip filters out most of the eroded soil.

terracette A small terrace, about 50 cm across, and closely spaced with
other terracettes. They rise above each other in steps of less than a metre.
Terracettes are held to be evidence of soil creep, but the cause of these
features is uncertain.

terrestrial magnetism See *geomagnetism.

terrestrial radiation The heat radiated from the earth. Short-wave
*solar radiation reaching the earth does not heat the atmosphere it passes

through, but does heat the earth's surface. In turn, and particularly on clear nights, much of this heat is radiated out from the earth. The earth also absorbs terrestrial radiation reflected from the overlying opaque *atmospheric layer. It is by long-wave terrestrial radiation that the *atmosphere is heated. Almost one-third of the solar radiation intercepted by earth is radiated back into space.

territorial justice The application of ideas of *social justice to an area of territory; that is, the identification by a government of areas of need, followed by a deliberate policy of redressing an imbalance. This implies higher government expenditure in areas of *deprivation, such as *depressed areas and *inner cities, than in affluent areas.

Ideas of social justice vary according to the *mode of production and the prevailing ideology, so that the nature of territorial justice is not the same world-wide.

territorial production complex, TPC A type of large scale industrial complex identified by Soviet planners. It is part of the industrial hierarchy of a *centrally planned economy, ranking in size and importance below the national economy and the major industrial unit but above specific industrial centres. Soviet planners used mathematical models to indicate the optimal location for TPCs, especially in the industrialization of Siberia. The term refers to planned, rather than spontaneously arising, industrial complexes.

territorial seas, territorial waters The coastal waters together with the sea bed beneath them and the air space above them, over which a state claims *sovereignty. Traditionally, this area included all the coastal waters up to three nautical miles from the coast. The definition of a landward baseline has been problematical for countries, such as Norway, with an indented coastline. In such cases, a baseline is drawn to link the major promontories.

The extent claimed from the baseline varies. Most countries claim twelve nautical miles. In 1983, the Law of the Sea Convention proposed a 200 nautical mile exclusive economic zone with rights over the sea and the resources of the sea bed. It has not been possible to demarcate such zones over most European waters since the nations are less than 400 nautical miles apart. In such cases, a median line is drawn between the baselines of the states concerned.

territorial social indicator A measure of *social well-being in a given area. Seven indicators may be used: wealth and employment, amenity, health, social problems, social belonging, and recreation and leisure. (The availability and selection of these data may vary from place to place.)

The data are analysed, using *principal components analysis, in order to detect spatial variations in social well-being, with the implication that governments can then develop appropriate strategies to help recovery in areas of deprivation. *See* *territorial justice, welfare geography.

territoriality The need by an individual or group to establish and hold an area of land. In animals, territoriality is an urge, fuelled by aggression, to define a territory for mating and food supply. In human beings, on the other hand, it is more an organization of space in order to make sense of it. The individual needs security and identity, and this is shown most clearly in relation to the home, which provides security of mind and body and a relatively threat-free environment. The community requires a suburb or small town with which to associate, providing an identity and the means of communicating that identity. The importance of territory extends to larger units; the reorganization of the counties of Britain always causes distress.

territory The living space of an animal which it will defend from the forays of other territorial animals. Animals need space in which to reproduce and their territory can be some or all of the following: a source of food, a source of mates, and a breeding area. When many individuals of a species divide an area into territories, the divisions may be spaces of relatively similar size. If all the available space is taken up, then the size of the population is at a maximum. The consequence of territoriality is to set a limit to population, but this consequence is a side-effect; territoriality is not a population control device.

Tertiary The earlier *period of the *Cenozoic era.

tertiary industry Economic activity concerned with the sale and use of economic goods and services, in other words, *service industry. Examples include retailing, wholesaling, and delivery. A high proportion of employment in services is characteristic of an economically advanced country; in 1987, 56.5% of Japanese, 59.2% of EC, and 68% of US employment was in services. See *post-industrial.

tessellation 1. (noun) Any infinitely repeatable pattern of a regular polygon. In *Geographic Information Systems, these may be square (*rasters), hexagonal, or triangular (see *triangulated irregular network)
   2. (verb) The partition of a two-dimensional plane, or a three-dimensional volume, into contiguous polygonal tiles or polyhedral blocks, respectively.

Tethys An ocean which developed during Paleozoic and Mezozoic times, running from the coast of southern Spain to south-east Asia. Great thicknesses of sediment were formed since the sea floor kept subsiding at the same rate as deposits were laid down. These sediments were compacted, subjected to *vulcanism, and then uplifted and deformed by the earth movements which formed the Alps (the Alpine orogeny).

thalweg The line of the fastest flow along the course of a river. This usually crosses and recrosses the stream channel.

thermal erosion See *thermo-erosion.

thermal expansion Also known as insolation weathering, this is the rupturing of rocks and minerals mainly as the result of large, daily temperature changes. The exterior of the rock expands more than the interior. Whether thermal expansion is effective in an environment with no water is open to question.

thermal low An intense, low pressure system caused by local heating of the earth's surface, and leading to the rising of air by *convection. Heavy rainfall will result if the air rises and cools enough for condensation to occur.

thermal pollution The contamination of cold water by adding warm water. Sources of heat include water used for cooling in electricity stations, the *urban heat island, and the construction of reservoirs. Many aquatic organisms cannot tolerate warm water.

thermal wind Not a real wind, but an expression of *wind shear for a given layer of *atmosphere; the *vector expressing the difference between the *geostrophic winds at the bottom and top of the layer. It is proportional to the thickness of the layer, and is directed along the *isotherms, with cold air to the left in the Northern Hemisphere, and to the right in the Southern.

However, the term is used to denote a wind developing as follows: the *pressure gradients which produce surface winds may be due to the presence of cold and warm *air masses. The fall in pressure with height is rapid in cold air, and much less rapid in warm air. Thus, at height, air pressure in the cold air will be less than that in the warm air. This creates a high-level pressure gradient and, therefore, a wind, often described as the 'thermal wind'. The strength of this wind is a function of its height and the temperature difference between air masses; the greater the difference, the stronger the wind. Since there is a marked *meridional temperature gradient in the *troposphere, influenced at height by a powerful westerly factor, thermal winds are very strong at the point where the temperature gradient is greatest; at the polar *front. The result is the polar front *jet. The force of a thermal wind may be strengthened by any pressure gradient at ground level.

thermocirque A large hollow on a hillside formed from the coalescence of *nivation hollows. The sides usually experience parallel retreat. Thermocirques are shallow because the centre of the hollow is protected from further erosion by a covering of snow.

thermodynamic diagram A type of graph, of which the *tephigram is an example, plotting the qualities of the atmosphere. Such diagrams are used as aids to weather forecasting.

thermo-electricity Power produced from a range of fuels such as coal, oil, peat, lignite, nuclear fuels, or *geothermal heat which drive steam turbines or internal combustion engines to turn over the generators.

thermo-erosion The combined thermal and mechanical activity of running water in *periglacial conditions. *Mass movement due to the melting of *permafrost is common. A river which is above freezing point may melt and undercut its bank to form a thermo-erosion niche. The undercut portions of the bank, frozen in the upper parts, collapse and melt. This leads to mass movement of the bank downslope.

thermokarst A landscape of irregular depressions, caused by the irregular heaving and melting of ground ice under *periglacial conditions. The exact form of the depressions depends on the original distribution of ice segregations, the subsurface movement of water during warmer periods, and the presence or absence of water in the hollows since water-filled hollows tend to perpetuate themselves. The term 'karst' is used to indicate the numerous features formed by subsidence, and does not imply the presence or development of a limestone landscape.

thermo-planation The *degradation of lowlands in a *periglacial zone due to *thermokarst processes. Sheet denudation is due to frost creep, frost heaving, nivation, solifluction, suffosion, and sheet wash.

thermistor A meteorological instrument used to measure temperatures. It contains a semiconductor with a large temperature coefficient of resistivity, and a linear relationship between current and applied electromotive force.

thermometer The mercury thermometer is ineffectual below −40 °C, the freezing point of mercury, so that alcohol thermometers are used for low temperatures, and in maximum−minimum thermometers.

thermosphere That part of the atmosphere, starting at about 85 km above the earth, the top of the mesosphere, extending to the uttermost fringe of the atmosphere. Here, temperatures increase with height. The warming of the thermosphere comes from the photo-dissociation of oxygen molecules, 50% of which dissociate into atomic oxygen, absorbing solar ultraviolet radiation.

Thiessen polygon A subdivision of a drainage basin, containing a rain gauge. Polygons are constructed by first siting the rain gauges. Their locations are plotted on a base map. These points are connected by drawing straight lines between the sites. The lines are bisected with perpendiculars which meet to form the polygons. The areas of the polygons are calculated and expressed as fractions of the total area. Each fraction is multiplied by the precipitation recorded by its rain gauge. The sum of these calculations represents total precipitation over the catchment area.

thinning The extraction of some of the young trees in a forest so that the remainder grow and develop fully. The aim is to remove as much timber as possible while maintaining output.

three-field system A farming system prevalent in Medieval lowland Britain whereby two of the three fields were cultivated while the third field was left fallow to recover its fertility. The crops were then rotated so that a different field was left fallow.

threshold population The minimum population needed to justify the provision of a certain good or service. This may be expressed crudely, in population numbers, so that M. G. Bradford and W. A. Kent (1977) give threshold populations of 7000 for a main post office and 15 000 for a library, but purchasing power may be a better yardstick for commercial goods or services.

The concept of the threshold can be demonstrated in terms of the provision of medical services, which will have an increasing threshold population as the sophistication of the service provided increases; from a single GP, to a group practice, to a hospital with consultants, and finally to a specialist hospital, such as Great Ormond Street Hospital for Sick Children.

Note that threshold populations may vary regionally and certainly vary nationally; the threshold population for a baker's shop is far lower in France than in the UK, for example. Each good or service may have two limits; the inner area containing the threshold population and the outer area bounded by the *range of the good or service. The actual evaluation of the threshold population for most goods and services is difficult.

throughfall *See* *interception.

throughflow The movement diagonally downslope of water through the soil, as opposed to the vertical movement known as *percolation. It may follow natural *percolines in the soil. Throughflow is a major factor in the hydrology of a drainage basin where the rocks underlying the soil are impermeable. *See* *interflow.

throw Of a fault, the vertical displacement of strata along a fault line.

thrust A movement causing the formation of a reverse *fault of a very low angle. The **thrust plane** is the low-angle fault face over which movement occurs.

thufur A low mound which forms part of a polygonal pattern in *periglacial, or cool areas, such as Spitzbergen. *See* *involution.

thunder When a stroke of *lightning passes through the atmosphere, the air becomes intensely hot, perhaps to 30 000 °C. The violent expansion thus caused makes a shock-wave heard as thunder.

thunderstorm A storm including strokes of *lightning, which cause the thunder, and draw off electrons earthward as part of the atmospheric electrical cycle.

tidal energy Energy based on the motions of the tide. Schemes to use tidal energy have been implemented at the Rance Barrage Tidal Scheme near St Malo and on the east coast of Canada.

tidal wave *See* *tsunami.

tide The twice daily rise and fall of sea level. Tides are the result of the pull exerted on the earth by the gravity of the moon and of the sun. This pull affects the land masses as well as the oceans but the reaction of the water is greater and more apparent. All of the earth is attracted by the moon's gravity but the greatest effect is exerted on each side of the earth as it faces the moon. The moon 'pulls out' two bulges of water from these sides. These bulges are fixed and the earth moves through them. This gives high water twice daily.

The sun also attracts water. When the effects of both sun and moon coincide, twice monthly in the second and fourth quarters of the moon, high spring tides occur. When the sun and moon seem to be at right angles to each other from the earth, the forces of moon and sun are opposed to each other, and lower, neap tides result.

The vertical distance between high and low tides is the **tidal range**. All places with a high tidal range have strong tidal currents, but swift currents can also occur in localities of low tidal range.

till The unsorted sediment deposited directly below a glacier, which exhibits a wide range of particle sizes, from fine clay to rock fragments and boulders. The lithological character of a till depends on the geology of the region the glacier has travelled over. Till is usually responsible for monotonous relief, sometimes diversified by the presence of *kettle holes, and sometimes overlain by *ablation moraine. Where sheets of till are old, they may form **till plateaux**, such as the smooth surface lying between Cambridge and St Neots, filling up stream valleys beneath.

Till results from melting at the surface and at the base of the glacier; the latter probably being of more importance. **Basal tills** are likely to be formed when the lower, debris-rich layers of a glacier are slowed down, perhaps by an obstacle. The material is then compressed, and water squeezed from it, by the weight of the ice above. A **till plain** blankets the ground, with only a few mounds and ridges poking through.

Other tills are classified by means of their origin: *ablation, or meltwater till, *lodgement till, and *sublimation till. **Flow till** is created when saturated debris, found at the top of the ice, flows into depressions within the ice and is then deposited.

till fabric analysis The study of the fabric of till to determine the movements of the glacier. The direction of ice advance is indicated by the orientation of the long axes of the pebbles incorporated in till, or by tracing the origin of pebbles.

time-space compression D. Harvey (1989) argued that, as *capitalism has developed, the pace of life has become faster and faster. The age-old barriers to action have been broken down so the world 'sometimes seems to collapse in on us'. This is not simply *time-space convergence, but the impact of new systems of transport and communications as experienced by the individual, and especially the emphasis given by modernity to 'the shock of the new', immediacy, and simultaneity.

time-space convergence Places are separated by absolute distance and by time. With improvements in communication systems and methods of transport, this time-distance diminishes. D. G. Janelle (*Prof. Geogr*, 1968) illustrated this with figures for the journey time between London and Edinburgh:

DATE	JOURNEY TIME IN MINUTES	MODE OF TRANSPORT
1658	20 000	stage coach
1770	6000	stage coach
1820	2700	stage coach
1860	700	railway
1950	200	aeroplane

and expressed the rate at which this takes place as:

$$\frac{TT_1 - TT_2}{Y_1 - Y_2}$$

where TT_1 and TT_2 are travel times in different years and where Y_1 and Y_2 are the relevant years.

In essence, time–space convergence means that the *friction of distance—a concept fundamental to conventional *central place theory, *diffusion theory, and *location theory—is lessening.

time–space geography An approach to geography developed at the University of Lund by Torsten Hägerstrand and his associates, (T. Carlstein, 1978, vol. 2).

Time and space provide the room needed for sequences of events and Hägerstrand expressed this as a web model. This is based on four propositions: that space and time are scarce resources which individuals draw on to achieve their aims; that achieving an aim is subject to *capability constraints, *coupling constraints, and *authority constraints; that these constraints interact to demarcate a series of probability boundaries; and that choices are made within these boundaries. A **time–space prism** is a representation of the constraints limiting the time within which the individual can act.

Time geography provides a method of mapping spatial movements through time.

time zone A division of the earth's surface, usually extending across 15° of longitude devised such that the standard time is the time at a meridian at the centre of the zone.

TIN *triangulated irregular network.

tithe A local tax first levied in England in the fourth century AD to pay for the church and its clergy. Tithes were at first paid in kind but subsequently commuted into money terms. The last tithes lapsed in 1996.

tolerance The ability of an organism to survive environmental conditions. The prefixes *eury-* and *steno-* refer to wide and narrow ranges of tolerance respectively. An organism can be widely tolerant of one factor, such as temperature (*eurythermal*), but narrowly tolerant of another, such as salinity (*stenohaline*).

tombolo A spit, resulting from *longshore drift which joins an offshore island to the mainland. In Tuscany, Italy, Monte Argentario is linked by three **tomboli** to the land.

topographic(al) map A map which indicates, to scale, the natural features of the earth's surface, as well as human features. The features are shown at the correct relationship to each other.

topological map A map designed to show only a selected feature, such as the stations on the London underground. Locations are shown as dots, with straight lines connecting them. Distance, scale, and relative orientation are not important.

topology The study of those properties of a geometric model, such as *connectivity, which are not dependent on position.

topophilia The feeling of affection which individuals have for particular places, a term introduced by Yi-Fu Tuan (1961). Places in this sense may vary in scale from a single room to a nation or continent. Topophilia is an important aspect of the symbolic meaning and significance of landscapes. *See* *iconography, *sense of place.

topple A form of *mass movement from a rock face where top-heavy rocks with vertical or forward-leaning bedding planes are separated from the bedrock and fall. As the rock peels from the top of the free face, it turns.

topsoil The cultivated soil; the surface soil as opposed to the *subsoil.

tor An upstanding mass of rocks or boulders which rises above the gentler slopes which surround it. Tors are thought to have been formed by *frost shattering in *periglacial conditions. This theory is disputed by those who maintain that tors form underground during the deep weathering typical of tropical areas. In this case, the tors are first formed before the *overburden is eroded away to reveal the boulders. According to these theories tors are examples of *relict landforms. Tors are found on granites from Dartmoor, UK, to the Asir Highlands of Saudi Arabia.

tornado A destructive, rotating storm under a funnel-shaped cloud which advances over the land along a narrow path. This storm is generated by powerful updraughts. The tangential speed of the whirling air may exceed 100 m s^{-1}, the core may perhaps be 200 m across, the duration of the storm about 20 minutes. Within a tornado, the central pressure is around 100 mb below that of the exterior; this may cause buildings to explode. This type of storm is very common in 'tornado alley', extending from northern Texas through Oklahoma, Kansas, and Missouri, with as many as 300 tornadoes a year.

The exact mechanism of its formation is not fully understood, but tornadoes are associated with intense local heating coupled with the meeting of warm, moist air from the Gulf of Mexico and cold air from the *basin and range area of the western United States. Tornadoes are often associated with hurricanes.

tourism Making a holiday involving an overnight stay away from the normal place of residence. This is in contrast to recreation which involves leisure activities lasting less than twenty-four hours. This holiday may be based on the cultural, historic, and social attractions of an urban centre, or on the appeal of a different environment. Urban tourism increases the importance of the *central place while tourism at the periphery can provide the income for economic development.

tower karst Limestone towers, from 30–200 m in height with nearly vertical walls and gently domed or serrated summits. The towers stand above large, flat *flood plains and swamps and show undercutting from rivers and swamps. Tower karsts are thought to represent the last remnant of a limestone outcrop.

Tower karst is well developed in south-east China, where the 200 m towers are formed of horizontally stratified, very pure limestone.

town A relatively small urban place. No limiting figures of population or areal extent are agreed upon. Towns may be regarded as central places providing goods and services to their surroundings but without the degree of economic specialization to be found in a city. In the USA, 'town' has a particular administrative connotation.

townscape In urban geography, the objective, visible scene of the urban area or the subjective *image of the city. The townscape has three separate, but closely related, parts; the street plan or layout, the architectural style, and the land use. Townscape is the urban counterpart of *landscape.

trace gas A gas naturally occurring in very small quantities, such as argon in the *atmosphere. A **trace element** is an element, such as zinc, copper, and cobalt, which is required in very small quantities to ensure normal development of an organism.

traction load *See* *bedload.

trade The movements of goods from producers to consumers. The classic explanation for trade is expressed in terms of *comparative advantage.

trade winds The tropical *easterlies, blowing towards the equator from the *subtropical anticyclones at a fairly constant speed. They are at their strongest around the equatorial flank of these *highs. The word 'trade' arises from the nautical expression 'to blow trade', i.e. to blow steadily, in a regular course. Trades are not true *zonal winds; their zonal character is weaker in summer because of the south-west *monsoon. They are most regular around 15°, and are associated with fine weather resulting from the *anticyclonic subsidence of the Hadley cell. Over the equator and the western oceans, **trade wind weather** is rainier.

trade wind inversion A temperature *inversion, found in the tropics at heights of between 3000 and 2000 m, caused by descending air of the *Hadley circulation. This inversion acts as a ceiling for pollution, an effect which is marked in cities with heavy motor traffic, such as Los Angeles. See *photochemical smog.

traffic The movement of people and vehicles along a routeway. **Traffic capacity** is the maximum number of vehicles which can pass over a route in a given time, while traffic density is the existing number of vehicles.

traffic calming The reduction of speed in road traffic, achieved by constructing speed ramps, by creating more pedestrian crossings, and by building low walls halfway across the road from alternate sides in order to create curves in a straight street.

traffic principle The basis of settlements about a *central place such that the number of services on straight-line routes is at a maximum. The number of settlements at progressively lower levels follows the sequence 1, 4, 16 . . . This is the k = 4 hierarchy as advanced by W. Christaller.

traffic segregation The subdivision of towns and cities into certain units where road traffic is restricted and pedestrians predominate. Each of these units is linked to the rest of the town by good roads which carry most of the traffic.

tragedy of the commons G. Hardin (*Science*, 1968) described an increase in the use of common land by a number of graziers, with each grazier continually adding to his stock of animals for as long as the marginal return from each animal is positive, even though the average return for each animal is falling, and even though the quality of the grazing deteriorates. Hardin used this metaphor to describe any situation where the interests of the individual do not coincide with the interests of the community, and where no organization has the power to regulate individual behaviour.

transactional analysis An analysis of the firms in an industry based on the proposition that capitalist enterprises are concerned above all with reducing transaction costs. This would explain the growth of *agglomerations.

transactionalism A perspective which selects the integrated system as a unit of study. 'The whole' is the proper unit of analysis, possessing properties which are not directly derived from its component parts, but these elements are thought to be independently definable and functioning. In this approach, the observers themselves are also aspects of any system, and the very fact of their observation will alter the system, or phenomenon, so that observers in different physical and psychological locations would yield different information about the same system.

transdisciplinary Describing a study which runs across traditional subject boundaries such as arts and science. Geography is often portrayed as a transdisciplinary subject since it has been concerned with the interplay between environment and humans, but many geographers argue that, with increasing specialization, the gulf between physical and human geographers has become very wide.

transfer costs Total transport costs involved in moving a cargo including extra costs such as tariffs and insurance. Transfer costs are highest for people, because of the very steep cost of insurance.

transfer price The price set by an organization for goods which are sold from one section of the organization to another. These prices can be used to minimize taxes.

transferability The capacity of a good to be transported. High transferability is linked with high value/low bulk, easily transported goods, such as Rolex watches; low transferability to low value/high bulk goods, like hay, or to low value yet fragile goods, like plate glass. E. L. Ullman (1954) believed transferability to be one of the three fundamental principles underlying *spatial interaction. The other two are *complementarity and intervening opportunity.

Transferability is largely determined by transport costs and movement will only take place if either the cost or *economic distance is not too great. As economic distance increases, so transferability decreases and any intervening source of goods will be used. Since economic distance and intervening opportunities vary, so transferability may change over time.

transform fault Faults which are parallel to the arc of sea floor spreading. They are *strike-slip faults, running transversely from the faults across the *oceanic ridge which they have displaced. The Pacific plate, for example, is separated from the American plate by the 600 km long San Andreas fault.

transformation The changing of data into another form. For example, absolute distance may be plotted as time-distance in a *cartogram.

transformation of data It is possible to make inferences about data when they show a *normal distribution. However, many data are skewed, with an asymmetrical distribution. It is possible to transform the distribution to make a 'normal' shape. This may be done if the distribution has only one *mode.

One of the most powerful transformations is to use log values for the data since logarithms 'shrink' the spread of data; this is used in plotting *rank-size relationships, for example. A milder transformation which may be applied is the use of square roots and square values. The mean and standard deviation may then be calculated. These transformations may bring about a more 'normal' shape, but the question of what is 'normal' is not always readily defined.

transhumance A seasonal movement of men and animals between different grazing grounds. Shepherds leave their lowland winter quarters, and move to upland, summer pastures. A farmer practising transhumance is not a nomad, since he has two fixed abodes.

transitional city A stage in the development of the capitalist city between the *pre-industrial and the Victorian. Workshops and factories are dispersed throughout the relatively compact city. Occupation is the main influence on residential location, and professionals are located near the city centre.

transitional zone The area of a city immediately surrounding the *CBD. It developed during the nineteenth century for residential purposes but is now an area of mixed use such as industry, shops and offices, poor housing, and multi-occupation of units.

translocation In soil science, the transfer of substances in solution or suspension from one *horizon to another.

transnational corporation Generally synonymous with *multinational corporation, although a transnational corporation may operate in only two national economies.

transpiration *See* *evapotranspiration.

transport costs Costs involved in relaying goods to and from a plant, including payments to transport firms for their services and any cost incurred by a plant in using and maintaining its own fleet of vehicles. Generally speaking, transport costs have fallen relatively as a result of improvements in transport technology and transport infrastructure. Early *location theory was based on transport costs, but *see* *time–space convergence and *tapering.

transport geography A branch of human geography concentrating on the movement of people and goods, the patterns of such movements, the volume of people and goods carried, the price of transport, and the role of transport in economic, political, and social development.

tree line The line beyond which trees will not grow. This occurs at high latitudes, as when *taiga gives way to *tundra, or at high altitudes.

trellised drainage *See* *drainage patterns.

trend surface map A three-dimensional diagram showing the uptake of an innovation through time and distance. Trend surface maps may be used to separate regular patterns of regional trends from localized anomalies which have no overall pattern, and in that respect act as filters which cut out short-wave irregularities but allow long-wave irregularities to pass through.

triangulated irregular network, TIN A type of tesseral model based on triangles. It is the standard method of representing terrain data, and continuous data, such as population density.

Triassic The oldest *period of *Mesozoic time stretching approximately from 225 to 190 million years BP.

tributary area *See* *umland.

trophic level An individual layer on the *pyramid of numbers which represents types of organisms living at parallel levels on *food chains. All herbivores live at one level, all primary carnivores on the next level, all secondary carnivores on the next level, and so on. The animals on each level are remarkably distinct in size from those on other levels; there is a clear jump in size between an insect and a bird, for example.

tropical cyclone *See* *hurricane.

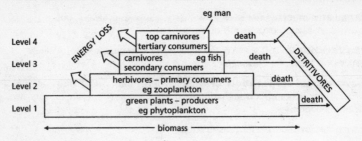

FIGURE 53: *Trophic level*

tropical forest Forested areas which often extend beyond the tropics and consist of *tropical rain forest and *mangrove forest.

tropical meteorology In this field, the boundaries of the tropics fluctuate being marked by the descending limbs of the *Hadley cells, and the centres of the *subtropical anticyclones. The tropical atmosphere is characterized by high temperatures, the transfer of energy in the form of *latent heat via the *Hadley cell, small temperature and *pressure gradients (so that there are no *fronts), high *humidity, easterly winds, and low values of the *Coriolis parameter, such that small pressure gradients will produce a stronger *geostrophic wind than in *mid-latitudes.

The following *synoptic scale phenomena are characteristic of tropical meteorology:

1. tropical wave disturbances. These have a wavelength of 2000–4000 km, travel across 6–7° of longitude a day, and last for about two weeks. Ahead of the trough is a ridge of high pressure bringing fine weather. With the approach of the trough, *cumulus cloud develops, wind *veers, and heavy showers fall.

2. tropical cyclones. *See* *hurricane.

3. *monsoon depressions.

4. subtropical cyclones. These occur when the cold upper air from high latitudes is cut off to form a wave some 300 km in width. They bring cloud and some rain.

5. tropical cloud clusters. *See* *squall line.

6. *easterly waves.

Small-scale variables such as relief, *local winds, and *ocean currents are vital. Cold ocean currents cause offshore fog, taking moisture from the winds. *Sea breezes then carry cool, dry air far inland. *See also* *inter-tropical convergence zone.

tropical rain forest Tropical forest of trees characterized by buttress roots, long, straight, lower trunks, and leathery leaves. The vegetation shows distinct layering: the canopy, or upper layer, at around 30 m; the intermediate layer at 20–25 m; and the lower layer at around 10–15 m. Undergrowth is poorly developed but *epiphytes and *lianas are common. Deciduous trees flower, fruit, and shed their leaves at random; there is no seasonality. The range of plant and animal species is immense and many plants yield important medicinal compounds. The felling of the rain forest causes soil erosion, the destruction of potentially useful species, and the reduction of oxygen from photosynthesis, while smoke from burning logs increases the quantity of *aerosols and carbon dioxide in the atmosphere. Some writers suggest that the hydrology of the earth may be altered if all the equatorial rain forest is destroyed.

tropical storm Within the *tropics, a weather system with strong *cyclonic circulation, cloud and rain, and wind speeds in excess of 19 m s^{-1}.

tropics The Tropic of Cancer lies approximately along latitude 23° 30′ N. Around 21–2 June, the sun's rays are perpendicular to the ground along this line and the sun exerts its maximum strength in the Northern Hemisphere. Conversely, the sun is overhead at the approximate latitude of 23° 30′ S, the Tropic of Capricorn, on 22–3 December when the sun's heat is at its maximum in the Southern Hemisphere. Between these two lines of latitude lie the tropics.

The term 'tropical' is used less exactly in climatology, where some areas outside the tropics are said to enjoy a 'tropical climate'.

tropopause The upper limit of the *troposphere, above which very few clouds, except for the *nacreous and *noctilucent, form.

troposphere The lower layer of the *atmosphere, extending to 16 km above ground level at the equator, 11 km at 50° N and S, and 9 km at the poles. Most clouds and *precipitation, and, indeed, weather events, occur within this layer. Increasingly, it is understood that air movements in the upper troposphere greatly influence weather systems in the lower troposphere. *See* *jet stream, *Rossby waves.

Within the troposphere, temperatures of rising air fall, at varying *lapse rates, because the air expands and, therefore, cools. Sinking parcels of air experience a corresponding heating. Most of the water vapour in the troposphere is concentrated in the lower, warmer zone; there is little where the temperature falls below about −40 °C because moist air, rising through *convection or *turbulence, condenses out as ice crystals form.

truncated spur A steep bluff on the side of a *glacial trough, protruding between *tributary, possibly *hanging, valleys. This landform is the result of mainly vertical glacial erosion.

tsunami A huge sea wave. Most are formed from earthquakes of 5.5 or more on the *Richter scale. Other causes include the eruption of submarine volcanoes, very large *landslides off coastal cliffs, or the *calving of very large icebergs from glaciers in *fiords. The most active source region of tsunamis between 1900 and 1983 was along the Japan–Taiwan island arc, where over a quarter of all tsunamis were generated.

A submarine earthquake off the north-east coast of Honshu generated the 1933 tsunami, producing a wave crest of up to 24 m. The death toll was 3008, with 1152 injured. While sea walls of up to 16 m offer some protection against tsunamis, the Japanese government have also offered subsidies for villagers to relocate on higher ground.

tufa A deposit of calcium carbonate found in deserts along a line of once-active springs. The presence of tufa in areas which are now arid points to a time of heavier rainfall; a *pluvial.

tumultuous bedding *See* *solifluction.

tundra The barren plains of northern Canada, the Alaska, and Eurasia. Temperatures and rainfall are low so vegetation is restricted to hardy shrubs, mosses, and lichens. The lower soil is permanently frozen, so that drainage is poor. Marshes and swamps are, therefore, common in summer.

tundra soil A dark soil with a thick, peat layer of poorly decomposed vegetation, which is usually underlain by a frozen layer of soil. *Translocation is limited and there is, therefore, little development of *horizons. Tundra soils range from *brown earths in the more humid areas to polar desert soils in arid areas.

tundra vegetation Characteristically, herbaceous perennials with scattered trees, mosses, lichens, and sedges. Tundra vegetation is restricted by the intense winter cold, insufficient summer heat, and waterlogged soil.

tunnel valley *See* *meltwater.

turbosphere That part of the *troposphere where the mixing of the atmosphere is achieved more by *convection than by *diffusion. Its upper limit is the turbopause, an ill-defined layer about 100 km above the earth's surface.

turbulence, turbulent flow A gustiness in the three-dimensional flow of a fluid, irregular in both space and time, and characterized by local, short-lived rotation currents known as *vortices. Turbulence is hierarchical; large *eddies produce smaller ones, and so on, down a series of smaller and smaller scales. In meteorology, it develops because of disturbances in air flows, the most important of which is *wind shear; air parcels caught in a wind shear tend to roll over and over. The *Reynolds number for turbulent flow is 2000 to 2500. In geomorphology, turbulent flow is classified according to the *Froude number of a stream. Compare with *laminar flow.

turbulent boundary layer *See* *boundary layer.

turnpike A toll road. Toll roads fell into disuse in Britain with the coming of the railways, but the term is still used for certain toll roads in the USA.

U

U-shaped valley Most U-shaped valleys—valleys with a parabolic *cross-section—are *glacial troughs. However, valleys with this form are also encountered in non-glaciated chalk topography.

ubac That side of a valley which receives less *insolation; the shaded side.

ubiquitous material In Weber's theory of industrial location, any material which is available everywhere, like water or air, and which therefore does not exert a locational pull.

ultisol A soil of the *US soil classification. *See* *ferruginous soil.

umland The area served by a city. The umland is also known as a *sphere of influence, *catchment area, *tributary area, or *urban field.

uncertainty The state of mind of an individual who is unable to make any estimate of future events. This differs from *risk in that the odds of an event occurring are known in a risk; uncertainty does not give any odds and all outcomes, expected or not, are possible. In the real world, decisions are often made under conditions of uncertainty since it may be difficult to predict the response of an individual to an event. When decision-makers are faced with uncertainty, they will react according to their nature. Some will assume the worst; others will hope for the best. In the former case, the individual is more concerned with possible future loss in a disaster than with actual gains. Most decision-makers are in the centre ground—partial optimists.

underbound(ed) city A city where the administrative boundary encloses an area smaller than that of the city itself; the city has 'burst' its bounds.

underdevelopment The original meaning of the term indicated that existing resources had not been exploited. The word is now close in meaning to 'poverty' although some oil-rich underdeveloped countries have high incomes which are enjoyed by the few. Indicators of underdevelopment include: high birth rates, high infant mortality, undernourishment, a large agricultural and small industrial sector, low per capita *GDP, high levels of illiteracy, and low life expectancy.

underfit stream Also known as a misfit stream, this is a stream which flows with narrower meander belts and shorter meander wavelengths than are appropriate to the valley. It is suggested that the valley was initially formed by a river of much greater *discharge than now obtains and that an underfit stream is evidence of climatic change.

underpopulation A situation where there are too few people to develop fully the economic potential of an area or nation; a larger population could be supported on the same resource base. Such a situation obtains in the Amazon Basin, but whether it would be wise to colonize such an area is disputable.

unemployment The state of being involuntarily out of work. The unemployment rate is the number unemployed as a percentage of the total population of working age. An unemployment level of under 3% is thought of as a natural rate as people change their jobs, their residences, and their state of health. **Structural unemployment** occurs when the labour market no longer requires a particular skill, as in the case of printing newspapers; new technology and new materials replace older working habits. Unemployment is more general and often reflects trade recession where no jobs are to be had. In the case of **fractional unemployment**, jobs are available but not taken up because of immobility or the lack of information.

uneven development The condition of an economy which has not benefited equally from development either in a regional sense and/or within classes in society. It may also occur between 'consumer goods' and 'capital goods' industries, and between sectors of the economy. It may be seen to be a result of capitalism, based as it is on competition and *accumulation, but it is not unique to capitalism. State socialism has often led to concentration on one sector at the expense of another.

unified field theory The theory that concepts within political geography can be linked together. Thus, ideas are explored until decisions are made which promote the movement of people, goods, and ideas. These movements take place within a field of circulation. Ultimately, an area emerges as an expression of the initial concept.

uniform delivered pricing Pricing a commodity at the same value regardless of the location of the customer. Demand will, therefore, not be affected by distance from the manufacturing point. Uniform delivered pricing is sometimes referred to as *c.i.f. pricing.

uniformitarianism The view that the interpretations of earth history can be based on the present-day evidence of natural processes. From this comes the maxim 'the present is the key to the past'. Although the processes may be the same, the rate of change may vary over geological time.

unit response graph For any *channelled flow, a *hydrograph which is produced by a storm of known rainfall amount.

unloading The removal, by erosion, of rock or, by ablation, of ice. With removal of rock or ice an exhumed landscape is revealed. The pressure release following unloading may cause the exhumed strata to 'burst' upwards.

upward transition region *See* *core-periphery model.

upwelling The rise of sea water from depths to the surface, bringing nutrients for plankton. Many of the world's best fishing grounds are located at such points. The effect called *el Niño is the failure of the Peruvian upwelling.

urban Of, living, or situated in a city or town. As no standard figures are given for the size of cities and towns, this concept can be rather vague. In Iceland, a settlement of 300 people is classed as urban; the figure is 10 000 in Spain. An area may be classified as urban by its role as a *central place for a tributary area, providing a range of shops, banks, and offices. A high density of population may also be used as an indicator but the city may include large areas of low-density housing.

urban blight A run-down area of the city. Some parts of the city become out-dated as buildings age and as variations occur in the type of demand. Certain activities, such as small-scale industry and warehousing, have an adverse effect on the urban environment and, as neighbourhoods decline, they become prone to vandalism. Erstwhile town houses are changed to multi-occupancy. Blighted urban areas of the inner city have now become a political concern and urban renewal schemes have become fashionable.

urban climates Built-up areas affect local climates in four major areas:
   1. The atmosphere. There are 10–20 times as many *aerosols in the urban atmosphere as in rural regions. Gases such as sulphur dioxide and nitrogen dioxide are at higher concentrations. *Pollution domes are common.
   2. Heat. Domestic heating, electricity production, and transport systems all give out heat. Thus, an urban heat island is created, with temperatures some 6–8 °C higher than those in the surrounding countryside. The amount of heating seems to be dependent on urban population densities rather than on city size.
   3. Air flows. Some buildings form wind-breaks; some streets form wind-tunnels. *Eddying is common as air 'bounces off' tall buildings.
   4. Moisture. *Run-off is rapid and plant life is relatively scarce. *Evapotranspiration is therefore lower and towns are less humid. However, rainfall in towns may be some 6–7% heavier than in the countryside.

urban containment The policy of limiting *urban sprawl by restricting out-of-town development.

urban density gradient In a city, the pattern of population density as it decreases with distance from the city centre. Population densities are higher near the centre where the poor, who need a central location in order to reduce the cost of the journey to work, live on small areas of valuable land, or where building densities are high in order to cover high land costs. Densities decline with distance from the centre as the rich, who can afford higher travel costs, locate at the periphery and use large areas of cheaper land. This is an Anglo-Saxon phenomenon.

urban development corporation, UDC A government-sponsored enterprise with the task of redeveloping derelict land in the *inner cities. These corporations are required to match public funding with privately raised capital (**leverage**) and work within the freedoms also granted to *enterprise zones. UDCs were responsible for much of the redevelopment of London's Docklands and for the Cardiff Bay scheme.

urban diseconomies Financial and social burdens arising from an urban location. These include constricted sites, high rates, traffic congestion, and pollution. *See* *decentralization.

urban ecology The application of the principles of *ecology to a study of urban environments. *See* *human ecology. Urban ecologists look at individual areas of the city—*natural areas—in the context of the whole city, and focus on the way a population organizes itself, and the way it adapts to change. Just as ecologists study the way in which an ecosystem seeks to re-establish equilibrium after a sudden alteration, so urban ecologists assume that people will try to re-establish equilibrium after sudden change.

Urban ecology posits that the urban realm is made up of four interrelated variables: a functionally integrated population, a self-sustaining system of relationships, an urban environment, and the technology and tools which sustain the community. A change in one will bring about a change in the other three.

Urban ecology has been criticized for focusing too much on competition at the expense of the cultural and subjective forces which shape the city. It flourished in the 1920s and 1930s, went through a period of neglect, was revived in the late 1950s and early 1960s, but is no longer seen as central.

urban field That area surrounding a city which is influenced by it. The inhabitants of the urban field depend on the city for services such as hospitals, higher education, employment, retailing, marketing, and finance and the city is served in its turn by labour.

The delimitation of urban fields poses problems. For example, the area served by a city newspaper may not be the same as the area served by the city's public transport, so that the boundary of a city's field is not demarcated by a single line; in fact there is a hierarchy of urban fields. The smaller fields of a number of towns may be 'nested' within the larger urban field of a city. The fields fall into three zones: a core area composed of the built-up area of the town, an outer area which uses the town for high-order goods and services, and a fringe area which uses the urban area rarely and then only for very high-order goods and services.

urban geography The study of the site, evolution, morphology, spatial pattern, and classification of towns. Historically, three themes may be distinguished: the quantitative, descriptive approach, establishing the spatial organization of the city; the behavioural method, emphasizing the decision-making process within the perceived environment; and the *radical tradition, which stresses not only the spatial inequalities within a city and the inequitable distribution of resources, but suggests strategies to remedy these inequalities. It also features the constraints imposed by society on individuals.

urban hydrology *Urbanization changes the hydrology of a drainage basin. Roads and artificial surfaces cut down infiltration and storage while storm sewers speed up the flow of water into rivers. It is suggested that urbanization increases the risk of flooding as rivers respond much more violently to a storm event.

urban land-value surface A three-dimensional representation of the variation in land values across a city. Most urban land value surfaces have a *peak land value intersection (PLVI) within the *CBD, where accessibility is at a maximum. Values fall with distance from the PLVI, but at varying gradients, and sub-peaks exist at more accessible locations: along ring roads, and especially where routes radiating from the city centre intersect ring roads. It is argued that the land value surface is of major importance in determining urban land use.

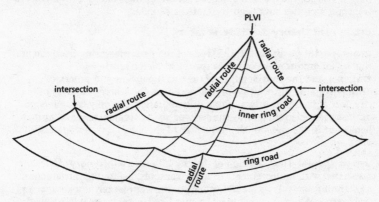

FIGURE 54: *Urban land-value surface*

urban managers and gatekeepers Those who allocate scarce urban resources and facilities. *Gatekeepers 'open the gate', usually for housing, to those who qualify and close it to those who do not. Gatekeepers include solicitors, estate agents, and financiers. Managers operate mainly in the public sector and include housing officers and local government councillors and planners.

urban morphology The form, function, and layout of the city, and the study of these features, including their development over time.

urban planning An attempt to manage the city, often in order to avoid, or alleviate, common urban problems such as *inner city decay, overcrowding, traffic and other forms of congestion. *See* *urban development corporation, *traffic calming.

urban renewal The attempt to reinvigorate a run-down urban area, such as the inner city. Slum clearance is *redevelopment after wholesale clearance of the site, but this has the effect of destroying communities and has become unacceptable; in Oxford, for example, the community of St Ebbe's was relocated at the edge of the city, in Blackbird Leys, leaving the population with a long journey to work and fewer amenities. By the 1970s, the emphasis had, therefore, shifted to refurbishment, and local authorities commonly gave improvement grants to upgrade existing housing. In many cases, this strategy lead to *gentrification, and a revival in the fortunes of a decaying area.

By the 1980s, and partly as a result of riots such as those in Toxteth, the British government was concerned to target regional aid, and set up *enterprise zones and *urban development corporations, while encouraging the growth of housing associations.

Many US federal programmes have been directed towards urban revitalization, not all of them successful, but the containerization of Boston's docks, the redevelopment of the waterfront, and improvements in transport facilities have been accounted a success.

urban rent theory *See* *bid-rent theory.

urban social geography The study of the patterns arising from the use that social groups make of urban space (*see* *neighbourhood, for example), and the consideration of the social patterns and processes arising from the distribution of, and access to, the scarce resources of the city. It is argued that urban spatial differentiation can only be understood against the background of the underlying social organization and human behaviour in the city, but the influences of space and distance are also vital.

urban sprawl The extension of the city into the countryside, particularly associated with improvements in mass transport. Before the introduction of planning controls in the UK, urban sprawl went largely unchecked, and *ribbon development along major routes like London's Great West Road, was rife. In an attempt to check further growth, *green belts have been established around Britain's cities.

urban system Any network of towns and cities, and their hinterlands, which can be seen as a system, since it depends on the movements of labour, goods and services, ideas, and *capital through the network. Crucial to the interactions within the system are efficient systems of transport and communication. With improved technology, it is possible to see urban systems which transcend national boundaries.

urban village A residential area of the city containing a cluster of individuals of similar culture and/or interests.

urbanism A way of life associated with urban dwelling. L. Wirth (R. Sennett, 1938) suggested that urban dwellers follow a distinctly different way of life from rural dwellers. Physical and social stimuli are high and urban residents may react by becoming aloof and indifferent in their relationships with others. Stress is higher than in the country, and this is said to account for higher levels of mental illness and crime in the city.

urbanization This is the migration of rural populations into towns and cities; an increasing proportion of the world's population resides in towns. Urbanization indicates a change of employment structure from agriculture and *cottage industries to mass production and *service industries. This backs up the view that urbanization results from, rather than causes, social change. This is most notable in the development of capitalism and its attendant industrialization. It is said that the development of the landless labourer and the concentration of wealth into a few hands encourages urbanization.

Others argue that urbanization is the inevitable result of economic growth, with the rise of specialized craftsmen, merchants, and administrators. A further view stresses the importance of *agglomeration economies. The city offers market, labour, and capital with a well-developed infrastructure. Urbanization is a relatively recent process in the Third World where it is even more rapid than population growth and where the largest agglomerations are growing most rapidly.

urbanization curve A model of the progress of urbanization, based on *empirical evidence from Europe. In a traditional society, urbanization is below 20%, and the rate of urbanization is slow, so the curve starts gently. With industrialization, and a rise in the importance of *manufacturing and *services, the pace of urbanization quickens, but the curve slackens after about 75%. While most developed countries have reached this third stage, the countries of the developing world are still on a rising curve of urbanization, often with a steeper gradient than is characteristic of advanced economies. Some geographers believe that, with increasing *decentralization, and an increase in home working, with the aid of computers, modems, and faxes, there may be a fourth stage, when urbanization actually decreases. *See* *counter-urbanization.

urbanization economies Advantages gained from an urban location. These include proximity to a market, labour supply, good communications, and financial and commercial services such as auditing, stockbroking, advertising, investment, industrial cleaning, and maintenance.

US soil classification The US Department of Agriculture recognized ten major soil groups in the Seventh Approximation. Alfisols are relatively young and acid soils with a clay B *horizon. Aridisols are semi-desert and *desert soils. Entisols are immature, mainly *azonal soils. Histosols are primarily organic in content, developing in marshes or peat bogs. Insectisols are young soils with weakly developed horizons. Mollisols are characteristic of grassland, high in *bases and with a thick, organically rich A horizon. Oxisols occur in tropical and subtropical areas. They are well-weathered and often have a layer of *plinthite near the surface. Spodosols have been *podzolized. Ultisols develop where summers are wet and winters are dry. They are quite deeply weathered and are often reddish-yellow in colour. Vertisols are clay soils characterized by deep, wide cracks in the dry season.

utility The satisfaction given to an individual by the goods and services used.

uvala A depression formed when two or more *dolines coalesce. The size of the hollow is not important in the recognition of a uvala.

V

vadose Referring to the zone immediately below the ground surface and above the *water-table in which the water content varies greatly in amount and position.

valley glacier A glacier situated in an upland valley or basin.

valley train A plain within a valley sloping down and away from the site of a glacier snout, composed of sands and gravels and containing pebbles and boulders.

valley wind An *anabatic air flow generated as a valley floor is heated by the sun. The warm air moves upslope. Valley winds are at their strongest in valleys of a north–south orientation.

vapour pressure In meteorology, the pressure exerted by the water vapour in the air. This is a partial pressure since pressure is also exerted by the air itself. *See* *saturation.

variability The extent to which a set of observations, such as annual rainfall totals over a number of years, spreads about the mean value.

variable A changing factor which may affect or be affected by another, such as river *discharge. Qualitative variables are plotted on a *nominal or *ordinal scale. Quantitative variables are usually plotted on a *ratio or *interval scale and the measurements may be *continuous or *discrete.

variable cost analysis A method of costing an industrial location in terms of the spatial variations in production and costs. In its simplest form:

$$TC_i = \sum_{j=1}^{n} Q_j \cdot U_{ij}$$

where TC_i = total cost of production at location i, Q_j = required quantity of input j, and U_{ij} = unit cost of input at location i.

variance, σ The average of the squares of the deviations from the arithmetic mean of a data set. It is a statistic which represents the extent to which a set of observations, for example, of annual rainfall over a series of years, spreads about the mean. Where the observations are closely grouped, the variance is low. *See* *standard deviation.

Varignon frame A string and pulley model for establishing the *least-cost location. Weights are used to represent the amounts of raw materials needed to make one unit of production. A weight is also used for the finished product. The weights are suspended below their point of origin on the strings and all the strings are tied together in the centre. The point of least costs will be the point at which the central knot stops.

varve A pair of coarse and fine deposits which reflect seasonal deposition of glacial debris deposited in *proglacial lakes. A **varve couplet** represents the total *fluvio-glacial deposition on a lake floor for one year. Summer deposition brings coarse sediments from meltwater streams; such sediments are from silt to sand in size. In winter the lake surface is frozen so that the water is calm. Under these still conditions, the fine deposits settle out in a thinner layer than that of the summer sediments. Analysis of a series of varves may help in the reconstruction of climatic changes during glaciation.

vector A force having both magnitude and direction, such as a westerly wind blowing at 30 m.p.h. *See also* *wind shear.

vector data In *Geographic Information Systems, positional data in the form of points, *lines, and polygons, expressed as *x*, *y* co-ordinates.

veering Of winds in the Northern Hemisphere, changing direction in a clockwise motion e.g. from westerly to northerly. The converse applies in the Southern Hemisphere. *See* *backing.

veld The wild grassland of the interior of South Africa. The veld has been greatly modified by fire, and experiments suggest that, when protected from farming or fire, much of the veld may develop into scrub or even forest.

vent In geomorphology, an opening in the *crust through which volcanic material flows. Some volcanoes have a single, central vent, others have a line of vents or side vents, also known as subsidiary vents.

vertex, vertices (pl.) Also known as a node. In *network analysis this is the place joined by two or more routes (*links).

vertical integration *See* *integration.

vertisols Soils of the *US soil classification found in regions of high temperature where bacteria destroy organic residues; hence the humus content is low. Alternate wet and dry seasons lead to the alternate swelling and shrinking of these soils. By these processes, horizons become mixed or inverted.

Victorian city A stage in the development of the capitalist city between the *transitional and the modern. The Victorian city is marked by residential segregation, especially of the emerging middle classes, and the rise of the *central business district. Industry is located outwards from the commercial core, following canals, railways, and roads, cutting the city into a series of wedges, and helping to define the residential areas.

village In Britain, just as there are no definitions for the population size which delimits a *town or *city, so there is no limiting size for a village.

virga Stringy trails of descending, and evaporating, water droplets, tapering down from the base of a cloud, made visible by back-lighting from the sun.

viscous Adhesive or glutinous. **Viscosity** is the resistance to flow exhibited
by a material.

vital rates Rates of those components, such as birth, marriage, fertility,
and death which indicate the nature and possible changes in a population.
Even when population numbers are stable, there may be changes in the
vital rates.

viticulture Vine cultivation, usually for wine-making.

V-notched weir An apparatus for measuring stream discharge. A
plywood dam is constructed across the stream. A 90° notch is cut into the
middle of the dam wall, such that water flows through the gap but not
over the wall. Discharge (Q) is then calculated from the equation $Q = 1.336 \times h2.43$ where h is the height of the water surface above the angle of the V-
notch.

volcanic ash Finely pulverized fragments of rock and lava which have
been thrown out during a volcanic eruption. The term 'ash' is a misnomer.

volcanic bomb A block of lava ejected into the air from a volcano. As it
is thrown out, it cools and spins, causing the block to be rounded or
decorated with spiral patterns.

volcano An opening in the crust out of which *magma, ash, and gases
erupt. The shape of the volcano depends very much on the type of lava.
*Cone volcanoes are associated with thick lava and much ash. *Shield
volcanoes are formed when less thick lava wells up and spreads over a
large area, creating a wide, gently sloping landform. Most volcanoes are
located at *destructive or *constructive plate margins.

von Thünen models J. H. von Thünen had two basic models. Both were
located in an isotropic plain where there was one market—the city—for
surplus agricultural production. One form of transport was available and
transport costs increased in direct proportion to distance. No external
trade took place, and farmers acted as *economic men. All farmers
received the same price for a particular crop at any one time.
    The first model postulates that the intensity of production of a
particular crop declines with distance from the market since transport
costs increase with distance from the market and the locational rent is
therefore lower. Intensive farming—which demands costly inputs—is only
profitable where locational rent is high to cover costs, so intensive farming
takes place only near the city.
    Von Thünen's second model is concerned with land use patterns.
Transport costs vary with the bulkiness and perishability of the product.
Product A is costly to transport but has a high market price and is
therefore farmed near the city. Product B sells for less but has lower
transport costs. At a certain distance, B becomes more profitable than A
because of its lower transport costs. Eventually, product C, with still lower
transport costs, becomes the most profitable product. The changing
pattern of the most profitable produce is therefore seen as a series of land

use rings around the city. This phenomenon may be illustrated by a graph showing the varying locational rent of three products, the most profitable product at each point, and the land use pattern which results.

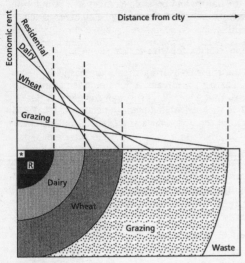

R Residential
* City centre

FIGURE 55: *von Thünen model*

vortex, vortices (pl.) *See* *turbulence.

vorticity In meteorology, a measure of the local spin of a part of the *atmosphere. The local spin of the atmosphere relative to the earth has opposite signs in *cyclones and *anticyclones; conventionally the cyclonic direction is taken as positive. Relative vorticity includes *curvature and *shear. A major principle governing the vorticity change of flowing air is the *conservation of angular momentum: as air spreads out horizontally, the rate of spin falls; as it contracts horizontally, its rate of spin rises.

vulcanian eruption (pl.) Also known as a Vesuvian eruption, this is marked by periodic lulls during which gas pressure builds up behind the lavas that clog the vent. This blockage is removed by an explosion which throws off *pyroclasts in large quantities.

vulcanicity The movement of magma, both into the earth's crust (*intrusion) and onto the earth's surface (*extrusion).

wadi In a hot desert, a steep-sided, flat-floored valley very occasionally occupied by an intermittent stream. Wadis were probably cut during *pluvials when rainfall was higher in deserts. *See* *raindrop erosion.

waning slope The low, *concave element at the foot of a hillslope.

warm front *See* *front.

warm glacier A glacier which has basal temperatures of around 0 °C. Warmth is imparted to the glacier by friction with the bedrock, by shearing within the ice mass, and from *geothermal heat. Rates of snow formation are rapid—between 25 and 40 years. Glacier flow is rapid because warm glaciers are lubricated at their base, and velocities of 20–30 m per day have been recorded. Warm glaciers are the most effective agents of *glacial erosion, especially when they exhibit *Blockschollen flow.

warm occlusion *See* *occlusion.

warp An alternative name for a *solifluction gravel.

wash board moraine Synonymous with De Geer *moraine.

water balance The balance at any location between the input—*precipitation, (P)—and the outputs: *evapotranspiration (E) and *run-off (R):

$$P = E + R.$$

If water balance is computed for a number of years, *groundwater storage (S) is held to be constant, but for studies of a single year, groundwater fluctuations are taken into account:

$$P = E + R \pm S$$

Storage and run-off tend to be higher in winter, and evapotranspiration is higher in summer. *See* *soil moisture budget.

waterfall A site on the *long profile of a river where water falls vertically. Waterfalls may be found at a band of more resistant rock, at a *knick point, or where *deposition has occurred. Perhaps the most famous are the two waterfalls that constitute Niagara Falls, with a drop of 55 m. The American Falls were retreating 0.6 m each year through *headward erosion; the Canadian Horseshoe Falls 1 m per year until the cementation associated with the construction of a power station reduced the rates to 2 and 6 cm per year respectively.

water-level weathering The development and enlargement of tidal pools by weathering and by the action of rock-grinding animals.

watershed The boundary between two river systems. The watershed marks the divide between *drainage basins, and usually runs along the highest points of the *interfluves.

water-table The level below which the ground is saturated. Any hole in the ground will fill with water when the water-table has been reached. This level often fluctuates with rainfall. The water-table is thus the upper surface of the *groundwater.

wave A ridge of water between two depressions. As waves approach a shore, they curl into an arc and break. The energy of surface waves is responsible for the *erosion of the coast. Waves also initiate currents which run along the coast and which are the moving force in *longshore drift. The height of a wave is generally proportional to the square of wind velocity.

wave-cut platform *See* *shore platform.

wave energy Energy generated by the force of ocean waves. The use of this energy is still in the experimental stage, but successful models have been built.

wave refraction The change in the approach angle of a wave as it moves towards the shore. As water becomes shallow, waves slow down. This

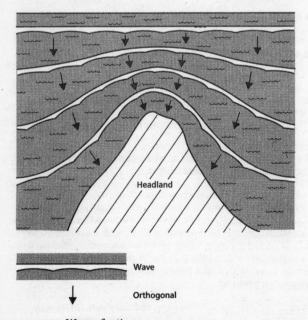

FIGURE 56: *Wave refraction*

change in speed causes the *orthogonals of a wave to 'bend' so that the
line of the wave mirrors the submarine contours. Refraction causes waves
to converge on *headlands and diverge in *bays. This means that the
energy of the waves is concentrated on the headlands rather than on the
beaches.

waxing slope The *convex element at the foot of a hillslope.

weather Current, rather than average, atmospheric conditions; the object
of study of *synoptic meteorology. Weather variables include humidity,
temperature, sunshine hours, cloud cover, visibility, and *precipitation
(fog, rain, snow, sleet, and frost).

weathering The breakdown, but not the removal, of rocks. Weathering
that causes chemical change is **chemical weathering** and includes the
processes of *hydration, *hydrolysis, *oxidation, *carbonation, and some
forms of *organic weathering. **Mechanical weathering** is the physical
disintegration of the rock, as in *pressure release, *crystal growth, *salt
weathering, *thermal expansion, and some forms of organic weathering
such as *chelation, and *bacterial reduction as in *gley soils.

weathering front The zone of contact of the *regolith with the
underlying rock.

weathering pit Depressions on a flat surface, usually on very soluble
rocks, and varying in shape and ranging in size from a few centimetres to
several metres in width. Most pits are initiated at a weak point in the rock.
*Erosion will enlarge the pits; often two or more pits combine. Honeycomb
weathering is a grouping of many, closely spaced pits. *Tafoni are
weathering pits which are cut into near vertical rock faces.

weathering rind The chemically altered 'skin' of rock around an
unaltered core.

Weber's theory of industrial location A model of industrial location
proposed by A. Weber (1909, trans. 1929), which assumes that industrialists
choose a *least-cost location for the development of new industry. The
theory is based on a number of assumptions, among them that markets are
fixed at certain specific points, that transport costs are proportional to the
weight of the goods and the distance covered by a raw material or a
finished product, that perfect competition exists, and that decisions are
made by *economic man.

   Weber postulated that raw materials and markets would exert a 'pull' on
the location of an industry through transport costs. Industries with a high
*material index would be pulled towards the raw material. Industries with
a low material index would be pulled towards the market.

   Once a least-cost location has been established, Weber goes on to
consider the deflecting effect of labour costs. To determine whether the
savings provided by moving to a location of cheaper or more efficient
labour would more than offset the increase in transport costs, *isodapanes
are constructed around the point of production at the point of minimum

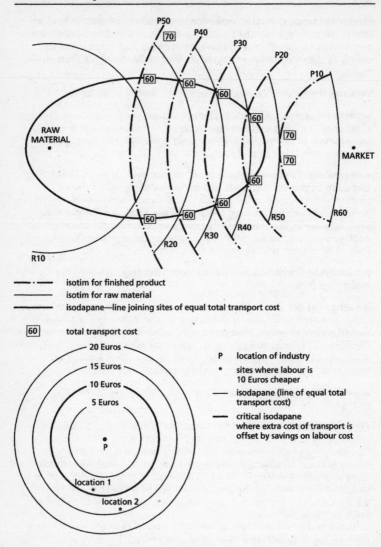

FIGURE 57: *Weber's theory of industrial location*

transport costs. The extra price of the wage bill is calculated for the point of production. If the source of cheap labour lies within the isodapane which has the value of the higher wages differential (the critical isodapane), it would be more profitable to choose the site with low labour costs rather than the least transport costs location.

Industrial location may be swayed by *agglomeration economies. The savings which would be made if, say, three firms were to locate together, are calculated for each plant. The isodapane with that value is drawn around the three least-cost locations. If these isodapanes overlap, it would be profitable for all three to locate together in the area of overlap.

weir Generally speaking, a small dam built across a river. In hydrology, weirs are erected to measure river flow. Water is impounded behind the dam and is fed through a notch. **Sharp-crested weirs** have a sharpened metal plate to dam the stream and a steep-sided notch. **Broad-crested weirs** are wider and lower. The rate of discharge of the river is calculated by different methods at each type of weir.

welfare While welfare might be equated with *well-being, within human geography it refers mostly to factors within the control of societies: environmental quality, security, and access to commodities and services. It therefore incorporates income, standard of living, housing, employment, and access to educational, health and social services. *See also* *welfare geography, *quality of life.

welfare geography An approach in human geography which features inequalities in *social well-being and *social justice, looking at the areal differentiation and spatial organization of human activity from the point of view of the welfare of the people involved: who gets what, where, and why?

Welfare geography developed, in part, in response to what was seen as the rather 'bloodless', narrowly economic preoccupations of the *quantitative revolution, and focuses on those factors which affect the quality of human life: crime/lack of crime, poverty/wealth, housing/homelessness, and the provision/lack of educational, health, leisure, and social services.

There are three broad lines of enquiry. The first is to identify inequality in the distribution of any of the welfare indicators listed above. This provides a base from which to evaluate the impact of past or proposed changes. From this, the second aim is an attempt to explain these inequalities. Current interpretations include explanations which link inequality with the *mode of production; it is argued that uneven development is an inevitable consequence of *capitalism, while recognizing that inequalities of a less extreme spatial expression also arise under socialism. Other explanations show how the location of services benefits some and disadvantages others. The third aim is to propose measures which will bring about a fairer distribution of resources and opportunities. *See* *territorial justice.

well-being A broad term, covering *quality of life, *welfare, *social well-being and standard of living.

westerlies, westerly winds Winds blowing from the west, most often occurring in mid *latitudes. The westerlies of the Northern Hemisphere blow from the south-west; those of the Southern Hemisphere blow from the north-west. These Southern Hemisphere westerlies are more constant

than those of the North because there are fewer land areas or relief
barriers in the South.

wet-bulb thermometer, wet-bulb hygrometer *See* *hygrometer.

wetland Any land which is intermittently or periodically waterlogged.
This includes *salt marshes, tidal *estuaries, marshes, and *bogs. Wetlands
are rapidly disappearing habitats; the Everglades National Park, Florida, is
a complex of coastal *mangroves, tropical saw-grass marshes, and forest on
the slightly raised areas, but flood-control measures to the north, and the
ever-increasing number of visitors cause intense pressure on the ecosystem.
Other wetlands are increasingly being reclaimed for agriculture, industry,
or housing.

wetted perimeter In a cross section of a river channel, the line of
contact between water and bed. *See* *hydraulic radius.

wetting front The lower limit, in a soil, of water infiltration from above.

wilderness An area which has generally been affected more by natural
forces than by human agency; a region little affected by people. Some
77 000 km² of the USA have been designated as **wilderness areas**, under the
Wilderness Act of 1964. These are, ideally, areas which have never been
subject to human manipulation of the ecology, whether deliberate or
unconscious, and which are set aside as nature reserves to which human
access is very severely restricted. Roads, motor vehicles, aircraft (except in
an emergency), and any economic use are all forbidden. However, pressure
to remove this protection did not slacken throughout the 1980s, with the
Administration arguing that the Federal Government owns too much
property, and the American Wilderness Society arguing for continued
public ownership.
  The motivation for establishing such areas includes scientific arguments
for *preservation rather than *conservation, the ethical view that not all
of nature should be exploited, and an appreciation of the spiritual quality
of the wilderness. The irony is that the designation of an area as a
wilderness often *increases* tourist interest.

wilting point The point at which a plant has to supply water from its
own tissues for *transpiration when the soil moisture is exhausted.

wind chill The power of the wind to remove the warm air close to the
surface of the skin. Cold air replaces air warmed by the body and, the
stronger the wind, the more heat is carried away. Wind chill hazard is a
normal feature of high latitude and high altitude areas.

wind energy Power generated by harnessing the wind, usually by
windmills. Early windmills were used to power millstones, pumps, and
forges. Future uses may include the generation of electricity but drawbacks
exist, such as the inconstant nature of the wind, the difficulty of
construction, and finding a suitable site. Wind farms have been, and are
being constructed in upland areas of the British Isles, such as Wales and
the Lake District, but they have been objected to because of visual and

noise pollution. Offshore locations have been suggested; here wind speeds are higher, and the impact of visual pollution is reduced.

wind gap A river-cut pass through a relief barrier which now contains no stream. It is postulated that such gaps were cut by rivers which were later *captured.

wind shear The local variation of the wind *vector, or any of its components, in a given direction. A change in wind speed and/or direction with height is the vertical shear. If the wind is *geostrophic, the vertical shear is given by the *thermal wind equation.

window In *meteorology, the ability of radiation of wavelength around 10 μm to escape absorption by the earth's *atmosphere. This wavelength nearly coincides with earth's peak radiation, and allows some of the outflow of terrestrial radiation to be lost to space, thus, in part, upholding the thermal equilibrium of the atmosphere, which is also achieved by *convection.

Wirthian theory of urbanism L. Wirth (*Amer. Jour. Soc.* 1938) suggested that key factors in *urbanization: the increased size of city populations, and their increased density and heterogeneity, had psychological and social consequences, most notably an impersonality in city dwellers, expressed, at its extreme, in loneliness, mental illness, and deviant behaviour. The spatial divorce between home, school, workplace, and relatives would, he suggested, lead to *anomie.

Evidence in support of Wirthian theory has been equivocal; while unhelpfulness, conflict, and crime do seem disproportionately prevalent in large cities, the urban individual's personal relationships and psychological morale seem as good as those of rural counterparts.

World Bank In 1945, at the Bretton Woods Conference in New Hampshire, USA, America, Canada, and the UK established both the *International Monetary Fund and the International Fund for Reconstruction and Development, now popularly known as the World Bank. Its original aim was to aid post-war reconstruction in Europe, but the Bank is now generally concerned with aid to the *developing world. Membership is now virtually world-wide.

The Bank acquires its funds partly through government subscription, but mainly through borrowing, and voting power is proportional to capital subscription, so the bank is effectively controlled by the rich countries. It has made loans of over US$200 billion. *See* debt crisis, *structural adjustment.

X

xeroll A soil of the *US soil classification. *See* *chestnut soil.

xerophyte A plant which is able to grow in very arid conditions because it has adapted to restrict any water loss. Such adaptations include dense hairs or waxy leaves and shedding leaves at the start of the arid season. Succulent xerophytes incorporate water into their structure.

xerosere A plant succession developed under dry conditions such as bare rock, or sand.

yardang In a *desert landscape, a long ridge which has been isolated by the removal of rocks on either side. Yardangs can be 100 m or more in height and can stretch for many kilometres.

yazoo A tributary stream which does not join the main stream directly but runs parallel to it for some distance, usually because it cannot breach the *levées which flank the main stream.

young fold mountain *See* *fold mountain.

youth *See* *cycle of erosion.

Z

Zelinsky *See* *mobility.

zero population growth The ending of population growth when birth and death rates are equal. This would require an average number of 2.3 children per family.

zero-sum game A *formal game whereby, on choosing a particular strategy, one competitor's gain is his opponent's loss, gain and loss summing to zero.

zeuge, zeugen (pl.) Also known as a mushroom rock, this is an upstanding rock in a *desert landscape, capped with a harder *stratum and undercut by wind at the base. It is indicative of *differential erosion, the base being of softer, more easily eroded rock. They are common in arid areas, such as the eastern province of Saudi Arabia.

zonal Referring to phenomena occurring in bands roughly parallel with lines of *latitude.

zonal index This indicates the strength of westerly winds in the upper air. *See* *Rossby waves.

zonal model Another term for *concentric zone model.

zonal soil A soil where differences in local rock formation and *lithology are largely masked by the over-riding effects of climate. The major zonal soils are *tundra soils, *podzols, *Mediterranean soils, *chernozems, *chestnut soils, and *ferallitic soils.

zonal winds Winds, such as the *trade winds or the *westerlies which are associated with particular *latitudinal zones.

zone of assimilation The area which increasingly develops the functions of the CBD; the CBD of the future, characterized by whole scale redevelopment of shops, offices, and hotels.

zone of discard That area, once a part of the CBD but now in decline and characterized by low status shops and warehouses, and vacant property.

zone of overlap An area served by more than one urban centre, i.e. within two or more different *urban fields.

zone in transition *See* *concentric zone model.

z-score A method of standardizing variables measured on interval or ratio scales. If different variables are measured in different units, they may be changed into standard scores—z-scores—by expressing the values in terms of the *standard deviation:

$$\text{z-score} = \frac{\text{deviation score}}{\text{standard deviation}}$$

where the deviation score is the difference between the value and the *arithmetic mean.

Factfinder

The countries in this factfinder are listed in order of their ranking in the Human Development Index. Most of the statistics come from the United Nations Development Programme (UNDP) *Human Development Report 1995* (OUP); those marked * come from the *Oxford Dictionary of the World*, 1995, and those marked † from the *Book of Vital World Statistics*. Sources, measurements, and dates of indicators are summarized below:

* capital
* population
* area
  human development index (1992)
  urban population as percent of total (1992)
† crude birth rate (1985–90): births per 1000 population per annum
† crude death rate (1985–90): deaths per 1000 population per annum
*† infant mortality rate: deaths per 1000 children under twelve months old
  life expectancy at birth (1992): in years
  real per capita GNP (1992): in $US
  percentage of labour force in agriculture, industry and services 1990–2
  commercial energy consumption per capita is expressed in kg. oil equivalent

Canada

capital Ottawa
population (1991) 27 296 860
urban population as percent of total 77
area 9 976 186 sq. km
human development index rating 1
† *crude birth rate* 14.1
† *crude death rate* 7.4
† *infant mortality rate* 7
life expectancy at birth 77.4
real GNP per capita $20 520
percentage of labour force in agriculture 5
percentage of labour force in industry 23
percentage of labour force in services 72
commercial energy consumption per capita 7912

USA

capital Washington, DC
population (1990) 248 709 870
urban population as percent of total 76
area 9 372 614 sq. km
human development index rating 2
† *crude birth rate* 15.1
† *crude death rate* 8.8
† *infant mortality rate* 10
life expectancy at birth 76.0
real GNP per capita $23 760
percentage of labour force in agriculture 3
percentage of labour force in industry 25
percentage of labour force in services 72
commercial energy consumption per capita 7622

Japan

capital Tokyo
population (1990) 123 612 000
urban population as percent of total 77
area 377 815 sq. km
human development index rating 3
† *crude birth rate* 11.4
† *crude death rate* 7.0
† *infant mortality rate* 5
life expectancy at birth 79.5
real GNP per capita $20 520
percentage of labour force in agriculture 7
percentage of labour force in industry 34
percentage of labour force in services 59
commercial energy consumption per capita 3586

Netherlands

capital Amsterdam
seat of government The Hague
population (1991) 15 010 445
urban population as percent of total 89
area 14 412 sq. km
human development index rating 4
† *crude birth rate* 11.8
† *crude death rate* 8.7
† *infant mortality rate* 8
life expectancy at birth 77.4
real GNP per capita $17 780
percentage of labour force in agriculture 5
percentage of labour force in industry 25
percentage of labour force in services 70
*commercial energy consumption per
  capita* 4560

Finland

capital Helsinki (Helsingfors)
population (1990) 4 998 500
urban population as percent of total 62
area 338 145 sq. km
human development index rating 5
† *crude birth rate* 12.5
† *crude death rate* 10.2
† *infant mortality rate* 6
life expectancy at birth 75.7
real GNP per capita $16 270
percentage of labour force in agriculture 9
percentage of labour force in industry 29
percentage of labour force in services 62
*commercial energy consumption per
  capita* 5560

France

capital Paris
population (est. 1991) 56 700 000
urban population as percent of total 73
area 547 026 sq. km
human development index rating 8
† *crude birth rate* 14.0
† *crude death rate* 10.4
† *infant mortality rate* 8
life expectancy at birth 76.9
real GNP per capita $19 510
percentage of labour force in agriculture 6
percentage of labour force in industry 29

percentage of labour force in services 65
*commercial energy consumption per
  capita* 4034

Spain

capital Madrid
population (est. 1991) 39 045 000
urban population as percent of total 76
area 504 750 sq. km
human development index rating 9
† *crude birth rate* 12.8
† *crude death rate* 9.1
† *infant mortality rate* 9
life expectancy at birth 77.6
real GNP per capita $13 400
percentage of labour force in agriculture 11
percentage of labour force in industry 33
percentage of labour force in services 56
*commercial energy consumption per
  capita* 2409

Sweden

capital Stockholm
population (1990) 8 590 630
urban population as percent of total 83
area 449 964 sq. km
human development index rating 10
† *crude birth rate* 11.2
† *crude death rate* 12.2
† *infant mortality rate* 6
life expectancy at birth 78.2
real GNP per capita $18 320
percentage of labour force in agriculture 3
percentage of labour force in industry 28
percentage of labour force in services 69
*commercial energy consumption per
  capita* 5395

Australia

capital Canberra
population (est. 1991) 17 500 000
urban population as percent of total 85
area 7 682 300 sq. km
human development index rating 11
† *crude birth rate* 15.0
† *crude death rate* 7.4
† *infant mortality rate* 9

455 **Factfinder**

life expectancy at birth 77.6
real GNP per capita $18 220
percentage of labour force in agriculture 6
percentage of labour force in industry 24
percentage of labour force in services 70
commercial energy consumption per capita 5263

Belgium

capital Brussels
population (1991) 9 978 700
urban population as percent of total 97
area 30 540 sq. km
human development index rating 12
† *crude birth rate* 11.7
† *crude death rate* 11.5
† *infant mortality rate* 10
life expectancy at birth 76.4
real GNP per capita $18 630
percentage of labour force in agriculture 3
percentage of labour force in industry 28
percentage of labour force in services 69
commercial energy consumption per capita 5100

Austria

capital Vienna
population (est. 1991) 7 700 000
urban population as percent of total 85
area 83 854 sq. km
human development index rating 11
† *crude birth rate* 11.6
† *crude death rate* 11.9
† *infant mortality rate* 8
life expectancy at birth 76.2
real GNP per capita $18 710
percentage of labour force in agriculture 7
percentage of labour force in industry 37
percentage of labour force in services 56
commercial energy consumption per capita 3266

Germany

capital Berlin
seat of government Bonn
population (1991) 78 700 000
urban population as percent of total 86

area 357 000 sq. km
human development index rating 15
† *crude birth rate* 10.4 (West Germany), 12.9 (East Germany)
† *crude death rate* 12.0 (West Germany), 12.8 (East Germany)
† *infant mortality rate* 8 (West Germany), 8 (East Germany)
life expectancy at birth 76.0
real GNP per capita $21 120
percentage of labour force in agriculture 3
percentage of labour force in industry 39
percentage of labour force in services 58
commercial energy consumption per capita 4538

Denmark

capital Copenhagen
population (est. 1991) 5 100 000
urban population as percent of total 85
area 43 075 sq. km
human development index rating 16
† *crude birth rate* 10.7
† *crude death rate* 11.3
† *infant mortality rate* 8
life expectancy at birth 75.3
real GNP per capita $19 080
percentage of labour force in agriculture 6
percentage of labour force in industry 28
percentage of labour force in services 66
commercial energy consumption per capita 3729

United Kingdom

capital London
population (1991) 55 700 000
urban population as percent of total 89
area 244 129 sq. km
human development index rating 17
† *crude birth rate* 13.4
† *crude death rate* 11.9
† *infant mortality rate* 9
life expectancy at birth 76.2
real GNP per capita $17 160
percentage of labour force in agriculture 2
percentage of labour force in industry 28
percentage of labour force in services 70
commercial energy consumption per capita 3743

Ireland

capital Dublin
population (1991) 5 093 370
urban population as percent of total 57
area 83 694 sq. km
human development index rating 57
† crude birth rate 18.1
† crude death rate 8.8
† infant mortality rate 12
life expectancy at birth 75.3
real GNP per capita $12 830
percentage of labour force in agriculture 14
percentage of labour force in industry 29
percentage of labour force in services 57
commercial energy consumption per
   capita 2881

Italy

capital Rome
population (1990) 57 746 160
urban population as percent of total 67
area 301 225 sq. km
human development index rating 20
† crude birth rate 10.8
† crude death rate 10.2
† infant mortality rate 10
life expectancy at birth 77.5
real GNP per capita $18 090
percentage of labour force in agriculture 9
percentage of labour force in industry 32
percentage of labour force in services 59
commercial energy consumption per
   capita 2755

Greece

capital Athens
population (1991) 10 269 000
urban population as percent of total 64
area 130 714 sq. km
human development index rating 22
† crude birth rate 11.9
† crude death rate 9.7
† infant mortality rate 13
life expectancy at birth 77.6
real GNP per capita $8310
percentage of labour force in agriculture 23
percentage of labour force in industry 27

percentage of labour force in services 50
commercial energy consumption per
   capita 2173

Luxembourg

capital Luxembourg
population (1990) 378 400
urban population as percent of total 88
area 2586 sq. km
human development index rating 27
† crude birth rate 11.5
† crude death rate 11.6
life expectancy at birth 75.7
real GNP per capita $21 520
percentage of labour force in agriculture 3
percentage of labour force in industry 31
percentage of labour force in services 66

Portugal

capital Lisbon
population (1991) 9 853 000
urban population as percent of total 34
area 92 072 sq. km
human development index rating 36
† crude birth rate 13.5
† crude death rate 10.1
† infant mortality rate 14
life expectancy at birth 74.6
real GNP per capita $9850
percentage of labour force in agriculture 17
percentage of labour force in industry 34
percentage of labour force in services 49
commercial energy consumption per
   capita 1816

Bahrain

capital Manama
population (1991) 518 250
urban population as percent of total 89
area 691 sq. km
human development index rating 44
crude birth rate 28.1
crude death rate 4.0
infant mortality rate 18
life expectancy at birth 71.6
real GNP per capita $14 590

percentage of labour force in agriculture 3
percentage of labour force in industry 14
percentage of labour force in services 82

United Arab Emirates

capital Abu Dhabi
population (est. 1991) 1 630 000
urban population as percent of total 82
area 77 700 sq. km
human development index rating 45
crude birth rate 23.2
crude death rate 3.7
infant mortality rate 19
life expectancy at birth 73.8
real GNP per capita $21 830
percentage of labour force in agriculture 21
percentage of labour force in industry 27
percentage of labour force in services 32
commercial energy consumption per capita 14 631

Poland

capital Warsaw
population (1990) 31 183 160
urban population as percent of total 63
area 304 463 sq. km
human development index rating 51
† *crude birth rate* 16.4
† *crude death rate* 9.9
† *infant mortality rate* 16
life expectancy at birth 71.1
real GNP per capita $4830
percentage of labour force in agriculture 27
percentage of labour force in industry 37
percentage of labour force in services 36
commercial energy consumption per capita 2407

Russian Federation

capital Moscow
population (est. 1991) 148 930 000
urban population as percent of total 75
area 17 075 400 sq. km
human development index rating 52
† *crude birth rate USSR figure* 18.4
† *crude death rate USSR figure* 10.6

† *infant mortality rate USSR figure* 25
life expectancy at birth 67.6
real GNP per capita $6140
percentage of labour force in agriculture 20
percentage of labour force in industry 46
percentage of labour force in services 34
commercial energy consumption per capita 5665

Malaysia

capital Kuala Lumpur
population (est. 1991) 18 294 000
urban population as percent of total 51
area 329 758 sq. km
human development index rating 59
crude birth rate 28.5
crude death rate 5.1
infant mortality rate 13
life expectancy at birth 70.8
real GNP per capita $7790
percentage of labour force in agriculture 26
percentage of labour force in industry 28
percentage of labour force in services 46
commercial energy consumption per capita 1445

Kuwait

capital Kuwait City
population (est. 1991) 1 200 000
urban population as percent of total 95
area 17 818 sq. km
human development index rating 61
crude birth rate 24.2
crude death rate 2.1
infant mortality rate 18
life expectancy at birth 74.9
real GNP per capita $8326
percentage of labour force in agriculture N/A
percentage of labour force in industry 26
percentage of labour force in services 73
commercial energy consumption per capita N/A

Brazil

capital Brasilia
population (1990) 155 566 100
urban population as percent of total 76

area 8 511 965 sq. km
human development index rating 63
crude birth rate 24.6
crude death rate 7.5
infant mortality rate 58
life expectancy at birth 66.3
real GNP per capita $5240
percentage of labour force in agriculture 25
percentage of labour force in industry 25
percentage of labour force in services 47
*commercial energy consumption per
    capita* 681

Kazakhstan

capital Almaty (Alma Ata)
population (1990) 16 691 000
urban population as percent of total 58
area 2 717 300 sq. km
human development index rating 64
crude birth rate USSR figure 18.4
crude death rate USSR figure 10.6
infant mortality rate USSR figure 25
life expectancy at birth 69.4
real GNP per capita $4270
percentage of labour force in agriculture 20
percentage of labour force in industry 22
percentage of labour force in services 58
*commercial energy consumption per
    capita* 4722

Cuba

capital Havana
population (est. 1990) 10 600 000
urban population as percent of total 75
area 110 860 sq. km
human development index rating 72
crude birth rate 16.9
crude death rate 6.8
infant mortality rate 12
life expectancy at birth 75.3
real GNP per capita $3412
percentage of labour force in agriculture 24
percentage of labour force in industry 29
percentage of labour force in services 47

Saudi Arabia

capital Riyadh (royal), Jeddah
    (administrative)

population (est. 1991) 15 431 000
urban population as percent of total 78
area 2 149 690 sq. km
human development index rating 76
crude birth rate 35.1
crude death rate 4.7
infant mortality rate 29
life expectancy at birth 69.7
real GNP per capita $9880
percentage of labour force in agriculture 48
percentage of labour force in industry 14
percentage of labour force in services 37
*commercial energy consumption per
    capita* 4463

Jamaica

capital Kingston
population (est. 1990) 2 500 000
urban population as percent of total 52
area 10 956 sq. km
human development index rating 88
crude birth rate 21.7
crude death rate 6.2
infant mortality rate 14
life expectancy at birth 73.6
real GNP per capita $3200
percentage of labour force in agriculture 26
percentage of labour force in industry 24
percentage of labour force in services 50
*commercial energy consumption per
    capita* 1075

Peru

capital Lima
population (est. 1991) 22 135 000
urban population as percent of total 71
area 1 285 216 sq. km
human development index rating 93
crude birth rate 27.3
crude death rate 6.9
infant mortality rate 64
life expectancy at birth 66.0
real GNP per capita $3300
percentage of labour force in agriculture 35
percentage of labour force in industry 12
percentage of labour force in services 53
*commercial energy consumption per
    capita* 330

Bolivia

capital La Paz *(administrative),* Sucre
  (judicial and legal)
population (est. 1991) 7 500 000
urban population as percent of total 58
area 1 098 580 sq. km
human development index rating 113
crude birth rate 35.7
crude death rate 10.2
infant mortality rate 75
life expectancy at birth 59.4
real GNP per capita $2 410
percentage of labour force in agriculture 47
percentage of labour force in industry 19
percentage of labour force in services 34
*commercial energy consumption per
  capita* 255

Pakistan

capital Islamabad
population (est. 1991) 115 588 000
urban population as percent of total 38
area 803 943 sq. km
human development index rating 128
crude birth rate 40.9
crude death rate 9.3
infant mortality rate 91
life expectancy at birth 61.5
real GNP per capita $2890
percentage of labour force in agriculture 47
percentage of labour force in industry 20
percentage of labour force in services 33
*commercial energy consumption per
  capita* 223

India

capital New Delhi
population (est. 1991) 859 200 000
urban population as percent of total 26
area 3 185 019 sq. km
human development index rating 134
crude birth rate 43.9
crude death rate 11.8
infant mortality rate 93
life expectancy at birth 60.4
real GNP per capita $1230
percentage of labour force in agriculture 62
percentage of labour force in industry 11

percentage of labour force in services 27
*commercial energy consumption per
  capita* 235

Nigeria

capital Abuja
population (1991) 88 514 500
urban population as percent of total 37
area 923 768 sq. km
human development index rating 141
crude birth rate 45.4
crude death rate 15.4
infant mortality rate 84
life expectancy at birth 50.4
real GNP per capita $1560
percentage of labour force in agriculture 48
percentage of labour force in industry 7
percentage of labour force in services 45
*commercial energy consumption per
  capita* 128

Haiti

capital Port-au-Prince
population (est. 1990) 6 500 000
urban population as percent of total 30
area 27 750 sq. km
human development index rating 148
crude birth rate 35.3
crude death rate 11.9
infant mortality rate 86
life expectancy at birth 56.6
real GNP per capita $1046
percentage of labour force in agriculture 68
percentage of labour force in industry 9
percentage of labour force in services 23

Malawi

capital Lilongwe
population (est. 1991) 8 796 000
urban population as percent of total 12
area 118 484 sq. km
human development index rating 157
crude birth rate 50.5
crude death rate 20.0
infant mortality rate 143
life expectancy at birth 45.6
real GNP per capita $820

percentage of labour force in agriculture 87
percentage of labour force in industry 5
percentage of labour force in services 8
*commercial energy consumption per
   capita* 40

Ethiopia

capital Addis Ababa
population (est. 1992) 45 892 000
urban population as percent of total 13
area 1 223 600 sq. km
human development index rating 171
crude birth rate 48.5
crude death rate 18.0
infant mortality rate 119
life expectancy at birth 47.5
real GNP per capita $330
percentage of labour force in agriculture 88
percentage of labour force in industry 2

percentage of labour force in services 10
commercial energy consumption per capita 21

Niger

capital Niamey
population (est. 1991) 7 909 000
urban population as percent of total 16
area 1 267 000 sq. km
human development index rating 174
crude birth rate 52.5
crude death rate 18.9
infant mortality rate 124
life expectancy at birth 46.5
real GNP per capita $820
percentage of labour force in agriculture 85
percentage of labour force in industry 3
percentage of labour force in services 12
commercial energy consumption per capita 39

OXFORD

MORE OXFORD PAPERBACKS

This book is just one of nearly 1000 Oxford Paperbacks currently in print. If you would like details of other Oxford Paperbacks, including titles in the World's Classics, Oxford Reference, Oxford Books, OPUS, Past Masters, Oxford Authors, and Oxford Shakespeare series, please write to:

UK and Europe: Oxford Paperbacks Publicity Manager, Arts and Reference Publicity Department, Oxford University Press, Walton Street, Oxford OX2 6DP.

Customers in UK and Europe will find Oxford Paperbacks available in all good bookshops. But in case of difficulty please send orders to the Cash-with-Order Department, Oxford University Press Distribution Services, Saxon Way West, Corby, Northants NN18 9ES. Tel: 01536 741519; Fax: 01536 746337. Please send a cheque for the total cost of the books, plus £1.75 postage and packing for orders under £20; £2.75 for orders over £20. Customers outside the UK should add 10% of the cost of the books for postage and packing.

USA: Oxford Paperbacks Marketing Manager, Oxford University Press, Inc., 200 Madison Avenue, New York, N.Y. 10016.

Canada: Trade Department, Oxford University Press, 70 Wynford Drive, Don Mills, Ontario M3C 1J9.

Australia: Trade Marketing Manager, Oxford University Press, G.P.O. Box 2784Y, Melbourne 3001, Victoria.

South Africa: Oxford University Press, P.O. Box 1141, Cape Town 8000.

Oxford
Paperback
Reference

OXFORD PAPERBACK REFERENCE

From *Art and Artists* to *Zoology*, the Oxford Paperback Reference series offers the very best subject reference books at the most affordable prices.

Authoritative, accessible, and up to date, the series features dictionaries in key student areas, as well as a range of fascinating books for a general readership. Included are such well-established titles as Fowler's *Modern English Usage*, Margaret Drabble's *Concise Companion to English Literature*, and the bestselling science and medical dictionaries.

The series has now been relaunched in handsome new covers. Highlights include new editions of some of the most popular titles, as well as brand new paperback reference books on *Politics*, *Philosophy*, and *Twentieth-Century Poetry*.

With new titles being constantly added, and existing titles regularly updated, Oxford Paperback Reference is unrivalled in its breadth of coverage and expansive publishing programme. New dictionaries of *Film*, *Economics*, *Linguistics*, *Architecture*, *Archaeology*, *Astronomy*, and *The Bible* are just a few of those coming in the future.

Oxford Paperback Reference

THE OXFORD DICTIONARY OF PHILOSOPHY

Edited by Simon Blackburn

* 2,500 entries covering the entire span of the subject including the most recent terms and concepts

* Biographical entries for nearly 500 philosophers

* Chronology of philosophical events

From Aristotle to Zen, this is the most comprehensive, authoritative, and up to date dictionary of philosophy available. Ideal for students or a general readership, it provides lively and accessible coverage of not only the Western philosophical tradition but also important themes from Chinese, Indian, Islamic, and Jewish philosophy. The paperback includes a new Chronology.

'an excellent source book and can be strongly recommended . . . there are generous and informative entries on the great philosophers . . . Overall the entries are written in an informed and judicious manner.'
Times Higher Education Supplement

Oxford
Paperback
Reference

THE CONCISE OXFORD DICTIONARY
OF POLITICS

Edited by Iain McLean

Written by an expert team of political scientists from Warwick University, this is the most authoritative and up-to-date dictionary of politics available.

* Over 1,500 entries provide truly international coverage of major political institutions, thinkers and concepts

* From Western to Chinese and Muslim political thought

* Covers new and thriving branches of the subject, including international political economy, voting theory, and feminism

* Appendix of political leaders

* Clear, no-nonsense definitions of terms such as veto and subsidiarity

Oxford
Paperback
Reference

THE CONCISE OXFORD COMPANION TO ENGLISH LITERATURE

Edited by Margaret Drabble and Jenny Stringer

Derived from the acclaimed *Oxford Companion to English Literature*, the concise maintains the wide coverage of its parent volume. It is an indispensable, compact guide to all aspects of English literature. For this revised edition, existing entries have been fully updated and revised with 60 new entries added on contemporary writers.

* Over 5,000 entries on the lives and works of authors, poets and playwrights

* The most comprehensive and authoritative paperback guide to English literature

* New entries include Peter Ackroyd, Martin Amis, Toni Morrison, and Jeanette Winterson

* New appendices list major literary prize-winners

From the reviews of its parent volume:

'It earns its place at the head of the best sellers: every home should have one'
Sunday Times

Oxford Paperback Reference

CONCISE SCIENCE DICTIONARY

New edition

Authoritative and up to date, this bestselling dictionary is ideal reference for both students and non-scientists. Fully revised for this third edition, with over 1,000 new entries, it provides coverage of biology (including human biology), chemistry, physics, the earth sciences, astronomy, maths and computing.

* 8,500 clear and concise entries

* Up-to-date coverage of areas such as molecular biology, genetics, particle physics, cosmology, and fullerene chemistry

* Appendices include the periodic table, tables of SI units, and classifications of the plant and animal kingdoms

'handy and readable . . . for scientists aged nine to ninety'
Nature

'The book will appeal not just to scientists and science students but also to the interested layperson. And it passes the most difficult test of any dictionary—it is well worth browsing through.'
New Scientist

Oxford
Paperback
Reference

THE CONCISE OXFORD DICTIONARY OF OPERA

New Edition

Edited by Ewan West and John Warrack

Derived from the full *Oxford Dictionary of Opera*, this is the most authoritative and up-to-date dictionary of opera available in paperback. Fully revised for this new edition, it is designed to be accessible to all those who enjoy opera, whether at the opera-house or at home.

* **Over 3,500 entries on operas, composers, and performers**

* **Plot summaries and separate entries for well-known roles, arias, and choruses**

* **Leading conductors, producers and designers**

From the reviews of its parent volume:

'the most authoritative single-volume work of its kind'
Independent on Sunday

'an invaluable reference work'
Gramophone

Oxford
Paperback
Reference

THE CONCISE OXFORD DICTIONARY OF MUSIC

New Edition

Edited by Michael Kennedy

Derived from the full *Oxford Dictionary of Music* this is the most authoritative and up-to-date dictionary of music available in paperback. Fully revised and updated for this new edition, it is a rich mine of information for lovers of music of all periods and styles.

* **14,000 entries on musical terms, works, composers, librettists, musicians, singers and orchestras.**

* **Comprehensive work-lists for major composers**

* **Generous coverage of living composers and performers**

'clearly the best around . . . the dictionary that everyone should have'
Literary Review

'indispensable'
Yorkshire Post

POLITICS

Kenneth Minogue

Since politics is both complex and controversial it is easy to miss the wood for the trees. In this Very Short Introduction Kenneth Minogue has brought the many dimensions of politics into a single focus: he discusses both the everyday grind of democracy and the attraction of grand ideals such as freedom and justice.

'Kenneth Minogue is a very lively stylist who does not distort difficult ideas.'
Maurice Cranston

'a dazzling but unpretentious display of great scholarship and humane reflection'
Professor Neil O'Sullivan, University of Hull

'Minogue is an admirable choice for showing us the nuts and bolts of the subject.'
Nicholas Lezard, *Guardian*

'This is a fascinating book which sketches, in a very short space, one view of the nature of politics . . . the reader is challenged, provoked and stimulated by Minogue's trenchant views.'
Talking Politics

ARCHAEOLOGY

Paul Bahn

'Archaeology starts, really, at the point when the first recognizable 'artefacts' appear—on current evidence, that was in East Africa about 2.5 million years ago—and stretches right up to the present day. What you threw in the garbage yesterday, no matter how useless, disgusting, or potentially embarrassing, has now become part of the recent archaeological record.'

This Very Short Introduction reflects the enduring popularity of archaeology—a subject which appeals as a pastime, career, and academic discipline, encompasses the whole globe, and surveys 2.5 million years. From deserts to jungles, from deep caves to mountain-tops, from pebble tools to satellite photographs, from excavation to abstract theory, archaeology interacts with nearly every other discipline in its attempts to reconstruct the past.

'very lively indeed and remarkably perceptive . . . a quite brilliant and level-headed look at the curious world of archaeology'
Professor Barry Cunliffe,
University of Oxford

POLITICS IN OXFORD PAPERBACKS

GOD SAVE ULSTER!

The Religion and Politics of Paisleyism

Steve Bruce

Ian Paisley is the only modern Western leader to have founded his own Church and political party, and his enduring popularity and success mirror the complicated issues which continue to plague Northern Ireland. This book is the first serious analysis of his religious and political careers and a unique insight into Unionist politics and religion in Northern Ireland today.

Since it was founded in 1951, the Free Presbyterian Church of Ulster has grown steadily; it now comprises some 14,000 members in fifty congregations in Ulster and ten branches overseas. The Democratic Unionist Party, formed in 1971, now speaks for about half of the Unionist voters in Northern Ireland, and the personal standing of the man who leads both these movements was confirmed in 1979 when Ian R. K. Paisley received more votes than any other member of the European Parliament. While not neglecting Paisley's 'charismatic' qualities, Steve Bruce argues that the key to his success has been his ability to embody and represent traditional evangelical Protestantism and traditional Ulster Unionism.

'original and profound . . . I cannot praise this book too highly.' Bernard Crick, *New Society*

POPULAR SCIENCE FROM OXFORD PAPERBACKS

THE SELFISH GENE

Second Edition

Richard Dawkins

Our genes made us. We animals exist for their preservation and are nothing more than their throwaway survival machines. The world of the selfish gene is one of savage competition, ruthless exploitation, and deceit. But what of the acts of apparent altruism found in nature—the bees who commit suicide when they sting to protect the hive, or the birds who risk their lives to warn the flock of an approaching hawk? Do they contravene the fundamental law of gene selfishness? By no means: Dawkins shows that the selfish gene is also the subtle gene. And he holds out the hope that our species—alone on earth—has the power to rebel against the designs of the selfish gene. This book is a call to arms. It is both manual and manifesto, and it grips like a thriller.

The Selfish Gene, Richard Dawkins's brilliant first book and still his most famous, is an international bestseller in thirteen languages. For this greatly expanded edition, endnotes have been added, giving fascinating reflections on the original text, and there are two major new chapters.

'learned, witty, and very well written . . . exhilaratingly good.' Sir Peter Medawar, *Spectator*

'Who should read this book? Everyone interested in the universe and their place in it.' Jeffrey R. Baylis, *Animal Behaviour*

'the sort of popular science writing that makes the reader feel like a genius' *New York Times*

POPULAR SCIENCE FROM OXFORD PAPERBACKS

THE AGES OF GAIA

A Biography of Our Living Earth

James Lovelock

In his first book, *Gaia: A New Look at Life on Earth*, James Lovelock proposed a startling new theory of life. Previously it was accepted that plants and animals evolve on, but are distinct from, an inanimate planet. Gaia maintained that the Earth, its rocks, oceans, and atmosphere, and all living things are part of one great organism, evolving over the vast span of geological time. Much scientific work has since confirmed Lovelock's ideas.

In *The Ages of Gaia*, Lovelock elaborates the basis of a new and unified view of the earth and life sciences, discussing recent scientific developments in detail: the greenhouse effect, acid rain, the depletion of the ozone layer and the effects of ultraviolet radiation, the emission of CFCs, and nuclear power. He demonstrates the geophysical interaction of atmosphere, oceans, climate, and the Earth's crust, regulated comfortably for life by living organisms using the energy of the sun.

'Open the cover and bathe in great draughts of air that excitingly argue the case that "the earth is alive".' David Bellamy, *Observer*

'Lovelock deserves to be described as a genius.' *New Scientist*